Series on Analysis, Applications and Computation – Vol. 1

ISAAC

Boundary Values
and Convolution in
Ultradistribution Spaces

Series on Analysis, Applications and Computation

Series Editors: Heinrich G W Begehr *(Freie Univ. Berlin, Germany)*
M. W. Wong *(York Univ., Canada)*
Robert Pertsch Gilbert *(Univ. Delaware, USA)*

Advisory Board Members:
Mikhail S Agranovich *(Moscow Inst. of Elec. & Math., Russia)*,
Ryuichi Ashino *(Osaka Kyoiku Univ., Japan)*,
Alain Bourgeat *(Univ. de Lyon, France)*,
Victor Burenkov *(Cardiff Univ., UK)*,
Jinyuan Du *(Wuhan Univ., China)*,
Antonio Fasano *(Univ. di Firenez, Italy)*,
Massimo Lanza de Cristoforis *(Univ. di Padova, Italy)*,
Bert-Wolfgang Schulze *(Univ. Potsdam, Germany)*,
Masahiro Yamamoto *(Univ. of Tokyo, Japan)* &
Armand Wirgin *(CNRS-Marseille, France)*

Published

Series on Analysis, Applications and Computation – Vol. 1

ISAAC

Boundary Values and Convolution in Ultradistribution Spaces

∘ Richard D Carmichael
Wake Forest University, USA

∘ Andrzej Kamiński
University of Rzeszów, Poland

∘ Stevan Pilipović
University of Novi Sad, Serbia

World Scientific

NEW JERSEY · LONDON · SINGAPORE · BEIJING · SHANGHAI · HONG KONG · TAIPEI · CHENNAI

Published by

World Scientific Publishing Co. Pte. Ltd.

5 Toh Tuck Link, Singapore 596224

USA office: 27 Warren Street, Suite 401-402, Hackensack, NJ 07601

UK office: 57 Shelton Street, Covent Garden, London WC2H 9HE

British Library Cataloguing-in-Publication Data
A catalogue record for this book is available from the British Library.

BOUNDARY VALUES AND CONVOLUTION IN ULTRADISTRIBUTION SPACES
Series on Analysis, Applications and Computation — Vol. 1

ISBN-13 978-981-270-769-7
ISBN-10 981-270-769-7

Printed in Singapore.

Preface

Numerous papers have been written concerning ultradistribution spaces (see [127], [7], [8], [82]-[86], [87], [56], [44], [11] and references therein). Such spaces are related to the solvability and the regularity problems of partial differential equations. Because of this relation, the study of the structural problems as well as problems of various operations and integral transformations in this setting is interesting in itself. The unpublished book of Komatsu [86] and in general papers of Komatsu are the basis for our approach. Important results in the framework of ultradistribution and hyperfunction spaces (see [129]-[130], [77]) which will not be included in this book were obtained by D. Kim, S. Y. Chung and their collaborators (see [38]-[40], [41]-[42], [79]), Matsuzawa (see [98]-[100]), Vogt, Meise, Taylor, Petzsche and their collaborators (see [104], [105], [110], [111]), the Italian school with Rodino, Gramchev (see [124], [54]) and many others. A list of papers with results on various problems within ultradistribution spaces is given in the references which is, however, far from being complete.

This book is intended to be an analysis of various spaces of ultradistributions considered as boundary values of analytic functions having appropriate growth estimates, and deals, in particular, with the Cauchy and Poisson integrals in the ultradistribution spaces $\mathcal{D}'(*, L^s)$, with the convolution of ultradistributions and tempered ultradistributions, and with the integral transforms of tempered ultradistributions.

The problems of characterizing analytic functions whose boundary values are elements of the spaces of distributions, ultradistributions, hyperfunctions, infra-hyperfunctions and, conversely, of finding boundary value representations of elements of the quoted spaces of generalized functions by analytic functions have a long history; for references see e.g. [88]-[89], [139]-[140], [147], [6], [152], [27] and references therein.

Carmichael and his co-workers ([16]-[27], [33]-[35]) have studied the Cauchy and Poisson kernels in appropriate tube domains. By considering the Cauchy and Poisson integrals of distributions in appropriate subspaces of the Schwartz space $\mathcal{D}'$, they obtained characterizations of these subspaces by the a priori estimates of the corresponding analytic or harmonic functions in tube domains.

Boundary value characterizations for the spaces $\mathcal{D}'((M_p), \Omega)$, $\mathcal{D}'(\{M_p\}), \Omega)$ of ultradistributions and the spaces $\mathcal{E}'((M_p), \Omega)$, $\mathcal{E}'(\{M_p\}, \Omega)$ of infra-hyperfunctions, related to a non-quasianalytic and quasianalytic sequence (M_p), respectively, are given in [127], [82], [111], [125].

The spaces $\mathcal{D}'((M_p), L^s)$ and $\mathcal{D}'((\{M_p\}), L^s)$ for $s \geq 1$ related to a non-quasianalytic sequence (M_p) are studied in papers by Carmichael and Pilipović. In this book, we investigate classes of analytic functions having boundary values in these spaces. For the analysis of Hardy type spaces of analytic functions, with bounds given by appropriate associated functions corresponding to the sequences (M_p), we apply the Cauchy and Poisson integrals as well as the Fourier transforms. The geometry of tube domains also is considered in this book. A complete boundary value characterization for the spaces $\mathcal{D}'(*, L^s)$ on $\mathbb{R}^n$, with $s \in (1, \infty)$, is given by means of almost analytic extensions, while in the cases $s = \infty$ and $s = 1$ only partial results are obtained.

One of the most important operations in the theory of generalized functions (Schwartz distributions, ultradistributions of Beurling and Roumieu type, hyperfunctions, Mikusiński operators) is the convolution. In the literature, the convolution of two generalized functions is usually defined and considered only in the case where one of them is of compact support and the definition has many applications. For instance, this definition of the convolution was used in the study of convolution equations in the space of ultradistributions by various authors. However, such a definition is not sufficient in many situations where the convolution of generalized functions should be defined without any restrictions on the supports of the generalized functions involved.

General definitions of the convolution of this kind for distributions (and tempered distributions) were considered by many authors (see [36, 132, 57, 156, 135, 149, 151, 2, 45, 69]), and it appeared that most of the definitions are equivalent (see [135, 45, 69]). Similar general definitions of the convolution for ultradistributions were first introduced in [119]. Then other analogues of the definitions of the convolution of distributions and tempered distributions were discussed for ultradistributions and tempered ultradistributions in [92, 70, 72, 73], and their equivalence was proved in [70] and in [73], respectively. Moreover, various sufficient conditions for the existence of the convolution of ultradistributions were studied in [71] in terms of the supports of ultradistributions involved (so-called compatibility conditions) and in terms of the weighted ultradistributional $\mathcal{D}'^{(M_p)}_{L^q}$ spaces. The results of the quoted papers are collected and improved in the book.

The book is organized as follows.

In this Preface we give historical comments below to the material presented in the book.

In Chapter 1 we define some notions connected with cones in $\mathbb{R}^n$ as well as the Cauchy and Poisson kernels corresponding to tube domains. We present there results which will be used later in proving boundary value representations.

Chapter 2 contains the definitions and main properties of the spaces of ultradifferentiable test functions of Beurling and Roumieu type as well as of the corresponding spaces of ultradistributions. We are mainly interested in the spaces $\mathcal{D}(*, L^s)$ and $\mathcal{S}^*$ and their strong duals. After presenting basic properties of the sequences (M_p) and ultradifferential operators generating the respective ultradistribution spaces, we prove structural theorems for these spaces. We also give the definitions of the Fourier and Laplace transforms.

Chapter 3 is devoted to characterizations of bounded sets in the spaces $\mathcal{D}'(*, L^t)$ of L^t ultradistributions of Beurling and Roumieu type for $t \in [1, \infty]$ and in the spaces $\mathcal{S}'^*$ of tempered ultradistributions of Beurling and Roumieu type. The characterizations are given in terms of representations of elements of these sets in the form of infinite series of derivatives of certain functions of the class L^t (of the class L^2) whose norms satisfy the respective estimates as well as in terms of images of ultradifferentiable operators of bounded sets in the respective spaces of functions.

In Chapter 4, the Cauchy and Poisson kernels are studied as elements of the ultradifferentiable spaces $\mathcal{D}(*, L^r)$ (Section 4.1). The Cauchy and Poisson integrals of ultradistributions in $\mathcal{D}'(*, L^s)$ are defined in Sections 4.2 and 4.3. For $s \geq 2$ the use of Cauchy integrals gives a complete boundary value characterization of elements in $\mathcal{D}'(*, L^s)$. Notice that the Poisson integral of an element of the space $\mathcal{D}'(*, L^s)$, $s > 1$, converges to this element in the corresponding general ultradistribution space.

In Chapter 5, we deal with the boundary values of analytic functions in appropriate tube domains. Section 5.1 concerns the Fourier transform and suitable generalizations of Hardy spaces within ultradistribution classes for $r \in (1, 2]$. In Section 5.2, we show that elements of such spaces have boundary values in $\mathcal{D}'((M_p), L^1)$; while appropriate L^s bounds for $s \geq 2$ lead to boundary values in $\mathcal{D}'((M_p), L^r)$ for $r \in (1, 2]$. The extension of the results of Section 5.2 to the case $r > 2$ is given in Section 5.3 for appropriate cones. By means of almost analytic extensions and Stokes' theorem, we give in Section 5.4 the complete boundary value characterization for the spaces $\mathcal{D}'((M_p), L^s)$ and $\mathcal{D}'(\{M_p\}, L^s)$ with $s > 1$. The results given in Section 5.4 for ultradistributions on the real line are true also in the multidimensional case. In Section 5.5 the cases $s = \infty$ and $s = 1$ are considered. Due to the method of Komatsu (see [82]) appropriate L^∞ and L^1 estimates are obtained for the corresponding boundary values in the respective ultradistribution spaces.

In Chapter 6, we develop the theory of convolution of ultradistributions and tempered ultradistributions. Various general definitions of the convolution in the spaces of ultradistributions and tempered ultradistributions of Beurling type are considered in an analogous way to the classical definitions of the convolution in the theory of distributions given by Schwartz in [132], Chevalley in [36], Vladimirov in [149], Dierolf and Voigt in [45], and Kamiński in [69]. As in the case of distributions, the respective definitions of the convolution of ultradistributions are equivalent both in $\mathcal{D}'^{(M_p)}$ and $\mathcal{S}'^{(M_p)}$, although the proofs of the equivalence require new methods. Also various sufficient conditions for the existence of convolution of two ultradistributions are given: in terms of their supports (appropriate compatibility conditions) and in terms of various subspaces of ultradistributions on which the convolution is defined as a bilinear mapping. In particular, the weighted ultradistributional $\mathcal{D}'^{(M_p)}_{L^q}$ spaces are studied.

In Chapter 7, different types of integral transforms in the spaces of tempered ultradistributions of Beurling and Roumieu type are defined and discussed. Various characterizations concerning the Fourier and Laplace transforms, Bargmann transform, the Wigner distribution and the Hilbert transform of tempered ultradistributions are given. Singular integral operators are studied in the spaces of tempered ultradistributions of Beurling and Roumieu type. Moreover, the Hermite expansions of elements of the spaces of test functions and their duals can be considered as a generalized integral transform.

Historically, the representation of distributions and other generalized functions as boundary values of analytic functions has its direct foundations in the two papers [88] and [89] by G. Köthe.

Motivated by suggestions of Köthe, H.-G. Tillmann obtained in [139] extensions and generalizations of the work of Köthe and did so with the analysis being in n dimensions. Namely, Tillmann gave in [139] a characterization of the analytic functions which have the distributions with compact support as boundary values and extended in [142] these results to vector valued distributions. In 1961, Tillmann published two additional classical papers, [140] and [141], in which the analytic functions with distributional boundary values in $\mathcal{D}'_{L^p}$ and $\mathcal{S}'$, respectively, were characterized. When stated in one dimension, the principal characterization results of [140] and [141] can be described in the following way.

Theorem 1. (see [140]) *Every distribution $U \in \mathcal{D}'_{L^p}$, $1 < p < \infty$, is the boundary value of an analytic function f of variable $z = x+iy$ with $\operatorname{Im} z \neq 0$ for which*

(a) $|f(z)| \leq M \max\{|y|^{-(p-1)/p}, |y|^{-m+1/p}\}$, $y = \operatorname{Im} z \neq 0$;

(b) $g_\varepsilon = f(\cdot + i\varepsilon) - f(\cdot - i\varepsilon) \in \mathcal{D}'_{L^p}$ *for $\varepsilon > 0$ and the set $\{g_\varepsilon : \varepsilon > 0\}$ is bounded in $\mathcal{D}'_{L^p}$;*

(c) $g_\varepsilon \to U$ *as $\varepsilon \to 0+$ in the strong topology of $\mathcal{D}'_{L^p}$.*

Conversely, every analytic function f of variable z with $\operatorname{Im} z \neq 0$ which satisfies conditions (a) and (b) has an element $U \in \mathcal{D}'_{L^p}$ as boundary value in the sense of (c).

The analytic function constructed from $U \in \mathcal{D}'_{L^p}$ in the proof of the sufficiency of Theorem 1 is the "indicatrix" of U or the Cauchy integral of U. In [96], Z. Łuszczki and Z. Zieleźny obtained results similar to those of [140] at about the same time. G. Bengel has extended in [6] the results for $\mathcal{D}'_{L^p}$ to vector valued distributions.

Theorem 2. (see [141]) *Every distribution $U \in \mathcal{S}'$ is the boundary value of an analytic function f of variable $z = x + iy$ with $\operatorname{Im} z \neq 0$ for which*

(a) $|f(z)| \leq M(1 + |z|^2)^m |y|^{-1/2-r}$, $y = \operatorname{Im} z \neq 0$;

(b) $g_\varepsilon = f(\cdot + i\varepsilon) - f(\cdot - i\varepsilon) \in \mathcal{S}'$, $\varepsilon > 0$;

(c) $g_\varepsilon \to U$ *as $\varepsilon \to 0+$ in the strong topology of $\mathcal{S}'$.*

Conversely, every analytic function f of variable z with $\operatorname{Im} z \neq 0$ which satisfies conditions (a) and (b) has as boundary value an element $U \in \mathcal{S}'$ in the sense of (c).

Associates of Tillmann, such as R. Meise (see [101], [102]), H. J. Petzsche (see [110], [111]), and D. Vogt (see [152]), have extended Theorem 2 and the related results. Several authors have continued the investigation of representing distributions and generalized functions as boundary values of analytic functions and associated analysis, such as recovery of the analytic functions from integrals of the boundary values. We mention the books of E. J. Beltrami and M. R. Wohlers [4], H. J. Bremermann [12], and B. W. Ross [126]. V. S. Vladimirov has extended much of the boundary value analysis to functions analytic in tube domains in n dimensional complex space and has discussed applications of this analysis to mathematical physics; we refer in particular to the two important books [147] and [151].

A more recent survey of distributional boundary value analysis and applications has been given in the book of R. D. Carmichael and D. Mitrović [30]. Here the analytic representation of distributions in $\mathcal{E}'$ and $\mathcal{O}'_\alpha$ in one dimension is given. Distributional Plemelj relations and representations of half plane analytic and meromorphic functions are obtained. The distributional boundary value results are applied to yield applications to boundary value problems and singular convolution equations. Analytic functions in tube domains in n dimensional complex space, as considered by Vladimirov, are studied with results being used to obtain the analytic representation of $\mathcal{E}'$ distributions, both in the scalar and vector valued case, $\mathcal{O}'_\alpha$ distributions, and $\mathcal{D}'_{L^p}$ distributions in terms of functions analytic in tube domains. Important in this analysis is the construction and properties of the Cauchy and Poisson kernel functions corresponding to tubes. Motivated by analysis of J. Sebastiõ e Silva (see [133], [134]) and H.-G. Tillmann (see [141]), the Cauchy integral of $\mathcal{S}'$ distributions is constructed and analyzed in n dimensions. The Hardy H^p functions in tubes are characterized in terms of the form of the boundary value as a subspace of those analytic functions which have $\mathcal{S}'$ boundary values.

A principal motivation of the present book is to present research concerning the representation of ultradistributions as boundary values of analytic functions and related topics that will aid in this study as the books [4], [12], [126], [147], [151], and [30] have done for distributions. In particular, the book [30] can be considered to be a companion book to the present one with the analysis there being for distributions as opposed to ultradistributions here.

Contents

Chapter 1

Cones in $\mathbb{R}^n$ and Kernels

1.1 Notation

We present the n-dimensional notation which will be used throughout. For the origin in $\mathbb{R}^n$, the n-dimensional Euclidean space, we use the standard symbol 0 and it follows easily from the context if 0 denotes the number or the vector. Thus $0 = (0, \ldots, 0) \in \mathbb{R}^n$. The operations on vectors in $\mathbb{R}^n$ (in particular, in $\mathbb{N}^n$ and $\mathbb{N}_0^n$) and inequalities between them are meant coordinatewise which, in particular, simplifies summation symbols involving indices $\alpha = (\alpha_1, \ldots, \alpha_n)$ and $\beta = (\beta_1, \ldots, \beta_n)$ in $\mathbb{N}_0^n$:

$$\sum_{0 \le \beta \le \alpha} a_\beta := \sum_{\beta_1=1}^{\alpha_1} \ldots \sum_{\beta_n=1}^{\alpha_n} a_{\beta_1,\ldots,\beta_n}.$$

Let $\alpha = (\alpha_1, \ldots, \alpha_n)$ be an n-tuple of arbitrary reals (in particular, arbitrary integers). If $t = (t_1, \ldots, t_n) \in \mathbb{R}^n$, we define $t^\alpha := t_1^{\alpha_1} \ldots t_n^{\alpha_n}$; in particular, $t^\alpha := t^{\alpha_1 + \ldots + \alpha_n}$ for $t \in \mathbb{R}$, whenever the symbols t^{α_j} make sense. The symbol z^α for $z \in \mathbb{C}^n$ is defined analogously. For $\alpha, \beta \in \mathbb{N}_0^n$ with $\alpha \le \beta$ we define $\bar{\alpha} := \alpha_1 + \ldots + \alpha_n$, $\alpha! := \alpha_1! \ldots \alpha_n!$ and

$$\binom{\alpha}{\beta} := \binom{\alpha_1}{\beta_1} \ldots \binom{\alpha_1}{\beta_1}.$$

Given two vectors $t = (t_1, \ldots, t_n)$ and $y = (y_1, \ldots, y_n)$ in $\mathbb{R}^n$ we use the symbol $\langle t, y \rangle$ for their scalar product, i.e.,

$$\langle t, y \rangle := t_1 y_1 + \ldots + t_n y_n.$$

A similar n-dimensional notation will be applied in the complex Euclidean space $\mathbb{C}^n$.

Let α denote an n-tuple of nonnegative integers, i.e., $\alpha := (\alpha_1, \ldots, \alpha_n) \in \mathbb{N}_0^n$. The symbol $D^\alpha = D_t^\alpha$ with $t = (t_1, \ldots, t_n) \in \mathbb{R}^n$

denotes the differential operator given by

$$D^\alpha = D_1^{\alpha_1} \ldots D_n^{\alpha_n} \quad \text{with} \quad D_j^{\alpha_j} := -\frac{1}{2\pi i}\frac{\partial^{\alpha_j}}{\partial t_j^{\alpha_j}} \qquad \text{for} \quad j = 1, \ldots, n.$$

$$(1.1)$$

On the other hand, the symbol $\partial^\alpha = \frac{\partial^\alpha}{\partial t^\alpha}$ with $t \in \mathbb{R}^n$ denotes the partial differential operator defined analogously as in (1.1), but with the constant 1 instead of $-(2\pi i)^{-1}$. We also write $\varphi^{(\alpha)}(t)$ instead of $\frac{\partial^\alpha \varphi(t)}{\partial t^\alpha}$ for functions φ on $\mathbb{R}^n$. A similar convention is applied to the symbols D_z^α, $\frac{\partial^\alpha}{\partial z^\alpha}$ and $\varphi^{(\alpha)}(z)$ for $z \in \mathbb{C}^n$ and functions φ on $\mathbb{C}^n$.

For $z = (z_1, \ldots, z_n) \in \mathbb{C}^n$ and, in particular, for $x = (x_1, \ldots, x_n) \in \mathbb{R}^n$, we denote

$$|z| := \Big(\sum_{j=1}^n |z_j|^2\Big)^{1/2}, \qquad |x| := \Big(\sum_{j=1}^n |x_j|^2\Big)^{1/2},$$

i.e., $|z|$ and $|x|$ denote the Euclidean norms of $z \in \mathbb{C}^n$ and $x \in \mathbb{R}^n$, respectively.

Functions and ultradistributions considered in the book can be treated as real- or complex-valued functions defined on (subsets of) $\mathbb{R}^n$ or $\mathbb{C}^n$. In general, we will try to distinguish a value of a function from the function itself, e.g. the symbols $\varphi(t), f(z) = f(x + iy)$ will mean the values of the functions φ, f at the points $t \in \mathbb{R}^n, z = x + iy \in T^C = \mathbb{R}^n + iC \subset \mathbb{C}^n$, where $C \subset \mathbb{R}^n$; consequently, for a given function $f\colon T^C \to \mathbb{C}$ and a fixed $y \in C$, the symbols $f(\cdot + iy), \|f(\cdot + iy)\|_{L^s}$ will mean the function $g_y\colon \mathbb{R}^n \to \mathbb{C}$ defined as $g_y(x) := f(x + iy)$ for $x \in \mathbb{R}^n$ and $\|g_y\|_{L^s}$, respectively.

Sometimes, however, it will be convenient to use the same symbol $f(x)$ for the function $f = f(\cdot)$ and its value $f(x)$ at the point x. For instance, in case of the Cauchy and Poisson kernels the traditional symbols $K(z - t)$ and $Q(z; t)$ will denote the values of the functions K and Q for concreate z and t, where $z = x + iy \in T^C = \mathbb{R}^n + iC$, for a given cone C in $\mathbb{R}^n$, and $t \in \mathbb{R}^n$, as well as the functions K and Q themselves. Similarly, if $\alpha = (\alpha_1, \ldots, \alpha_n) \in \mathbb{N}_0^n$, then the symbol x^α will mean both the power function which assigns to each $x = (x_1, \ldots, x_n) \in \mathbb{R}^n$ the number $x^\alpha = x_1^{\alpha_1} \cdot \ldots \cdot x_n^{\alpha_n}$ and its value at the given point x. Moreover, for arbitrary $\beta = (\beta_1, \ldots, \beta_n) \in \mathbb{N}_0^n$, we denote by $<x>^\beta$ both the function on $\mathbb{R}^n$ and its value at the point $x = (x_1, \ldots, x_n) \in \mathbb{R}^n$ defined by the formula:

$$<x>^\beta := \prod_{j=1}^n [1 + (x_j)^2]^{\beta_j/2}.$$

$$(1.2)$$

We shall apply the following very convenient notation for exponents with two variables z, $\zeta \in \mathbb{C}^n$:

$$E_z(\zeta) := \exp\left[2\pi i \langle z, \zeta \rangle\right], \qquad z \in \mathbb{C}^n, \quad \zeta \in \mathbb{C}^n. \tag{1.3}$$

We also denote

$$e_y(t) := \exp\left[-2\pi \langle y, t \rangle\right], \qquad y \in \mathbb{R}^n, \quad t \in \mathbb{R}^n, \tag{1.4}$$

i.e., we have $E_{iy} = e_y$ for $y \in \mathbb{R}^n$. In particular, $e_y(t) = \exp(-2\pi yt)$ for $y, t \in \mathbb{R}$. The symbols $\tilde{g}$ and $\tilde{T}$ for a given real- or complex-valued function g and ultradistribution T in $\mathbb{R}^n$ are meant as follows:

$$\tilde{g}(x) := g(-x), \qquad x \in \mathbb{R}^n, \tag{1.5}$$

and

$$\langle \tilde{T}, \varphi \rangle := \langle T, \tilde{\varphi} \rangle \tag{1.6}$$

for every function φ from the respective space of test functions (see Section 2.3), where $\tilde{\varphi}$ is defined in (1.5).

The *Fourier transform* of a real- or complex-valued L^1 function φ, denoted by $\mathcal{F}[\varphi]$ or by $\hat{\varphi}$, is defined by (and used in Chapters 1-5)

$$\mathcal{F}[\varphi](x) = \hat{\varphi}(x) := \int_{\mathbb{R}^n} \varphi(t) e^{2\pi i \langle x, t \rangle} \, dt = \int_{\mathbb{R}^n} \varphi(t) E_x(t) \, dt \tag{1.7}$$

and the *inverse Fourier transform* of an L^1 function φ, denoted by $\mathcal{F}^{-1}[\varphi]$ or by $\check{\varphi}$ is defined by

$$\mathcal{F}^{-1}[\varphi](x) = \check{\varphi}(x) := \int_{\mathbb{R}^n} \varphi(t) e^{-2\pi i \langle x, t \rangle} \, dt = \int_{\mathbb{R}^n} \varphi(t) E_{-x}(t) \, dt. \tag{1.8}$$

In Chapter 7, another version of the Fourier and inverse Fourier transforms, $\mathcal{F}_0$ and $\mathcal{F}_0^{-1}$, will be convenient to be considered instead of $\mathcal{F}$ and $\mathcal{F}^{-1}$ defined in (1.7) and (1.8). This will not cause any misinterpretation, because both versions of the definitions differ from each other by constants, and all results remain true in both cases.

We assume familiarity on the part of the reader with properties of the Fourier transform on L^r, $1 \le r \le 2$, the corresponding inverse Fourier transform, and the associated Plancherel theory for the Fourier transform.

The *inclusion* of two sets will be denoted by means of the symbol $A \subseteq B$ and the *proper inclusion* by the symbol $A \subset B$. The *closure* of the set $A \subseteq \mathbb{R}^n$ (in the sense of the Euclidean topology) will be denoted by $\bar{A}$.

By the *support* of $1°$ a given function in $\mathbb{R}^n$; $2°$ a given ultradistribution T in $\mathbb{R}^n$, denoted by $1°$ supp g, $2°$ supp T we will mean $1°$ the set

$$\text{supp} \, g := \overline{\{t \in \mathbb{R}^n : \ g(t) \ne 0\}}; \tag{1.9}$$

$2°$ the smallest closed set $A \subseteq \mathbb{R}^n$ such that $\langle T, \varphi \rangle = 0$ for all functions φ from the respective space of test functions such that supp $\varphi \subseteq A^c$.

1.2   Cones in $\mathbb{R}^n$

We introduce the definitions and notation associated with cones in $\mathbb{R}^n$ and tubes in $\mathbb{C}^n$ (cf. [147], [151]).

A set $C \subseteq \mathbb{R}^n$ is a *cone* (with vertex at zero) if $y \in C$ implies $\lambda y \in C$ for all positive reals λ. The intersection of the cone C with the unit sphere $\{y \in \mathbb{R}^n : |y| = 1\}$ is called the *projection* of C and is denoted by $\operatorname{pr}(C)$. If C_1 and C_2 are cones such that $\operatorname{pr}(\overline{C_1}) \subset \operatorname{pr}(C_2)$, the cone C_1 will be called a *compact subcone* of C_2 and we will write then $C_1 \subset\subset C_2$. An open convex cone C such that $\overline{C}$ does not contain any entire straight line will be called a *regular cone*. The set

$$C^* := \{t \in \mathbb{R}^n : \langle t, y \rangle \geq 0 \quad \text{for all } y \in C\}$$

is the *dual cone* of the cone C. A cone is called *self dual* if $C^* = \overline{C}$. For every cone C, the dual cone C^* is closed and convex. We have $C^* = \overline{C}^* = (O(C))^*$ and $C^{**} = \overline{O(C)}$, where $O(C)$ denotes the convex hull of C. The function u_C defined by

$$u_C(t) := \sup_{y \in \operatorname{pr}(C)} (-\langle t, y \rangle), \qquad t \in \mathbb{R}^n$$

is said to be the *indicatrix* of the cone C.

We have $C^* = \{t \in \mathbb{R}^n : u_C(t) \leq 0\}$. Moreover, $u_C(t) \leq u_{O(C)}(t)$ for all $t \in \mathbb{R}^n$ and $u_C(t) = u_{O(C)}(t)$ for $t \in C^*$.

Given a cone C, put $C_* := \mathbb{R}^n \setminus C^*$. The number

$$\rho_C := \sup_{t \in C_*} u_{O(C)}(t)/u_C(t)$$

characterizes the convexity of C. Notice that a cone C is convex if and only if $\rho_C = 1$. Further, if a cone is open and consists of a finite number of components, then $\rho_C < +\infty$.

We give some examples of cones and their dual cones. If $C = (0, \infty)$, then $C^* = [0, \infty)$, $u_C(t) = -t$ and $\rho_C = 1$. The case $C = (-\infty, 0)$ is analogous. If $C = \mathbb{R}^n$, then $C^* = \{0\}$, $u_C(t) = |t|$ and $\rho_C = 1$.

Let Θ be the set of all n-tuples whose entries are -1 or 1, i.e.,

$$\Theta := \{u = (u_1, \ldots, u_n) \in \mathbb{R}^n : u_j \in \{-1, 1\} \text{ for } j = 1, \ldots, n\}. \quad (1.10)$$

Fix $u = (u_1, \ldots, u_n) \in \Theta$, an arbitrary element of the 2^n elements of Θ. Then

$$C_u := \{y \in \mathbb{R}^n : u_j y_j > 0 \text{ for } j = 1, \ldots, n\} \quad (1.11)$$

is a self dual cone in $\mathbb{R}^n$ for every $u \in \Theta$. Each of the 2^n sets C_u with $u \in \Theta$ will be called an n-*rant* in $\mathbb{R}^n$.

Each n-rant C_u in $\mathbb{R}^n$ ($u \in \Theta$) defined in (1.11) is an example of a regular cone. The forward and backward light cones, defined by

$$\Gamma^+ := \{y \in \mathbb{R}^n: \ y_1 > (y_2^2 + \ldots + y_n^2)^{1/2}\},$$
$$\Gamma^- := \{y \in \mathbb{R}^n: \ y_1 < -(y_2^2 + \ldots + y_n^2)^{1/2}\},$$

respectively, are important self dual cones in mathematical physics.

For an arbitrary cone C in $\mathbb{R}^n$ the set

$$T^C := \mathbb{R}^n + iC = \{z = x + iy: \ x \in \mathbb{R}^n, \ y \in C\}$$

will be called a *tube* in $\mathbb{C}^n$. The set $\{z = x + iy: \ x \in \mathbb{R}^n, \ y = 0\}$ is called the *distinguished boundary* of the tube T^C, while $\mathbb{R}^n + i\partial C$, with ∂C denoting the boundary of C, is the *topological boundary* of T^C.

We now present two important lemmas concerning cones and dual cones which will be of particular use in the construction and analysis of the Cauchy and Poisson kernel functions below. The lemmas are proved in [147], Section 25; we give here a separate proof of the second lemma.

Lemma 1.2.1. *Let C be an open connected cone in $\mathbb{R}^n$. The closure $\overline{O(C)}$ of $O(C)$ contains an entire straight line if and only if the dual cone C^* lies in some $(n-1)$-dimensional plane.*

Lemma 1.2.2. *Let C be an open (not necessarily connected) cone in $\mathbb{R}^n$. For every $y \in O(C)$ there exists a positive δ (depending on y) such that*

$$\langle y, t \rangle \geq \delta \, |y| \, |t|, \qquad t \in C^*. \tag{1.12}$$

Further, if C' is an arbitrary compact subcone of $O(C)$, then there exists a $\delta > 0$ (depending only on C' and not on $y \in C'$) such that (1.12) holds for all $y \in C'$ and all $t \in C^*$.

Proof. Since $u_C(t) = u_{O(C)}(t)$ for $t \in C^*$, we have $\langle y, t \rangle \geq 0$ for all $y \in O(C)$ and all $t \in C^*$. For an arbitrary $y \in O(C)$, we have

$$\tilde{y} := y/|y| \in \text{pr}(O(C)) \subset O(C),$$

since $O(C)$ is a cone. Moreover, $O(C)$ is open, because C is open. Thus there exists a $\delta = \delta_y > 0$ such that

$$N(\tilde{y}, 2\delta) := \{y': \ |y' - \tilde{y}| < 2\delta\} \subset O(C).$$

Hence

$$\tilde{y} - (t/|t|)\,\delta \in N(\tilde{y}, 2\delta) \subset O(C)$$

and thus
$$\langle \tilde{y} - (t/|t|)\, \delta, t\rangle \geq 0$$
for every $t \in C^*$, but this implies (1.12). Now, let C' be an arbitrary compact subcone of $O(C)$. Let d be the distance from $\mathrm{pr}(C')$ to the complement of $O(C)$ in $\mathbb{R}^n$, that is,
$$d := \inf\{|y_1 - y_2|: \ y_1 \in \mathrm{pr}(C'), \ y_2 \notin O(C)\}.$$
Obviously, d is positive and depends only on C' and not on $y \in C'$. Define now $\delta = d/2$. The preceding considerations show that (1.12) holds for all $y \in C'$ and $t \in C^*$. The proof is complete. $\square$

For C being an open connected cone in $\mathbb{R}^n$, we denote the distance from $y \in C$ to the topological boundary ∂C of C by
$$d(y) := \inf\{|y - y_1|: \ y_1 \in \partial C\}.$$
It has been shown in [151], p. 159, that
$$d(y) = \inf_{t \in \mathrm{pr}(C^*)} \langle t, y\rangle, \qquad y \in C. \tag{1.13}$$
Let C' be an arbitrary compact subcone of C. It follows from Lemma 1.2.2 and (1.13) that there exists a $\delta = \delta(C') > 0$, depending only on C' and not on $y \in C'$, such that
$$0 < \delta|y| \leq d(y) \leq |y|, \qquad y \in C' \subset\subset C. \tag{1.14}$$
Let C be an open connected cone in $\mathbb{R}^n$. We make the following convention concerning the notation $y \to 0$, $y \in C$, which normally means that y varies arbitrarily within C while $y \to 0$. But frequently the above symbol will mean that $y \to 0, y \in C'$ for every compact subcone C' of C. We shall distinguish between these two convergences only when necessary; in most relevant situations the analysis clearly shows which of the interpretations of the symbol $y \to 0, y \in C$, is used in a given case.

Let V be an ultradistribution and let f be a function of the variable $z = x + iy \in T^C$ for a given cone C. By $f(\cdot + iy) \to V$ in the weak topology of the ultradistribution space as $y \to 0, y \in C$, we mean the convergence
$$\langle f(\cdot + iy), \varphi\rangle \to \langle V, \varphi\rangle$$
as $y \to 0, y \in C$, for each fixed element φ in the corresponding test function space. By $f(\cdot + iy) \to V$ in the strong topology of the ultradistribution space as $y \to 0, y \in C$, we mean
$$\langle f(\cdot + iy), \varphi\rangle \to \langle V, \varphi\rangle$$
as $y \to 0, \ y \in C$, where the convergence is uniform for an arbitrary bounded set of functions φ in the corresponding test function space. Then V is called the weak or strong, respectively, ultradistributional boundary value of f and is defined on the distinguished boundary of the tube T^C.

1.3 Cauchy and Poisson kernels

Let C be a regular cone in $\mathbb{R}^n$, that is C is an open convex cone such that $\overline{C}$ does not contain any entire straight line. The *Cauchy kernel* corresponding to the tube T^C, denoted traditionally by $K(z - t)$, is defined as a function of variables $z = x + iy \in T^C$ and $t \in \mathbb{R}^n$ by the formula

$$K(z - t) := \int_{C^*} E_{z-t}(u)\, du = \int_{C^*} \exp\left[2\pi i \langle z - t, u \rangle\right] du, \quad z \in T^C, \ t \in \mathbb{R}^n.$$

$$(1.15)$$

Note that the Cauchy kernel $K(z - t)$ is well defined for $z = x + iy \in T^C$ and $t \in \mathbb{R}^n$, because $\langle y, u \rangle \geq 0$ for $y \in C$ and $u \in C^*$ and

$$|E_{z-t}(u)| = |\exp\left[2\pi i \langle x - t, u \rangle\right] \exp\left[-2\pi \langle y, u \rangle\right]| = \exp\left[-2\pi \langle y, u \rangle\right].$$

Moreover, denoting by I_{C^*} the characteristic function of C^* and using the definition (1.8) of the inverse Fourier transform $\mathcal{F}^{-1}$, we can write formula (1.15) in the form:

$$K(z - t) := \int_{C^*} \exp\left[-2\pi i t\right] E_z(u)\, du = \mathcal{F}^{-1}[I_{C^*} E_z](t), \quad z \in T^C, \ t \in \mathbb{R}^n.$$

$$(1.16)$$

In case $C = C_u$ is any of the 2^n n-rants in $\mathbb{R}^n$, the Cauchy kernel $K(z - t) = K_u(z - t)$ takes the classical form

$$K(z - t) := \frac{(-1)^u}{(2\pi i)^n} \prod_{j=1}^{n} (t_j - z_j)^{-1}, \quad z \in \mathbb{R}^n + iC_u, \ t \in \mathbb{R}^n,$$

since $C_u^* = \overline{C}_u$ in this case.

The *Poisson kernel* Q corresponding to the tube T^C is the function of variables $z \in T^C$ and $t \in \mathbb{R}^n$ given by

$$Q(z; t) := \frac{K(z - t)\overline{K(z - t)}}{K(2iy)}, \quad z = x + iy \in T^C, \ t \in \mathbb{R}^n. \quad (1.17)$$

In case $C = C_u$ is any of the n-rants, the Poisson kernel $Q(z; t) = Q_u(z; t)$ reduces to the classical form

$$Q(z; t) = \frac{(-1)^u}{\pi^n} \prod_{j=1}^{n} \frac{y_j}{(t_j - x_j)^2 + y_j^2}, \quad z = x + iy \in \mathbb{R}^n + iC_u, \ t \in \mathbb{R}^n.$$

If the cone C above had been assumed to be open and connected but not necessarily convex, we would have defined the kernels $K(z - t)$ and $Q(z; t)$ for $z \in T^{O(C)}$ and would obtain all the properties concerning the kernels for $z \in T^{O(C)}$. Thus we have assumed that C is convex without loss of generality. From Lemma 1.2.1, the dual cone C^* will lie in an $(n-1)$-dimensional plane if $\overline{C}$ contains an entire straight line. In this case the Lebesgue measure of C^* would be zero. Hence the Cauchy kernel $K(z-t)$ would be zero and the Poisson kernel $Q(z;t)$ would be undefined. To avoid this situation we have to assume that $\overline{C}$ does not contain any entire straight line. Therefore we consider regular cones unless explicitly stated otherwise.

We conclude this section with several technical lemmas which will be used in our analysis concerning the Cauchy and Poisson kernels.

Lemma 1.3.1. *Let C be an open connected cone in $\mathbb{R}^n$.*

I. Fix arbitrarily $z \in T^C = \mathbb{R}^n + iC$ and denote by I_{C^*} the characteristic function of C^*. Then $E_z I_{C^*} \in L^p$ for all p, $1 \le p \le \infty$.

II. Assume that g is a continuous function on $\mathbb{R}^n$ with support in C^* such that, for arbitrary $m > 0$ and compact subcone C' of C,

$$|g(t)| \le M(C',m)\exp[2\pi(\langle w,t\rangle + \sigma|w|)], \qquad t \in \mathbb{R}^n, \tag{1.18}$$

whenever $\sigma > 0$ and $w \in C' \setminus (C' \cap \overline{N}(0,m)))$, where $\overline{N}(0,m)$ is the closure of the ball with center at 0 and radius m and $M(C',m)$ is a constant. Then, for an arbitrary y in C, $y \ne 0$, we have $e_y g \in L^p$, whenever $1 \le p < \infty$.

Proof. To prove part I fix z in T^C and let $y = \operatorname{Im} z$. Applying Lemma 1.2.2, we find a $\delta = \delta_y > 0$ such that

$$|E_z(t)|\, I_{C^*}(t) = e_y(t)\, I_{C^*}(t) \le e_{\delta|y|}(|t|)\, I_{C^*}(t) \le 1 \tag{1.19}$$

for all $z = x + iy \in T^C$ and all $t \in \mathbb{R}^n$, since $I_{C^*}(t) = 0$ for $t \notin C^*$. Part I of the lemma for $p = \infty$ follows from (1.19). For $1 \le p < \infty$, we use (1.19) and integration by parts $n-1$ times (or the gamma function after the change of variable for $v = 2\pi\delta p|y|r$) to get

$$\int_{\mathbb{R}^n} |E_z(t)I_{C^*}(t)|^p\, dt \le \int_{\mathbb{R}^n} e_{p\delta|y|}(|t|)\, dt$$

$$= \Omega_n \int_0^\infty r^{n-1} e_{p\delta|y|}(r)\, dr = (n-1)!\,\Omega_n (2\pi\delta p|y|)^{-n}, \tag{1.20}$$

where Ω_n is the surface area of the unit sphere in $\mathbb{R}^n$. The estimate in (1.20) proves part I of the lemma for $1 \le p < \infty$.

To prove part II fix a point y in C. Since C is open, there exists a compact subcone C' of C and a $m > 0$ such that $y \in C' \setminus (C' \cap \overline{N}(0,m))$. Since $y \notin \overline{N}(0,m))$, we have $|y| > m$. Choose $w := \lambda y$, where λ is an arbitrary number such that $m/|y| < \lambda < 1$. Since C' is a cone, $y \in C'$ and $\lambda|y| > m$, we have $w = \lambda y \in C' \setminus (C' \cap \overline{N}(0, m))$, i.e., the estimate given by (1.18) is true for w just chosen. Since $C' \subset\subset C$, it follows from Lemma 1.2.2 that there is a $\delta = \delta(C') > 0$, not depending on $y \in C'$, such that (1.12) holds for all $t \in C^*$. Hence, denoting $A(\sigma, \lambda, y) := M(C', m) \exp[2\pi\sigma\lambda|y|]$, we have

$$|e_y(t)g(t)| \le A(\sigma, \lambda, y)e_{(1-\lambda)y}(t) \le A(\sigma, \lambda, y)e_{(1-\lambda)\delta|y|}(|t|)$$

for $t \in C^*$. Integrating by parts (or using the gamma function) yields

$$\int_{\mathbb{R}^n} |E_{iy}(t)g(t)|^p \, dt \le \int_{C^*} e_{p(1-\lambda)\delta|y|}(|t|) \, dt$$

$$= \Omega_n [A(\sigma, \lambda, y)]^p \int_0^\infty r^{n-1} e_{p(1-\lambda)\delta|y|}(r) \, dr$$

$$= (n-1)! \, \Omega_n [A(\sigma, \lambda, y)]^p [2\pi p(1-\lambda)\delta|y|]^{-n} < \infty,$$

since $\operatorname{supp} g \subseteq C^*$. This completes the proof of part II and the lemma. $\square$

Lemma 1.3.2. *Let C be a regular cone. The Cauchy kernel $K(z - t)$ is an analytic function of the variable $z \in T^C$ for each fixed $t \in \mathbb{R}^n$.*

Proof. Let I_{C^*} denote the characteristic function of C^*. By the proof of Lemma 1.3.1, $I_{C^*} E_{z-t} \in L^1$ for fixed $z \in T^C$ and $t \in \mathbb{R}^n$. Let K be an arbitrary compact subset of T^C and let $z \in K \subset T^C$. There exists a compact subcone C' of C such that $y = \operatorname{Im} z \in C'$ and y has positive distance (say k) from 0. By Lemma 1.2.2, there is a $\delta = \delta(C') > 0$, depending only on C', such that

$$|I_{C^*}(u)E_{z-t}(u)| = I_{C^*}(u) \exp\left[-2\pi\langle y, u \rangle\right] \le I_{C^*}(u) \exp\left[-2\pi\delta k|u|\right] \quad (1.21)$$

for $t \in \mathbb{R}^n$ and $u \in C^*$. The right side of (1.21) is an L^1 function of variable $u \in \mathbb{R}^n$ for arbitrary $z \in K$ and $t \in \mathbb{R}^n$, according to the proof of Lemma 1.3.1, and the function $z \mapsto I_{C^*}(u)E_{z-t}(u)$ is analytic in $z \in T^C$ for each fixed $t \in \mathbb{R}^n$ and $u \in \mathbb{R}^n$. To conclude the assertion it remains to use a well known theorem concerning integrals involving a parameter (see e.g. [15], pp. 295-296). $\square$

Lemma 1.3.3. *Let C be a regular cone and fix $w = u + iv \in T^C$. The function*

$$K(z + w) := \int_{C^*} E_{z+w}(u) \, du, \qquad z \in T^C,$$

is analytic in $z \in T^C$ and

$$|K(z+w)| \le M_v < \infty, \qquad z \in T^C,$$

where M_v is a constant which depends only on $v = \operatorname{Im} w$.

Proof. The proof that $K(z+w)$ is analytic in $z \in T^C$ is the same as in the proof of Lemma 1.3.2. We have $\langle y, u \rangle \ge 0$ for $y \in C$ and $u \in C^*$. By Lemma 1.2.2, there is a $\delta = \delta_v > 0$ such that $\langle v, u \rangle \ge \delta |v| |u|$ for $v \in C$ and $u \in C^*$. The assertion now follows by similar analysis as in (1.20). $\square$

Lemma 1.3.4. *Let $h \in L^p$, $1 \le p \le 2$ and let $g := \mathcal{F}^{-1}[h]$ in the sense of the space L^p. Assume that $gE_z \in L^1$ for $z \in T^C$ and $\operatorname{supp} g \subseteq C^*$ almost everywhere. We have*

$$\int_{C^*} g(u) E_z(u) \, du = \int_{\mathbb{R}^n} h(t) K(z-t) \, dt, \qquad z \in T^C. \qquad (1.22)$$

Proof. Let $z \in T^C$. Let $1 \le p \le 2$ and $1/p + 1/q = 1$. As a result of the remarks below, $K(z-t)$ as a function of $t \in \mathbb{R}^n$ belongs to L^q, i.e. $K(z - \cdot) \in L^q$, for every $z \in T^C$. Therefore the integral on the right side of (1.22) is well defined. First consider $p = 1$. By Lemma 1.3.3, Fubini's theorem and definition (1.8) of the inverse Fourier transform $\mathcal{F}^{-1}[h]$ of h, we have

$$\int_{\mathbb{R}^n} h(t) K(z-t) \, dt = \int_{\mathbb{R}^n} h(t) \, dt \int_{C^*} e^{2\pi i \langle z-t, u \rangle} \, du$$

$$= \int_{C^*} e^{2\pi i \langle z, u \rangle} \, du \int_{\mathbb{R}^n} h(t) E_{-u}(t) \, dt = \int_{C^*} g(u) E_z(u) \, du, \qquad (1.23)$$

which proves (1.22) for $p = 1$. In case $1 < p \le 2$, the function g is the limit in the L^q norm of the sequence (g_k) of the functions

$$g_k(u) := \int_{|t| \le k} h(t) E_{-u}(t) \, dt \qquad (k \in \mathbb{N})$$

and so, by Hölder's inequality,

$$\int_{C^*} |gE_z - g_k E_z| \, du \le \|g - g_k\|_{L^q} \|e_y\|_{L^p} \to 0$$

as $k \to \infty$ for every $z \in T^C$ (i.e. $y \in C$). Consequently, applying Fubini's theorem, we conclude from (1.23) that

$$\int_{C^*} g(u) E_z(u) \, du = \lim_{k \to \infty} \int_{C^*} g_k(u) E_z(u) \, du$$

$$= \lim_{k \to \infty} \int_{|t| \le k} h(t) \, dt \int_{C^*} E_{z-t}(u) \, du = \int_{\mathbb{R}^n} h(t) K(z-t) \, dt$$

for $z \in T^C$, which shows (1.22) in the cases $1 < p \leq 2$ and Lemma 1.3.4 is thus proved. $\square$

The Poisson kernel defined in (1.17) has been known for some time to be an approximate identity. We state this in the following lemma (see [137], p. 105):

Lemma 1.3.5. *Let C be a regular cone, let $z \in T^C$ and $t \in \mathbb{R}^n$. The Poisson kernel $Q(z;t)$ has the following properties:*

(i) $\quad Q(z;t) \geq 0, \qquad z \in T^C, \;\; t \in \mathbb{R}^n;$

(ii) $\quad \int\limits_{\mathbb{R}^n} Q(z;t)\,dt = 1, \qquad z \in T^C;$

(iii) $\quad \lim\limits_{z \to t_0, z \in T^C} \int\limits_{|t-t_0|>\delta} Q(z;t)\,dt = 0, \qquad \delta > 0$

uniformly for all $t_0 \in \mathbb{R}^n$.

We shall prove later that the Cauchy and Poisson kernels are in certain ultradifferentiable function spaces.

Chapter 2

Ultradifferentiable Functions and Ultradistributions

2.1 Sequences (M_p)

We define subspaces of some of the Schwartz test spaces through the use of sequences of positive real numbers which satisfy certain conditions. The corresponding dual spaces then contain the generalized functions of Schwartz.

By $(M_p) = (M_p)_{p \in \mathbb{N}_0}$ we will denote a sequence of positive numbers which satisfies some of the following conditions:

$(M.1)$ $\quad M_p^2 \leq M_{p-1} M_{p+1}, \qquad p \in \mathbb{N};$

$(M.2)$ $\quad$ there are positive constants A and H such that

$$M_p \leq A H^p \min_{0 \leq q \leq p} M_q M_{p-q}, \qquad p \in \mathbb{N}_0;$$

$(M.3)$ $\quad$ there is a constant $A > 0$ such that

$$\sum_{q=p+1}^{\infty} M_{q-1}/M_q \leq A p M_p/M_{p+1}, \qquad p \in \mathbb{N}.$$

Sometimes $(M.2)$ and $(M.3)$ will be replaced by the following weaker conditions:

$(M.2')$ $\quad$ there are constants A and H such that

$$M_{p+1} \leq A H^p M_p, \qquad p \in \mathbb{N}_0;$$

$(M.3')$ $\quad \sum_{p=1}^{\infty} M_{p-1}/M_p < \infty .$

It is worth noting that constants A and H in conditions $(M.2)$, $(M.2')$ and $(M.3)$ can be assumed to be greater than 1 and condition $(M.2)$ has the following equivalent symmetric form:

$(M.2)$ there are positive (greater than 1) constants A and H such that

$$M_{p+q} \leq AH^{p+q} M_p M_q, \qquad p, q \in \mathbb{N}_0. \tag{2.1}$$

In Lemma 2.1.1 below, we will show that condition $(M.1)$ implies

$$M_p M_q \leq M_0 M_{p+q}, \qquad p, q \in \mathbb{N}_0; \tag{2.2}$$

that is, the inequality (2.2) is in the converse order of that in (2.1) (with M_0 assuming the role of the constant A and the constant 1 assuming the role of H in (2.1)). Conversely, the inequality (2.2) together with condition $(M.2)$ imply the following weaker version of condition $(M.1)$:

$(M.1')$ there are positive (greater than 1) constants A and H such that

$$M_p^2 \leq AH^p M_{p-1} M_{p+1}, \qquad p \in \mathbb{N}.$$

Sequences (M_p) satisfying some or all of these properties are the basis for the ultradistributions to be studied here. The paper of Komatsu [82] serves as a basic reference for these sequences. If $s > 1$, the Gevrey sequences (M_p) given by $M_p = (p!)^s$, $M_p = p^{ps}$ and $M_p = \Gamma(1 + ps)$, where Γ denotes the gamma function, are basic examples of sequences satisfying some of the above stated conditions.

We will prove some properties of sequences (M_p). It will be convenient to consider together with a given sequence (M_p) also its multi-dimensional variant M_α with multi-indices $\alpha = (\alpha_1, \ldots, \alpha_n) \in \mathbb{N}_0^n$ defined by

$$M_\alpha := M_{\alpha_1 + \ldots + \alpha_n}, \qquad \alpha = (\alpha_1, \ldots, \alpha_n) \in \mathbb{N}_0^n.$$

Lemma 2.1.1. *Let (M_p) be an arbitrary sequence of positive numbers.*
   (i)   If the sequence (M_p) satisfies condition $(M.1)$, then

$$M_p M_q \leq M_0 M_{p+q}, \qquad p, q \in \mathbb{N}_0. \tag{2.3}$$

   (ii)   If the sequence (M_p) satisfies condition $(M.2)$, then

$$M_p \geq \frac{M_{p+q}}{(AM_1)^q H^{p+1} \ldots H^{p+q}}, \qquad p \in \mathbb{N}_0, \ q \in \mathbb{N}, \tag{2.4}$$

where the positive constants A and H are from $(M.2)$, and

$$M_\alpha \leq BE^\alpha M_{\alpha_1} \ldots M_{\alpha_n} \tag{2.5}$$

for every $\alpha = (\alpha_1, \ldots, \alpha_n) \in \mathbb{N}_0^n$, where B and E are positive constants.

Proof. Applying $(M.1)$ repeatedly, we get

$$\frac{M_p}{M_{p+1}} \leq \frac{M_{p-1}}{M_p} \leq \cdots \leq \frac{M_0}{M_1}.$$

Using this and similar arguments, we have

$$\frac{M_p}{M_{p+q}} = \frac{M_p}{M_{p+1}} \frac{M_{p+1}}{M_{p+2}} \cdots \frac{M_{p+q-1}}{M_{p+q}} \leq \frac{M_0}{M_1} \frac{M_1}{M_2} \cdots \frac{M_{q-1}}{M_q} = \frac{M_0}{M_q},$$

from which (2.3) follows. Inequalities (2.4) and (2.5) follow by repeated applications of $(M.2)$. $\square$

Let (M_p) and (N_p) be sequences of positive numbers which satisfy $(M.1)$. Following Komatsu [82], Definition 3.1, p. 52, we write

$$M_p \subset N_p$$

if there exist constants $L > 0$ and $B > 0$, independent of p, such that

$$M_p \leq BL^p N_p, \qquad p \in \mathbb{N}_0. \tag{2.6}$$

Following [82], Definition 3.9, p. 53, we write

$$M_p \prec N_p \tag{2.7}$$

if for each $L > 0$ there is a constant $B > 0$, independent of p, such that (2.6) holds.

Komatsu has proved in [82], p. 74, that $p! \prec M_p$ for every sequence (M_p) satisfying $(M.1)$ and $(M.3')$. This and Stirling's formula imply $p^p \prec M_p$ (with the convention $0^0 = 1$ and the assumption $M_0 = 1$), as noticed by Pilipović in [113], p. 209. Moreover, $\Gamma(s+p) \prec M_p$, $s > 0$, by Lemma 4.1 in [82], p. 56, and analysis as in [82], p. 74. We summarize these facts in the following lemma.

Lemma 2.1.2. *Let the sequence (M_p) satisfy conditions $(M.1)$ and $(M.3')$. We have $p! \prec M_p$, $p^p \prec M_p$ and $\Gamma(s+p) \prec M_p$ for $s > 0$.*

For a sequence (M_p) the *associated functions* M and M^* of Komatsu, are defined by

$$M(\rho) := \sup_{p \in \mathbb{N}_0} \log_+ (\rho^p M_0/M_p), \qquad 0 < \rho < \infty; \tag{2.8}$$

$$M^*(\rho) := \sup_{p \in \mathbb{N}_0} \log_+ (\rho^p p! M_0/M_p), \qquad 0 < \rho < \infty, \tag{2.9}$$

where $\log_+ \rho := \max (\log \rho, 0)$. Some properties of the associated function M are collected in the following lemma.

Lemma 2.1.3. *If the sequence (M_p) satisfies $(M.1)$, then*

$$M\Big(\sum_{j=1}^{k} \rho_j\Big) \leq \sum_{j=1}^{k} M(k\rho_j), \qquad \rho_1 > 0, \ \ldots, \ \rho_k > 0, \qquad (2.10)$$

for arbitrary $k \in \mathbb{N}$. If the sequence (M_p) satisfies $(M.1)$ and $(M.2)$, then

$$2M(\rho) \leq M(H\rho) + \log_+(AM_0), \qquad \rho > 0, \qquad (2.11)$$

where A and H are the constants in $(M.2)$; if $L \geq 1$, then there is a constant $K > 0$ such that

$$M(L\rho) \leq (3/2)LM(\rho) + K, \qquad \rho > 0; \qquad (2.12)$$

if $L \geq 1$, then there is a constant $B > 0$ and a constant $E_L > 0$ depending on L such that

$$LM(\rho) \leq M(B^{L-1}\rho) + E_L, \qquad \rho > 0. \qquad (2.13)$$

Proof. Inequality (2.10) for $k = 2$ was proved by Petzsche in [111], p. 142 (Lemma 1.10), under the assumption that (M_p) satisfies $(M.1)$. However, we will need inequality (2.10) in the general form for arbitrary $k \in \mathbb{N}$ (see (5.89) in the proof of Lemma 5.2.9). Therefore we give here the proof of the general case. Since $\log_+$ is a nondecreasing function on $(0,\infty)$, we have

$$\log_+ \frac{(\sum_{j=1}^{k} t_j)^p M_0}{M_p} \leq \log_+ \frac{(k\max_{1\leq j\leq k} t_j)^p M_0}{M_p} \leq \sum_{j=1}^{k} \log_+ \frac{(kt_j)^p M_0}{M_p}$$

$$\leq \sum_{j=1}^{k} \sup_{q\in\mathbb{N}_0} \log_+ \frac{(kt_j)^q M_0}{M_q} \leq \sum_{j=1}^{k} M(kt_j)$$

for arbitrary $t_1,\ldots,t_k \in (0,\infty)$ and $p \in \mathbb{N}$. Hence

$$M\Big(\sum_{j=1}^{k} t_j\Big) = \sup_{p\in\mathbb{N}_0} \log_+ \frac{(\sum_{j=1}^{k} t_j)^p M_0}{M_p} \leq \sum_{j=1}^{k} M(kt_j).$$

Inequality (2.11) is shown by Komatsu in [82], p. 51 (Proposition 3.6) and by Petzsche in [111], p. 138 (Lemma 1.4), under conditions $(M.1)$ and $(M.2)$ on the sequence (M_p).

Inequalities (2.12) and (2.13) are proved in [111], Lemma 1.7, p. 140. $\square$

2.2 Ultradifferential operators

We denote by $\mathcal{R}$ the family of all sequences (r_p) of positive numbers which increase (not necessarily strictly) to infinity. This set is partially ordered and directed by the relation $(r_p) \preceq (s_p)$, which means that there exists p_0 such that $r_p \leq s_p$ for every $p > p_0$.

An operator of the form $P(D) = \sum_{\alpha \in \mathbb{N}_0^n} c_\alpha D^\alpha$, $c_\alpha \in \mathbb{C}$, is called an *ultradifferential operator* of class (M_p) (respectively, of class $\{M_p\}$) if there are constants $A > 0$, $h > 0$ (respectively, for every $h > 0$ there is an $A > 0$) such that

$$|c_\alpha| \leq \frac{Ah^\alpha}{M_\alpha}, \qquad \alpha \in \mathbb{N}_0^n. \tag{2.14}$$

Special classes of entire functions will be needed. We recall some facts from [82], [86].

Let $r > 0$ and $m_p := M_p/M_{p-1}$ for $p \in \mathbb{N}$ and let n' be a fixed integer greater than $n/2$. Put

$$P_r(\zeta) := (1 + \zeta_1^2 + \ldots + \zeta_n^2)^{n'} \prod_{j=1}^{\infty} \left(1 + \frac{\zeta_1^2 + \ldots + \zeta_n^2}{r^2 m_j^2} \right), \qquad \zeta \in \mathbb{C}^n, \tag{2.15}$$

where $\zeta = (\zeta_1, \ldots, \zeta_n)$ and n' is a fixed integer greater than $n/2$. If conditions $(M.1)$, $(M.2)$ and $(M.3)$ hold, then $P_r(D)$ is an ultradifferential operator of class (M_p); it maps $\mathcal{D}((M_p), \mathbb{R}^n)$ (cf. the next paragraph) into itself and

$$\mathcal{F}[P_r(D)\phi](\xi) = P_r(\xi)\widehat{\varphi}(\xi), \qquad \xi \in \mathbb{R}^n, \tag{2.16}$$

for $\varphi \in \mathcal{D}((M_p), \mathbb{R}^n)$. Put, for a given (r_p),

$$P_{(r_p)}(\zeta) := (1 + \zeta_1^2 + \ldots + \zeta_n^2)^{n'} \prod_{j=1}^{\infty} \left(1 + \frac{\zeta_1^2 + \ldots + \zeta_n^2}{r_j^2 m_j^2} \right), \qquad \zeta \in \mathbb{C}^n. \tag{2.17}$$

If conditions $(M.1)$, $(M.2)$ and $(M.3)$ are satisfied, the function $P_{(r_p)}$ is of class $\{M_p\}$. For elements of $\mathcal{D}(\{M_p\}, \mathbb{R}^n)$ and the ultradifferential operator $P_{(r_p)}(D)$ equation (2.16) holds as well.

For a given sequence (M_p) and $(r_p) \in \mathcal{R}$ we consider the corresponding sequence (N_p), defined by

$$N_p := M_p \prod_{j=1}^{p} r_j, \qquad p \in \mathbb{N}.$$

If the associated function corresponding to the sequence (M_p), given by (2.8), is denoted by M, then the associated function corresponding to the sequence (N_p) defined above is denoted by N and, according to the definition in (2.8), is given by the formula

$$N(\rho) = \sup\{\log_+(\rho^p N_0/N_p)\colon p \in \mathbb{N}_0\}, \qquad \rho > 0.$$

If an element of $\mathcal{R}$ is denoted by $(\tilde{r}_p)$, the corresponding associated function is denoted by $\tilde{N}$. It follows from the definition that for every $(r_p) \in \mathcal{R}$ and constants $C > 0$ and $c > 0$ there are $(\tilde{r}_p) \in \mathcal{R}$, and $\rho_0 > 0$ such that

$$CN(c\rho) \leq \tilde{N}(\rho), \qquad \rho > \rho_0. \tag{2.18}$$

Assume the conditions $(M.1)$, $(M.2)$ and $(M.3)$ are satisfied. From [82], Proposition 4.5 and p. 91, it follows that there exist constants $D > 0$ and $c > 0$ such that

$$D \exp\left[-N(c|\xi|)\right] \leq |1/P_{(r_p)}(\xi)| \leq \exp\left[-N(\xi)\right], \qquad \xi \in \mathbb{R}^n. \tag{2.19}$$

Using the Cauchy formula

$$\partial^k(1/P_{(r_p)}(\xi)) = \frac{k!}{(2\pi i)^n} \int_{\Gamma_1} \cdots \int_{\Gamma_n} \frac{[1/P_{(r_p)}(\zeta)]\,d\zeta_1 \ldots d\zeta_n}{(\zeta_1 - \xi_1)^{k_1+1} \ldots (\zeta_n - \xi_n)^{k_n+1}},$$

for $k \in \mathbb{N}_0^n$ and $\xi \in \mathbb{R}^n$, where $\Gamma_j := \{\zeta_j\colon |\zeta_j - \xi_j| = d\}$ with $d > 0$ for $j = 1,\ldots,n$, we see that there exists a $C > 0$ such that

$$|\partial^k(1/P_{(r_p)}(\xi))| \leq Ck!d^{-k} \exp\left[-N(\xi)/C\right], \qquad \xi \in \mathbb{R}^n. \tag{2.20}$$

The two-dimensional version of Lemma 3.4 from [86] will be needed.

Lemma 2.2.1. *Let $a_{p,q} > 0$ for $p, q \in \mathbb{N}_0$.*

(i)   There exists a constant $h > 0$ such that

$$\sup\left\{\frac{a_{p,q}}{h^{p+q}}\colon p, q \in \mathbb{N}_0\right\} < \infty \tag{2.21}$$

if and only if

$$\sup\left\{\frac{a_{p,q}}{R_p S_q}\colon p, q \in \mathbb{N}_0\right\} < \infty \tag{2.22}$$

for arbitrary sequences $(r_j), (s_j)$ in $\mathcal{R}$, where

$$R_0 = S_0 = 1, \quad R_p := \prod_{j=1}^{p} r_j, \quad S_q := \prod_{j=1}^{q} s_j, \qquad p, q \in \mathbb{N}. \tag{2.23}$$

(ii)   There exist sequences $(r_j), (s_j) \in \mathcal{R}$ such that

$$\sup\left\{R_p S_q\, a_{p,q}\colon p, q \in \mathbb{N}_0\right\} < \infty, \tag{2.24}$$

where R_p and S_q are given by (2.23), if and only if

$$\sup\left\{h^{p+q}\, a_{p,q}\colon p, q \in \mathbb{N}_0\right\} < \infty \tag{2.25}$$

for every $h > 0$.

Proof. One has only to prove the if parts.

(i) Assume that (2.22) holds for arbitrary $(r_j), (s_j) \in \mathcal{R}$, but (2.21) does not hold for any $h > 0$, i.e. there exists a sequence (p_k, q_k) in $\mathbb{N}_0^2$ such that

$$p_{k+1} + q_{k+1} > p_k + q_k, \qquad h^{-(p_k+q_k)} a_{p_k,q_k} > k \qquad (k \in \mathbb{N}_0).$$

According as the sequences $(p_k), (q_k)$ are bounded or not, the following cases may appear:

(a) there is a $j_0 \in \mathbb{N}_0$ and a subsequence $(\bar{q}_m), \bar{q}_m := q_{k_m}$, of (q_k) such that $(p_{j_0}, \bar{q}_m)$ is a subsequence of (p_k, q_k) and $(\bar{q}_m)$ is strictly increasing;

(b) the symmetric case to the previous one;

(c) there are subsequences $(\bar{p}_m), \bar{p}_m := p_{k_m}$, and $(\bar{q}_m), \bar{q}_m := q_{k_m}$, of (p_k) and (q_k) such that $(\bar{p}_m, \bar{q}_m)$ is a subsequence of the sequence (p_k, q_k) and both $(\bar{p}_m)$ and $(\bar{q}_m)$ are strictly increasing sequences.

Assume (c) and let (h_k) be a sequence strictly increasing to ∞. Define

$$r_j := h_1, 1 \le j \le \bar{p}_1, r_j := (h_m^{\bar{p}_m} h_{m-1}^{-\bar{p}_{m-1}})^{1/(\bar{p}_m - \bar{p}_{m-1})}, \bar{p}_{m-1} < j \le \bar{p}_m;$$

$$s_j := h_1, 1 \le j \le \bar{q}_1, s_j := (h_m^{\bar{q}_m} h_{m-1}^{-\bar{q}_{m-1}})^{1/(\bar{q}_m - \bar{q}_{m-1})}, \bar{q}_{m-1} < j < \bar{q}_m,$$

where $m = 2, 3, \ldots$. The constructed sequences (r_j) and (s_j) increase to ∞, i.e. are elements of $\mathcal{R}$, but they do not satisfy (2.22), since if we put $i := \bar{p}_k, j := \bar{q}_k$ for arbitrary $k \in \mathbb{N}$, we have

$$R_i = R_{\bar{p}_k} = h_1^{\bar{p}_1} \prod_{j=2}^{k} h_j^{\bar{p}_j} h_{j-1}^{-\bar{p}_{j-1}} = h_k^{\bar{p}_k}; \quad S_j = S_{\bar{q}_k} = h_k^{\bar{q}_k},$$

i.e.

$$\frac{a_{i,j}}{R_i S_j} = \frac{a_{\bar{p}_k, \bar{q}_k}}{h_k^{\bar{p}_k + \bar{q}_k}} > k,$$

which contradicts the assumption in case (c). The proof in cases (a) and (b) is similar.

(ii) Let us assume now that condition (2.25) is satisfied for every $h > 0$. Denote $N_k := \{(p, q): p, q \in \mathbb{N}_0, p + q = k\}$ and

$$b_k := \sup\{a_{p,q}: (p, q) \in N_k\}; \qquad C_{h,k} := \sup\{h^k a_{p,q}: (p, q) \in N_k\}; \qquad (2.26)$$

$$\tilde{C}_h := \sup\{h^k b_k: k \in \mathbb{N}_0\}; \qquad C_h := \sup\{h^{p+q} a_{p,q}: p, q \in \mathbb{N}_0\} \qquad (2.27)$$

for $k \in \mathbb{N}_0$ and $h > 0$. Obviously, $C_{1,k} = b_k$ and $C_{h,k} = h^k b_k$. Moreover let

$$H_k := \sup\{h^k C_h^{-1}: h \ge 1\}, \qquad k \in \mathbb{N}_0. \qquad (2.28)$$

By assumption, $C_h < \infty$ for every $h > 0$. According to (2.26) and (2.27),

$$C_{h,k} \leq C_h, \qquad h > 0, \quad k \in \mathbb{N}_0, \tag{2.29}$$

which implies that $b_k = C_{1,k} < \infty$ for $k \in \mathbb{N}_0$. Since

$$h^{p+q} a_{p,q} \leq C_{h,k} = h^k b_k \leq \tilde{C}_h \tag{2.30}$$

for any fixed $p, q \in \mathbb{N}_0$ with $k := p + q$, we see that $C_h \leq \tilde{C}_h$, by definition of C_h in (2.27). As a matter of fact, we have $\tilde{C}_h = C_h$, because

$$\tilde{C}_h = \sup \{C_{h,k} : \ k \in \mathbb{N}_0\} \leq C_h,$$

by definition of $\tilde{C}_h$ in (2.27), equality in (2.30) and inequality (2.29). Hence

$$H_{p+q} a_{p,q} \leq \sup \left\{ \frac{h^k a_{p,q}}{C_h} : \ h \geq 1 \right\} \leq \sup \left\{ \frac{C_{h,k}}{C_h} : \ h \geq 1 \right\} \leq 1 \tag{2.31}$$

for arbitrary $p, q \in \mathbb{N}_0$, where $k := p + q$, due to (2.28), (2.26) and (2.29).

Put $r_j = s_j := H_j / H_{j-1}$ for $j \in \mathbb{N}$. Since

$$\left(\frac{h^j}{C_h} \right)^2 \leq H_{j-1} H_{j+1}$$

for every $h \geq 1$, we get $H_j / H_{j-1} \leq H_{j+1} / H_j$ for $j \in \mathbb{N}$, i.e. the sequences (r_j) and (s_j) are increasing. Moreover, for every $h > 0$, we have

$$\frac{H_j}{h^j} \geq C_{h_0}^{-1} \left(\frac{h_0}{h} \right)^j \to \infty \qquad \text{as } j \to \infty,$$

where h_0 is fixed so that $h_0 > \max(h, 1)$. Hence it follows that the sequences (r_j) and (s_j) are unbounded and so increase to infinity, i.e. $(r_j), (s_j) \in \mathcal{R}$.

Since the sequence (s_j) is increasing, we have

$$R_p S_q\, a_{p,q} = \prod_{j=1}^{p} r_j \prod_{j=1}^{q} s_j\, a_{p,q} \leq \prod_{j=1}^{p+q} r_j\, a_{p,q} = H_{p+q}\, a_{p,q} \leq 1,$$

in view of (2.31), i.e. condition (2.25) holds for the sequences (r_j), (s_j), constructed above, so the assertion is proved. $\square$

2.3 Functions and ultradistributions of Beurling and Roumieu type

In this section, we introduce ultradifferentiable functions and various types of spaces of test functions and ultradistributions meant as elements of the respective dual spaces.

Let (M_p), $p \in \mathbb{N}_0$, be a sequence of positive numbers. We define $\mathcal{D}((M_p), \Omega)$ (respectively, $\mathcal{D}(\{M_p\}, \Omega)$), where Ω is an open set in $\mathbb{R}^n$, to be the set of all complex valued infinitely differentiable functions φ with compact support in Ω such that there exists an $N > 0$ for which

$$\sup_{t \in R^n} |\partial^\alpha \varphi(t)| \le N h^\alpha M_\alpha, \qquad \alpha \in \mathbb{N}_0^n, \tag{2.32}$$

for all $h > 0$ (respectively, for some $h > 0$). Here the positive constants N and h depend only on φ; they do not depend on α. The topologies of $\mathcal{D}((M_p), \Omega)$ and $\mathcal{D}(\{M_p\}, \Omega)$ are given in Komatsu [82], p. 44, which is a good source of information concerning these spaces. Let $\mathcal{D}(h, K)$ denote the space of smooth functions supported by a compact set K for which (2.32) holds and $\mathcal{D}((M_p), K)$ and $\mathcal{D}(\{M_p\}, K)$ denote subspaces of $\mathcal{D}((M_p), \Omega)$ and $\mathcal{D}(\{M_p\}, \Omega)$ consisting of elements supported by K, respectively. Recall that

$$\begin{aligned}
\mathcal{D}^{(M_p)}(\Omega) = \mathcal{D}((M_p), \Omega) &:= \mathop{\text{ind lim}}_{K \subset\subset \Omega} \ \mathop{\text{proj lim}}_{h \to 0} \ \mathcal{D}(h, K) \\
&= \mathop{\text{ind lim}}_{K \subset\subset \Omega} \ \mathcal{D}((M_p), K);
\end{aligned}$$

$$\begin{aligned}
\mathcal{D}^{\{M_p\}}(\Omega) = \mathcal{D}(\{M_p\}, \Omega) &:= \mathop{\text{ind lim}}_{K \subset\subset \Omega} \ \mathop{\text{ind lim}}_{h \to 0} \ \mathcal{D}(h, K) \\
&= \mathop{\text{ind lim}}_{K \subset\subset \Omega} \ \mathcal{D}(\{M_p\}, K).
\end{aligned}$$

The notation $\mathcal{D}_K^{(M_p)} := \mathcal{D}((M_p), K)$ and $\mathcal{D}_K^{\{M_p\}} := \mathcal{D}(\{M_p\}, K)$ is also used. The strong duals of the above spaces, denoted by $\mathcal{D}'^{(M_p)}(\Omega) = \mathcal{D}'((M_p), \Omega)$ and $\mathcal{D}'^{\{M_p\}}(\Omega) = \mathcal{D}'(\{M_p\}, \Omega)$ are called the spaces of Beurling and Roumieu ultradistributions, respectively.

In case $\Omega = \mathbb{R}^n$, we shall use simpler symbols $\mathcal{D}^{(M_p)}$, $\mathcal{D}'^{(M_p)}$ and $\mathcal{D}^{\{M_p\}}$, $\mathcal{D}'^{\{M_p\}}$ instead of $\mathcal{D}^{(M_p)}(\mathbb{R}^n) = \mathcal{D}((M_p), \mathbb{R}^n)$, $\mathcal{D}'^{(M_p)}(\mathbb{R}^n) = \mathcal{D}'((M_p), \mathbb{R}^n)$ and $\mathcal{D}^{\{M_p\}}(\mathbb{R}^n) = \mathcal{D}(\{M_p\}, \mathbb{R}^n)$, $\mathcal{D}'^{\{M_p\}}(\mathbb{R}^n) = \mathcal{D}'(\{M_p\}, \mathbb{R}^n)$, respectively.

The spaces of test functions and ultradistributions which correspond to the spaces $\mathcal{D}_{L^s} = \mathcal{D}_{L^s}((\mathbb{R}^n))$ and $\mathcal{D}'_{L^s} = \mathcal{D}'_{L^s}((\mathbb{R}^n))$ of L. Schwartz (see [131], pp. 199-205) will be basic for our work. The space $\mathcal{D}((M_p), L^s)$ (respectively, $\mathcal{D}(\{M_p\}, L^s)$), $1 \le s \le \infty$, is defined to be the set of all complex valued infinitely differentiable functions φ such that there is a constant $N > 0$ for which

$$\|\partial^\alpha \varphi\|_{L^s} \le N h^\alpha M_\alpha, \qquad \alpha \in \mathbb{N}_0^n \tag{2.33}$$

for all $h > 0$ (respectively, for some $h > 0$). We have

$$\mathcal{D}((M_p), L^s) \subset \mathcal{D}(\{M_p\}, L^s), \qquad 1 \le s \le \infty.$$

Further,

$$\mathcal{D}^{(M_p)} \subset \mathcal{D}((M_p), L^s), \qquad 1 \le s \le \infty,$$

and

$$\mathcal{D}^{\{M_p\}} \subset \mathcal{D}(\{M_p\}, L^s), \qquad 1 \le s \le \infty.$$

A natural topology is defined on $\mathcal{D}((M_p), L^s)$, $1 \le s \le \infty$, as follows. First put

$$\|\varphi\|_{s,h} := \sup_\alpha \frac{\|\partial^\alpha \varphi\|_{L^s}}{h^\alpha M_\alpha}, \qquad h > 0, \tag{2.34}$$

and

$$\mathcal{D}((M_p), h, L^s) := \{\varphi \in C^\infty : \|\varphi\|_{s,h} < \infty\}, \qquad h > 0. \tag{2.35}$$

Now, since $\mathcal{D}((M_p), h_1, L^s) \subset \mathcal{D}((M_p), h_2, L^s)$, whenever $0 < h_1 < h_2$, we may equip the set $\mathcal{D}((M_p), L^s)$ with the projective limit topology by putting

$$\mathcal{D}((M_p), L^s) := \operatorname*{proj\,lim}_{h \to 0} \ \mathcal{D}((M_p), h, L^s). \tag{2.36}$$

A net (φ_λ) of elements of $\mathcal{D}((M_p), L^s)$ converges to $\varphi \in \mathcal{D}((M_p), L^s)$ as $\lambda \to \infty$ in this topology, and we write $\varphi_\lambda \to \varphi$ in $\mathcal{D}((M_p), L^s)$ as $\lambda \to \infty$, if

$$\lim_{\lambda \to \infty} \|\partial^\alpha (\varphi_\lambda - \varphi)\|_{L^s} = 0, \qquad \alpha \in \mathbb{N}_0^n, \tag{2.37}$$

and, in addition, there is a constant $N > 0$, independent of λ, such that

$$\|\partial^\alpha (\varphi_\lambda - \varphi)\|_{L^s} \le N h^\alpha M_\alpha, \qquad \alpha \in \mathbb{N}_0^n, \tag{2.38}$$

for all $h > 0$.

In $\mathcal{D}(\{M_p\}, L^s)$, $1 \le s \le \infty$, we define the inductive limit topology in the following way:

$$\mathcal{D}(\{M_p\}, L^s) := \operatorname*{ind\,lim}_{h \to \infty} \ \mathcal{D}((M_p), h, L^s),$$

where the spaces $\mathcal{D}((M_p), h, L^s)$ are defined in (2.35) with the topology given by the family of norms defined in (2.34). In this topology a net (φ_λ) of elements of $\mathcal{D}(\{M_p\}, L^s)$ converges to $\varphi \in \mathcal{D}(\{M_p\}, L^s)$ as $\lambda \to \infty$, and we write $\varphi_\lambda \to \varphi$ in $\mathcal{D}(\{M_p\}, L^s)$ as $\lambda \to \infty$, if (2.37) holds and, in addition, there are constants $N > 0$ and $h > 0$, independent of λ and α, such that (2.38) holds.

Throughout we assume that the sequence (M_p) will satisfy at least conditions $(M.1)$ and $(M.3')$ so that $\mathcal{D}((M_p), L^s)$ and $\mathcal{D}(\{M_p\}, L^s)$ contain sufficiently many functions (see Komatsu [82], p. 26).

Note that the spaces $\mathcal{D}((M_p), L^s)$ defined above in (2.36) and the spaces $\mathcal{D}_{L^s}^{(M_p)}(\mathbb{R}^n)$, defined by Pilipović in [112], §3, coincide for $1 \leq s \leq \infty$; it is easy to verify that the norms in (2.34) are equivalent to the norms $\gamma_{s,h}(\varphi)$ in the sense of [112], §3. Various important properties of the spaces $\mathcal{D}((M_p), L^s)$ are proved in [112], §3; among them the fact that $\mathcal{D}^{(M_p)}$ is dense in $\mathcal{D}((M_p), L^s)$, whenever $1 \leq s \leq \infty$. An additional function space is $\dot{\mathcal{B}}(*, \mathbb{R}^n)$, corresponding to the Schwartz space $\dot{\mathcal{B}}$. The space $\dot{\mathcal{B}}(*, \mathbb{R}^n)$ is defined to be the completion of $\mathcal{D}(*, \mathbb{R}^n)$ in $\mathcal{D}(*, L^\infty)$. We denote by $\mathcal{D}'((M_p), L^s)$ and $\mathcal{D}'(\{M_p\}, L^s)$ the spaces of continuous linear forms on $\mathcal{D}((M_p), L^s)$ and $\mathcal{D}(\{M_p\}, L^s)$, respectively. Following several authors, we call $\mathcal{D}'((M_p), L^s)$ (respectively, $\mathcal{D}'(\{M_p\}, L^s)$) the space of *ultradistributions* of class (M_p) or of *Beurling type* (respectively, of class $\{M_p\}$ or of *Roumieu type*). Following Komatsu (see [82], pp. 47 and 61), we use the notation $\mathcal{D}(*, L^s)$ and $\mathcal{D}'(*, L^s)$, where $*$ is the common notation for the symbols (M_p) and $\{M_p\}$

We now present characterization results for $\mathcal{D}'(*, L^s)$. We prove the result for $\mathcal{D}'(\{M_p\}, L^s)$ here. The proof for $\mathcal{D}'((M_p), L^s)$ is similar and can be found in Pilipović [112] (Theorem 5).

Theorem 2.3.1. *Let* $1 \leq s < \infty$. *Let* $\{g_\alpha\}_{\alpha \in \mathbb{N}_0^n}$ *be a sequence of functions in* L^r, $1/r + 1/s = 1$, *such that for all* $k > 0$

$$\|g_\alpha\|_{L^r} = \mathcal{O}\left(\frac{1}{k^\alpha M_\alpha}\right) \qquad as \quad |\alpha| \to \infty. \tag{2.39}$$

Then

$$V = \sum_{\alpha \in \mathbb{N}_0^n} \partial^\alpha g_\alpha \tag{2.40}$$

is an element of $\mathcal{D}'(\{M_p\}, L^s)$. *Conversely, if* $V \in \mathcal{D}'(\{M_p\}, L^s)$, *then* V *has the form (2.40) where* $\{g_\alpha\}_{\alpha \in \mathbb{N}_0^n}$ *is a sequence of functions in* L^r *satisfying (2.39) for all* $k > 0$.

Proof. Let $\varphi \in \mathcal{D}(\{M_p\}, L^s)$ and let V be given by (2.40) with $\{g_\alpha\}$ satisfying (2.39).

First we shall show that the series

$$\sum_{\alpha \in \mathbb{N}_0^n} \int_{R^n} g_\alpha(t) \partial^\alpha \varphi(t)\, dt \tag{2.41}$$

converges absolutely and $\langle V, \varphi \rangle$ is a well defined complex number. Indeed, by the definition of $\mathcal{D}(\{M_p\}, L^s)$, there exist constants $N > 0$ and $H > 0$

such that

$$\sum_{\alpha \in \mathbb{N}_0^n} \left| \int_{R^n} g_\alpha(t) \partial^\alpha \varphi(t) \, dt \right| \leq \sum_{\alpha \in \mathbb{N}_0^n} \|g_\alpha\|_{L^r} \|\partial^\alpha \varphi\|_{L^s}$$

$$\leq \sum_{\alpha \in \mathbb{N}_0^n} N H^\alpha M_\alpha \|g_\alpha\|_{L^r}. \tag{2.42}$$

By (2.39), there exist $a > 0$ and $\alpha_0 \in \mathbb{N}_0^n$ such that

$$\|g_\alpha\|_{L^r} \leq a/(k^\alpha M_\alpha), \qquad \alpha \notin A_0,$$

for all $k > 0$, where $A_0 := \{\beta \in \mathbb{N}_0^n : \beta \leq \alpha_0\}$.

Choosing $k := 2H$ we have

$$\sum_{\alpha \notin A_0} H^\alpha M_\alpha \|g_\alpha\|_{L^r} \leq a \sum_{\alpha \notin A_0} (1/2)^\alpha < \infty, \tag{2.43}$$

which proves that the series on the right of (2.42) converges and the series in (2.41) converges absolutely.

Let (φ_j) be a sequence in $\mathcal{D}(\{M_p\}, L^s)$ such that $\varphi_j \to 0$ in $\mathcal{D}(\{M_p\}, L^s)$ as $j \to \infty$. We have

$$|\langle V, \varphi_j \rangle| \leq \sum_{\alpha \in \mathbb{N}_0^n} \|g_\alpha\|_{L^r} \|\partial^\alpha \varphi_j\|_{L^s}. \tag{2.44}$$

From the convergence in $\mathcal{D}(\{M_p\}, L^r)$ it follows that there exist $N > 0$ and $H > 0$, which are independent of α and j, such that

$$\|\partial^\alpha \varphi_j\|_{L^s} \leq N H^\alpha M_\alpha, \qquad \alpha \in \mathbb{N}_0^n.$$

This together with (2.43) and (2.39) shows that the series on the right of (2.44) converges uniformly in j. This in turn yields $|\langle V, \varphi_j \rangle| \to 0$ as $j \to \infty$, since $\varphi_j \to 0$ in $\mathcal{D}(\{M_p\}, L^s)$ as $j \to \infty$. This proves that V is continuous on $\mathcal{D}(\{M_p\}, L^s)$. The linearity of V on $\mathcal{D}(\{M_p\}, L^s)$ is obvious. Consequently, $V \in \mathcal{D}'(\{M_p\}, L^s)$.

We now prove the converse. In the Roumieu case we put $\mathcal{F} := L^s$ and consider the space $E(L^s, \{M_p\})$ of Roumieu (see [127], p. 43). Put

$$\Phi(\{M_p\}, L^s) := \{\{(-1)^\alpha \partial^\alpha \varphi\}_{\alpha \in \mathbb{N}_0^n} : \varphi \in \mathcal{D}(\{M_p\}, L^s)\},$$

where $1 \leq s < \infty$. From the defining properties of $\mathcal{D}(\{M_p\}, L^s)$ we conclude

$$\Phi(\{M_p\}, L^s) \subset E(L^s, \{M_p\}), \qquad 1 \leq s < \infty,$$

and the topology of the subspace $\Phi(\{M_p\}, L^s)$ is induced by the topology of $E(L^s, \{M_p\})$. Let $V \in \mathcal{D}'(\{M_p\}, L^s)$. We define now an element $V_1 \in \Phi'(\{M_p\}, L^s)$, corresponding to V, by

$$\langle V_1, \{(-1)^\alpha \partial^\alpha \varphi\}_{\alpha \in \mathbb{N}_0^n} \rangle := \langle V, \varphi \rangle, \qquad \varphi \in \mathcal{D}(\{M_p\}, L^s). \tag{2.45}$$

By the Hahn - Banach theorem there is an element $V_2 \in E'(L^s, \{M_p\})$ such that $V_1 = V_2$ on $\Phi(\{M_p\}, L^s)$. Thus, by the characterization of $E'(L^s, \{M_p\})$ given in Roumieu (see [127], Proposition 3, p. 45), we can find a sequence $\{g_\alpha\}_{\alpha \in \mathbb{N}_0^n}$ such that $g_\alpha \in L^r$ with $1/r + 1/s = 1$ for $\alpha \in \mathbb{N}_0^n$ and (2.39) holds for all $k > 0$ such that

$$\langle V_1, \{(-1)^\alpha \partial^\alpha \varphi\}_{\alpha \in \mathbb{N}_0^n} \rangle = \sum_{\alpha \in \mathbb{N}_0^n} \langle g_\alpha, (-1)^\alpha \partial^\alpha \varphi \rangle \tag{2.46}$$

for $\varphi \in \mathcal{D}(\{M_p\}, L^s)$. Notice that (2.45) and (2.46) yield (2.40). The proof is thus complete. $\square$

As previously noted, a similar characterization result is true for $\mathcal{D}'((M_p), L^s)$ and we now present it. The proof is similar to that of Theorem 2.3.1 and can be found in Pilipović [112] (Theorem 5).

Theorem 2.3.2. *Let $1 < s < \infty$. Let $\{g_\alpha\}_{\alpha \in \mathbb{N}_0^n}$ be a sequence of functions in L^r with $1/r + 1/s = 1$ such that, for some $k > 0$,*

$$\|g_\alpha\|_{L^r} = \mathcal{O}\left(\frac{1}{k^\alpha M_\alpha}\right) \qquad as \ |\alpha| \to \infty. \tag{2.47}$$

Then

$$V = \sum_{\alpha \in \mathbb{N}_0^n} \partial^\alpha g_\alpha \tag{2.48}$$

is an element of $\mathcal{D}'((M_p), L^s)$. Conversely, if $V \in \mathcal{D}'((M_p), L^s)$, then V has the form (2.48), where $\{g_\alpha\}_{\alpha \in \mathbb{N}_0^n}$ is a sequence of functions in L^r satisfying (2.47) for some $k > 0$.

Condition (2.47) on the sequence $\{g_\alpha\}$ is equivalent to

$$\sup_\alpha (k^\alpha M_\alpha \|g_\alpha\|_{L^r}) < \infty \tag{2.49}$$

for some $k > 0$. The derivatives in (2.40) and (2.48) are to be taken in the usual ultradistribution sense.

Notice that $\mathcal{D}'((M_p), L^s)$ and $\mathcal{D}'(\{M_p\}, L^s)$ are not distribution spaces in the sense of Schwartz but are ultradistribution spaces in the sense of Komatsu and Roumieu. These spaces are generalizations of the Schwartz spaces $\mathcal{D}'_{L^s}$. Theorems 2.3.1 and 2.3.2 show the difference between distributions in $\mathcal{D}'_{L^s}$ which are finite sums of distributional derivatives of L^r functions and ultradistributions in $\mathcal{D}'(*, L^s)$ which are infinite sums of ultradistribution derivatives of L^r functions satisfying conditions (2.39) or (2.47), respectively.

2.4   Fourier transform on $\mathcal{D}(*, L^s)$ and $\mathcal{D}'(*, L^s)$

We now consider the Fourier transform acting on $\mathcal{D}(*, L^s)$ and study the resulting spaces. Using this analysis we are able to define an inverse Fourier transform on the dual spaces of these Fourier transform spaces which will map the dual spaces to $\mathcal{D}'(*, L^s)$. We use these results in some of our ultradistributional boundary value analysis presented later.

Consider the spaces $\mathcal{D}(*, L^r)$, $1 \le r \le 2$, where $*$ is either (M_p) or $\{M_p\}$. Put

$$\mathcal{FD}(*, L^r) := \{\hat{\varphi} : \ \varphi \in \mathcal{D}(*, L^r)\}, \qquad 1 \le r \le 2,$$

where $\hat{\varphi} = \mathcal{F}[\varphi]$ denotes the Fourier transform of the function φ defined by formula (1.7). We have

$$\mathcal{FD}(*, L^r) \subset L^s, \qquad 1/r + 1/s = 1,$$

and the Fourier transform is a one-to-one mapping of $\mathcal{D}(*, L^r)$ onto $\mathcal{FD}(*, L^r)$. To determine a topology on $\mathcal{FD}(*, L^r)$ let $\varphi \in \mathcal{D}(*, L^r)$ and recall from the Fourier transform theory that

$$\mathcal{F}[D^\alpha \varphi] = x^\alpha \hat{\varphi} \in L^s$$

with $1/r + 1/s = 1$ for every n-tuple α of nonnegative integers.

For an arbitrary $\varphi \in \mathcal{D}((M_p), L^r)$ (respectively, $\varphi \in \mathcal{D}(\{M_p\}, L^r)$) and $\psi = \hat{\varphi} \in \mathcal{FD}((M_p), L^r)$, (respectively, $\psi = \hat{\varphi} \in \mathcal{D}(\{M_p\}, L^r)$), we have

$$\sup_\alpha \frac{\|x^\alpha \psi\|_{L^s}}{h^\alpha M_\alpha} = \sup_\alpha \frac{\|\mathcal{F}[D^\alpha \varphi]\|_{L^s}}{h^\alpha M_\alpha} \le \sup_\alpha \frac{\|D^\alpha \varphi\|_{L^r}}{h^\alpha M_\alpha} < \infty \qquad (2.50)$$

for all (respectively, for some) $h > 0$, in view of the Parseval inequality and (2.16).

On the space $\mathcal{FD}((M_p), L^r)$ (respectively, on the space $\mathcal{FD}(\{M_p\}, L^r)$) define the family $\{\tau_h^s\}_{h>0}$ of norms as follows:

$$\tau_h^s(\psi) := \sup_\alpha \frac{\|x^\alpha \psi\|_{L^s}}{h^\alpha M_\alpha} \qquad (2.51)$$

for all $\psi \in \mathcal{FD}((M_p), L^r)$ (respectively, for all $\psi \in \mathcal{FD}(\{M_p\}, L^r)$). We endow the space $\mathcal{FD}((M_p), L^r)$ (respectively, the space $\mathcal{FD}(\{M_p\}, L^r)$) with the projective (respectively, inductive) limit topology with respect to this family of norms.

A net (ψ_λ) of functions in $\mathcal{FD}((M_p), L^r)$ (respectively, in $\mathcal{FD}(\{M_p\}, L^r)$) converges to zero as $\lambda \to \infty$ in this topology in $\mathcal{FD}((M_p), L^r)$ (respectively, in $\mathcal{FD}(\{M_p\}, L^r)$) if

$$\lim_{\lambda \to \infty} \|x^\alpha \psi_\lambda\|_{L^s} = 0$$

for all n-tuples α of nonnegative integers and for every $h > 0$ there is a constant $N > 0$ which is independent of α and λ (respectively, there are constants $N > 0$ and $h > 0$ which are independent of α and λ) such that

$$\sup_\alpha \frac{\|x^\alpha \psi_\lambda\|_{L^s}}{h^\alpha M_\alpha} \leq N$$

for all $h > 0$ (respectively, for the given $h > 0$).

Using this meaning of convergence, we have the following lemma.

Lemma 2.4.1. *The Fourier transform is an isomorphism from* $\mathcal{D}(*, L^r)$ *onto* $\mathcal{F}\mathcal{D}(*, L^r)$ *for* $1 \leq r \leq 2$.

Proof. We have previously noted that the Fourier transform is a one-to-one mapping of $\mathcal{D}(*, L^r)$ onto $\mathcal{F}\mathcal{D}(*, L^r)$, $1 \leq r \leq 2$. Now let (φ_λ) be a net in $\mathcal{D}(*, L^r)$ which converges to zero in $\mathcal{D}(*, L^r)$ as $\lambda \to \infty$. Since

$$\mathcal{F}[D^\alpha \varphi] = x^\alpha \hat{\varphi} \in L^s,$$

with $1/r + 1/s = 1$, for every $\varphi \in \mathcal{D}(*, L^r)$ and $\alpha \in \mathbb{N}_0^n$, we conclude from the Parseval inequality, (2.50), and the definition of convergence in $\mathcal{D}(*, L^r)$, given in Section 2.3, that $\psi_\lambda = \hat{\varphi}_\lambda$ converges to zero in $\mathcal{F}\mathcal{D}(*, L^r)$ as $\lambda \to \infty$. Consequently, the Fourier transform is a continuous mapping from $\mathcal{D}(*, L^r)$ to $\mathcal{F}\mathcal{D}(*, L^r)$. The proof of Lemma 2.4.1 is complete. $\square$

Let $\mathcal{F}'\mathcal{D}(*, L^r)$ for $1 \leq r \leq 2$ denote the space of all continuous linear forms on $\mathcal{F}\mathcal{D}(*, L^r)$. We now define the inverse Fourier transform $\mathcal{F}^{-1}$ on the space $\mathcal{F}'\mathcal{D}(*, L^r)$ in case $1 \leq r \leq 2$.

For $V \in \mathcal{F}'\mathcal{D}(*, L^r)$, we define $\mathcal{F}^{-1}[V] =: \check{V}$ by the Parseval formula:

$$\langle \mathcal{F}^{-1}[V], \varphi \rangle = \langle \check{V}, \varphi \rangle := \langle V, \tilde{\hat{\varphi}} \rangle, \qquad \varphi \in \mathcal{D}(*, L^r), \qquad (2.52)$$

where $\hat{\phi} = \mathcal{F}[\phi]$ denotes the Fourier transform of ϕ defined by formula (1.7) and $\tilde{\phi}(x) := \phi(-x)$ for $x \in \mathbb{R}^n$ according to the notation introduced in (1.5). If we denote $v := \mathcal{F}^{-1}[V]$ and $\psi := \hat{\varphi}$, the definition (2.52) of $\mathcal{F}^{-1}[V]$ can be written in the form:

$$\langle v, \varphi \rangle := \langle V, \tilde{\psi} \rangle, \qquad \varphi \in \mathcal{D}(*, L^r), \ \psi = \hat{\varphi} \in \mathcal{F}\mathcal{D}(*, L^r).$$

For $V \in \mathcal{F}'\mathcal{D}(*, L^r)$, we have $U = \mathcal{F}^{-1}[V] \in \mathcal{D}'(*, L^r)$, i.e., U is a continuous linear form on $\mathcal{D}(*, L^r)$. Linearity is obvious and continuity of U on $\mathcal{D}(*, L^r)$ follows, because the convergence of a given net (φ_λ) to zero in $\mathcal{D}(*, L^r)$ implies the convergence of the respective net (ψ_λ) to zero in $\mathcal{F}\mathcal{D}(*, L^r)$, according to the inequality in (2.50). In this way, we have proved the following assertion:

Lemma 2.4.2. *The inverse Fourier transform defined on* $\mathcal{F}'\mathcal{D}(*, L^r)$ *by formula (2.52) maps* $\mathcal{F}'\mathcal{D}(*, L^r)$ *to* $\mathcal{D}'(*, L^r)$ *for* $1 \leq r \leq 2$.

We shall use the construction (2.52) in boundary value results subsequently.

2.5 Ultradifferentiable functions of ultrapolynomial growth

We assume that $(M.1)$ and $(M.3')$ hold. The spaces of Gelfand-Shilov type S whose elements are ultradifferentiable functions of ultrapolynomial growth are the test spaces for spaces of tempered ultradistributions. These spaces are studied in [56], [115], [92], [70], [38], [39], and many other papers. Here we follow the preprint [75].

Definition 2.5.1. Let $m > 0$ and $r \in [1, \infty)$ be given. Denote by $\mathcal{S}_r^{(M_p),m} = \mathcal{S}_r^{(M_p),m}(\mathbb{R}^n)$ the space of smooth functions φ on $\mathbb{R}^n$ such that

$$\sigma_{m,r}(\varphi) := \left[\sum_{\alpha,\beta \in \mathbb{N}_0^n} \int_{\mathbb{R}^n} \left| \frac{m^{\alpha+\beta}}{M_\alpha M_\beta} < x >^\beta \varphi^{(\alpha)}(x) \right|^r dx \right]^{1/r} < \infty,$$

that is,

$$\sigma_{m,r}(\varphi) := \left[\sum_{\alpha,\beta \in \mathbb{N}_0^n} \left(\frac{m^{\alpha+\beta}}{M_\alpha M_\beta} \| < x >^\beta \varphi^{(\alpha)} \|_{L^r} \right)^r \right]^{1/r} < \infty. \tag{2.53}$$

Denote by $\mathcal{S}_\infty^{(M_p),m} = \mathcal{S}_\infty^{(M_p),m}(\mathbb{R}^n)$ the space of smooth functions φ on $\mathbb{R}^n$ such that

$$\sigma_{m,\infty}(\varphi) := \sup_{\alpha,\beta \in \mathbb{N}_0^n} \frac{m^{\alpha+\beta}}{M_\alpha M_\beta} \| < x >^\beta \varphi^{(\alpha)} \|_{L^\infty} < \infty. \tag{2.54}$$

The spaces $\mathcal{S}_r^{(M_p),m}$ and $\mathcal{S}_\infty^{(M_p),m}$ are equipped with the topologies induced by the norms $\sigma_{m,r}$ and $\sigma_{m,\infty}$, respectively.

The spaces $S_r^{(M_p),m}$ are Banach spaces for $r \in [1, \infty]$. In particular, the space $S_2^{(M_p),m}$ is a Hilbert space with the scalar product defined by

$$(\varphi, \psi) := \sum_{\alpha,\beta \in N_0} \left(\frac{m^{\alpha+\beta}}{M_\alpha M_\beta} \right)^2 \int_{\mathbb{R}^n} \langle x \rangle^{2\beta} \varphi^{(\alpha)}(x) \overline{\psi^{(\alpha)}(x)} \, dx,$$

for $\varphi, \psi \in S_2^{(M_p),m}$.

Now we give the fundamental definition for our considerations of the spaces $\mathcal{S}^{(M_p)}$ and $\mathcal{S}^{\{M_p\}}$ and the dual spaces $\mathcal{S}'^{(M_p)}$ and $\mathcal{S}'^{\{M_p\}}$ of tempered ultradistributions of Beurling and Roumieu type, respectively.

Definition 2.5.2. Let

$$\mathcal{S}^{(M_p)} = \mathcal{S}^{(M_p)}(\mathbb{R}^n) := \underset{m \to \infty}{\text{proj lim}} \; \mathcal{S}_2^{(M_p),m}(\mathbb{R}^n) \tag{2.55}$$

and

$$\mathcal{S}^{\{M_p\}} = \mathcal{S}^{\{M_p\}}(\mathbb{R}^n) := \underset{m \to \infty}{\text{ind lim}} \; \mathcal{S}_2^{(M_p),m}(\mathbb{R}^n) \tag{2.56}$$

The structure of the test spaces is described in the following two theorems. A simple consequence will be, if $(M.2')$ is satisfied, that $\mathcal{S}^{(M_p)}$ and $\mathcal{S}^{\{M_p\}}$ are the projective (as $m \to \infty$) and the inductive (as $m \to 0$) limits not only of the spaces of the spaces $\mathcal{S}_2^{(M_p),m}$ but also of the spaces $\mathcal{S}_r^{(M_p),m}$ respectively, where $r \in [1, \infty]$.

In the theorem below and further on we shall use a convention, analogous to that applied for the one-dimensional case in (2.23), which will simplify the notation of products of subsequent elements of sequences belonging to the family $\mathcal{R}$ described in Section 2.2. Namely for a given sequence $(a_p) \in \mathcal{R}$ and $\alpha = (\alpha_1, \ldots, \alpha_n) \in \mathbb{N}_0$ we define the associated sequence $(A_\alpha)_{\alpha \in \mathbb{N}_0^n}$ of products of subsequent elements of (a_p) as follows:

$$A_0 = 1; \qquad A_\alpha := \prod_{j=1}^{\bar{\alpha}} a_j, \qquad \alpha \in \mathbb{N}_0^n, \; \alpha \neq 0, \tag{2.57}$$

where $\bar{\alpha} := \alpha_1 + \ldots + \alpha_n$. Analogous products of subsequent elements of sequences $(b_j), (r_j), (s_j)$ etc. in $\mathcal{R}$ will be denoted by $B_\alpha, R_\alpha, S_\alpha$ etc., respectively, for a given $\alpha \in \mathbb{N}_0$.

Theorem 2.5.1. *Let (M_p) satisfy $(M.1)$ and $(M.3')$. Then*

$$\mathcal{S}^{\{M_p\}} = \underset{(a_j),(b_j) \in \mathcal{R}}{\text{proj lim}} \; \mathcal{S}_{(a_j),(b_j)}^{(M_p)},$$

where $\mathcal{S}_{(a_j),(b_j)}^{(M_p)}$ is the space of functions $\varphi \in C^\infty$ such that

$$\sigma_{(a_j),(b_j),2}(\varphi) := \sup \left\{ \frac{\| <x>^\beta \varphi^{(\alpha)} \|_{L^2}}{M_\alpha A_\alpha M_\beta B_\beta} : \alpha, \beta \in \mathbb{N}_0^n \right\} < \infty,$$

where $<x>^\beta$ means the function given by $<x>^\beta = (1 + |x^2|)^{\beta/2}$ (see (1.2)).

Proof. From Lemma 2.2.1 it follows that $\varphi \in C^\infty(\mathbb{R}^n)$ belongs to $\mathcal{S}^{\{M_p\}}$ if and only if $\sigma_{(a_j),(b_j),2}(\varphi) < \infty$ for arbitrary $(a_j), (b_j) \in \mathcal{R}$.

Every norm $\sigma_{(a_j),(b_j),2}$, where $(a_j), (b_j) \in \mathcal{R}$, is continuous on the space $\mathcal{S}_h^{(M_p)}$, $h > 0$, and so on the space $\mathcal{S}^{\{M_p\}}$.

Since $\mathcal{S}^{\{M_p\}}$ is reflexive, every continuous seminorm p is bounded by the seminorm p^B, where B is a bounded set in $\mathcal{S}'^{\{M_p\}}$, defined by

$$p^B(\varphi) := \sup\{|<f,\varphi>|:\ f \in B\}.$$

We have

$$p^B(\varphi) \le \sup_{f\in B} \sum_{\alpha,\beta\in\mathbb{N}_0^n} \| <x>^\beta \varphi^{(\alpha)} \|_{L^2}.$$

By Lemma 2.2.1, it follows that there exist (a_j) and (b_j) from $\mathcal{R}$ such that for some $C > 0$ we have

$$p^B(\varphi) \le C\sigma_{(a_j),(b_j)}(\varphi).$$

The proof of Theorem 2.5.1 is completed. $\square$

Definition 2.5.3. Let $(a_p), (b_p) \in \mathcal{R}$ and let $\mathcal{S}^{(M_p)}_{(a_p),(b_p),\infty}$ be the space of smooth functions φ on $\mathbb{R}^n$ such that

$$\sigma_{(a_p),(b_p),\infty}(\varphi) := \sup_{\alpha,\beta\in\mathbb{N}_0^n} \frac{\|\langle x\rangle^\beta \varphi^{(\alpha)}\|_{L^\infty}}{M_\alpha A_\alpha M_\beta B_\beta} < \infty, \qquad (2.58)$$

equipped with the topology induced by the norm $\sigma_{(a_p),(b_p),\infty}$.

In addition, consider in $\mathcal{S}^{(M_p)}_{(a_p),(b_p),\infty}$ the norms $\sigma_{m,r}(m > 0, r \in [1,\infty])$ defined in (2.53) and (2.54) as well as the following ones:

$$\sigma'_{m,r}(\varphi) := \sum_{\alpha,\beta\in\mathbb{N}_0} \frac{m^{\alpha+\beta}}{M_\alpha M_\beta} \|x^\beta \varphi^{(\alpha)}\|_{L^r}; \qquad (2.59)$$

$$\sigma'_{m,\infty}(\varphi) := \sup_{\alpha,\beta\in\mathbb{N}_0} \frac{m^{\alpha+\beta}}{M_\alpha M_\beta} \|x^\beta \varphi^{(\alpha)}\|_{L^\infty},; \qquad (2.60)$$

$$\tau_{m,r}(\varphi) := \sup_{\alpha\in\mathbb{N}_0} \frac{m^\alpha}{M_\alpha} \|\varphi^{(\alpha)} \exp[M(m|\cdot|)]\|_{L^r}; \qquad (2.61)$$

$$\tau_{m,\infty}(\varphi) := \sup_{\alpha\in\mathbb{N}_0} \frac{m^\alpha}{M_\alpha} \|\varphi^{(\alpha)} \exp[M(m|\cdot|)]\|_{L^\infty}; \qquad (2.62)$$

for arbitrary $m > 0$ and $r \in [1, \infty)$ and

$$\sigma_{(a_p),(b_p),r}(\varphi) := \sum_{\alpha,\beta\in\mathbb{N}_0} \frac{\| <x>^\beta \varphi^{(\alpha)} \|_{L^r}}{M_\alpha A_\alpha M_\beta B_\beta} ; \qquad (2.63)$$

$$\sigma'_{(a_p),(b_p),r}(\varphi) := \sum_{\alpha,\beta\in\mathbb{N}_0} \frac{\| x^\beta \varphi^{(\alpha)} \|_{L^r}}{M_\alpha A_\alpha M_\beta B_\beta} ; \qquad (2.64)$$

$$\sigma'_{(a_p),(b_p),\infty}(\varphi) := \sup_{\alpha,\beta\in\mathbb{N}_0} \frac{\| x^\beta \varphi^{(\alpha)} \|_{L^\infty}}{M_\alpha A_\alpha M_\beta B_\beta} ; \qquad (2.65)$$

$$\tau_{(a_p),(b_p),r}(\varphi) := \sup_{\alpha,\beta\in\mathbb{N}_0} \frac{\| \varphi^{(\alpha)} \exp[N_{(b_p)}(|\cdot|)] \|_{L^r}}{M_\alpha A_\alpha} ; \qquad (2.66)$$

$$\tau_{(a_p),(b_p),\infty}(\varphi) := \sup_{\alpha,\beta\in\mathbb{N}_0} \frac{\| \varphi^{(\alpha)} \exp[N_{(b_p)}(|\cdot|)] \|_{L^\infty}}{M_\alpha A_\alpha} ; \qquad (2.67)$$

for arbitrary $(a_p), (b_p) \in \mathcal{R}$ and $r \in [1, \infty)$.

Definition 2.5.4. It will be convenient to denote various families of norms in the following way:

$$\mathbf{S}_r := \{\sigma_{m,r} : m > 0\}, \qquad \tilde{\mathbf{S}}_r := \{\sigma_{(a_p),(b_p),r} : (a_p), (b_p) \in \mathcal{R}\},$$

$$\mathbf{S}'_r := \{\sigma'_{m,r} : m > 0\}, \qquad \tilde{\mathbf{S}}'_r := \{\sigma'_{(a_p),(b_p),r} : (a_p), (b_p) \in \mathcal{R}\},$$

$$\mathbf{T}_r := \{\tau_{m,r} : m > 0\}, \qquad \tilde{\mathbf{T}}_r := \{\tau_{(a_p),(b_p),r} : (a_p), (b_p) \in \mathcal{R}\}$$

for $r \in [1, \infty]$.

From the theorem below it follows that the space $\mathcal{S}^{(M_p)}_{(a_p),(b_p),\infty}$ endowed with the topology of any of the above families of norms coincides with the space $\mathcal{S}^{\{M_p\}}$. In Section 7.2 we will consider additional families of norms in this space.

Theorem 2.5.2. *The above defined families of norms have the following properties:*

(1) *The families $\mathbf{S}_\infty$ and $\mathbf{S}'_\infty$ (respectively, $\tilde{\mathbf{S}}_\infty$ and $\tilde{\mathbf{S}}'_\infty$) of norms in the space $\mathcal{S}^{(M_p)}$ (respectively, $\mathcal{S}^{\{M_p\}}$) are equivalent;*

(2) *If condition (M.2') holds, then for every $r \in [1, \infty]$ the families $\mathbf{S}_r$, $\mathbf{S}'_r$ and $\mathbf{S}$ (respectively, $\tilde{\mathbf{S}}_r$, $\tilde{\mathbf{S}}'_r$ and $\tilde{\mathbf{S}}$) of norms in the space $\mathcal{S}^{(M_p)}$ (respectively, $\mathcal{S}^{\{M_p\}}$) are equivalent.*

Proof. For the sake of simplicity we will prove the assertions in the case $n = 1$. Parts of respective assertions given in parentheses can be proved in a similar way.

Proof of Part (1). Obviously, $\sigma'_{m,\infty}(\varphi) \leq \sigma_{m,\infty}(\varphi)$ for every smooth function φ and $m > 0$. Condition $(M.3')$ implies that, for every $m > 0$,

$$\frac{m^k k!}{M_k} \longrightarrow 0 \qquad \text{as } k \to \infty \tag{2.68}$$

(see [82], (4.8)), so there is a constant $C_m \geq 1$ such that $m^k/M_k \leq C_m$ for all $k \in \mathbb{N}_0$. Since

$$\langle x \rangle^\beta \leq 2^{\beta/2} \max\left(1, |x|^\beta\right) \leq 2^\beta \max\left(1, |x|^\beta\right), \qquad x \in \mathbb{R}, \ \beta \in \mathbb{N}_0,$$

for every $m > 0$ there exists a $C_m > 0$ such that, for every smooth function φ and $\alpha, \beta \in \mathbb{N}_0$, we have

$$\frac{m^{\alpha+\beta}}{M_\alpha M_\beta} \|\langle x \rangle^\beta \varphi^{(\alpha)}\|_{L^\infty} \leq \frac{m^{\alpha+\beta}}{M_\alpha M_\beta} 2^\beta \max\left(\|\varphi^{(\alpha)}\|_{L^\infty}, \|x^\beta \varphi^{(\alpha)}\|_{L^\infty}\right)$$

$$\leq \max\left(C_m \frac{(2m)^\alpha}{M_\alpha} \|\varphi^{(\alpha)}\|_{L^\infty}, \frac{(2m)^{\alpha+\beta}}{M_\alpha M_\beta} \|x^\beta \varphi^{(\alpha)}\|_{L^\infty}\right)$$

$$\leq C_m \sup_{\beta \in \mathbb{N}_0} \frac{(2m)^{\alpha+\beta}}{M_\alpha M_\beta} \|x^\beta \varphi^{(\alpha)}\|_{L^\infty} = C_m \sigma'_{2m,\infty}(\varphi).$$

Therefore, for every $m > 0$ there exists a $C_m > 0$ such that $\sigma_{m,\infty}(\varphi) \leq C_m \sigma'_{2m,\infty}(\varphi)$ for every smooth function φ. Thus the families $\mathbf{S}_\infty$ and $\mathbf{S}'_\infty$ of norms are equivalent. The equivalence of the families $\tilde{\mathbf{S}}_\infty$ and $\tilde{\mathbf{S}}'_\infty$ of norms follows, by Lemma 2.2.1.

Proof of Part (2). We start by introducing additional notation for arbitrary $t \in [1, \infty]$. For a given function ψ denote by $\|\psi\|'_{L^t}$, $\|\psi\|''_{L^t}$ and $\|\psi\|_{L^t}$ its L^t norms on the sets $[-1, 1]$, $\mathbb{R} \setminus [-1, 1]$ and $\mathbb{R}$, respectively.

Let $\alpha, \beta, \gamma \in \mathbb{N}_0$ and $t \in [1, \infty)$. For a given smooth function φ denote

$$a_{\alpha,\beta}(\varphi) := \|x^\beta \varphi^{(\alpha)}\|'_{L^\infty}; \, b_{\alpha,\beta}(\varphi) := \|x^\beta \varphi^{(\alpha)}\|''_{L^\infty};$$

$$c_{\alpha,\beta}(\varphi) := \|x^\beta \varphi^{(\alpha)}\|_{L^\infty};$$

$$A^t_{\alpha,\beta}(\varphi) := \|x^\beta \varphi^{(\alpha)}\|'_{L^t}; \, B^t_{\alpha,\beta}(\varphi) := \|x^\beta \varphi^{(\alpha)}\|''_{L^t};$$

$$C^t_{\alpha,\beta}(\varphi) := \|x^\beta \varphi^{(\alpha)}\|_{L^t}.$$

Moreover, denote by $I_{\gamma,t}$ the L^t norm of the function $\tau(x) = x^{-\gamma}$ on $\mathbb{R} \setminus [-1, 1]$ and $\mathbb{R}$.

Due to $(M.2')$, for every $m > 0$ there exists a constant $D > 0$ such that

$$\sigma'_{m,t}(\varphi) \leq \sum_{\alpha,\beta \in \mathbb{N}_0} \frac{m^{\alpha+\beta}}{M_\alpha M_\beta} \left[a_{\alpha,\beta}(\varphi) + b_{\alpha,\beta+\gamma}(\varphi) I_{\gamma,1} \right]$$

$$\leq \sum_{\alpha,\beta \in \mathbb{N}_0} \frac{m^{\alpha+\beta}}{M_\alpha M_\beta} c_{\alpha,\beta}(\varphi) + D \sum_{\alpha,\beta \in \mathbb{N}_0} \frac{m^{\alpha+\beta} H^{\sigma\beta}}{M_\alpha M_{\beta+\gamma}} c_{\alpha,\beta+\gamma}(\varphi)$$

$$\leq D\sigma'_{m(1+H^\gamma),\infty}(\varphi) \tag{2.69}$$

for every smooth function φ.

Clearly,

$$|x^\beta \varphi^{(\alpha)}(x)| \leq \beta C_{\alpha,\beta,1}(\varphi) + C_{\alpha+1,\beta,1}(\varphi)$$

for α, $\beta \in \mathbb{N}_0$ and $x \in \mathbb{R}$. Hence, by condition $(M.2')$, for every $m > 0$ there exists a $D > 0$ such that

$$\sigma'_{m,\infty}(\varphi) \leq D \sup_{\alpha,\beta \in \mathbb{N}_0} \frac{H^\alpha m^{\alpha+\beta}}{M_{\alpha+1} M_\beta} \left(2^\beta C_{\alpha,\beta,1}(\varphi) + H m^{\alpha+2} C_{\alpha+1,\beta,1}(\varphi) \right)$$

$$\leq D\sigma'_{2m(1+H),1}(\varphi) \tag{2.70}$$

for every smooth function φ.

Now let $t \in (1,\infty)$, $q := t/(t-1)$ and $\gamma := [1/q] + 1$. The Hölder inequality, (2.68) and $(M.2')$ imply that for every $m > 0$ there exists a $D > 0$ such that, for every smooth function φ,

$$\sigma'_{m,1}(\varphi) = \sum_{\alpha,\beta \in \mathbb{N}_0} \frac{m^{\alpha+\beta}}{M_\alpha M_\beta} \left(A_{\alpha,0,1}(\varphi) + B_{\alpha,\beta,1}(\varphi) \right)$$

$$\leq \sum_{\alpha,\beta \in \mathbb{N}_0} \frac{m^{\alpha+\beta}}{M_\alpha M_\beta} \left[D A_{\alpha,0,t}(\varphi) + B_{\alpha,\beta+\gamma,t}(\varphi) I_{\gamma,q} \right]$$

$$\leq D \sum_{\alpha,\beta \in \mathbb{N}_0} \frac{m^{\alpha+\beta}}{M_\alpha M_\beta} \left(\|\varphi^{(\alpha)}\|_{L^t} + \|x^{\beta+\gamma} \varphi^{(\alpha)}\|_{L^t} \right)$$

$$\leq D \left[\sum_{\alpha \in \mathbb{N}_0} \frac{m^\alpha}{M_\alpha} \|\varphi^{(\alpha)}\|_{L^t} + \sum_{\alpha,\beta \in \mathbb{N}_0} \frac{m^{\alpha+\beta} H^{\gamma\beta}}{M_\alpha M_{\beta+\gamma}} \|x^{\beta+\gamma} \varphi^{(\alpha)}\|_{L^t} \right]$$

$$\leq C\sigma'_{m(1+H^\gamma),t}(\varphi). \tag{2.71}$$

The equivalence of the families $\{\sigma'_{m,r} : m > 0\}$ and $\{\sigma'_{m,p} : m > 0\}$ for $r, p \in [1,\infty]$ follows from (2.69), (2.70) and (2.71).

The proof of the equivalence of $\{\sigma_{m,p} : m > 0\}$ and $\{\sigma_{m,r} : m > 0\}$, where $r, p \in [1,\infty]$, is analogous.

Condition $(M.2')$ implies that for every $\varphi \in \mathcal{S}^{(M_p)}$ and every $m > 0$ there exists $D > 0$ such that for every $\alpha, \beta \in \mathbb{N}_0$ and $|x| > k > 1$, we have

$$\frac{m^{\alpha+\beta}}{M_\alpha M_\beta} |x^\beta \varphi^{(\alpha)}(x)| \leq D \frac{m^\alpha (mH)^{\beta+1}}{M_\alpha M_{\beta+1}} |x^\beta \varphi^{(\alpha)}(x)|$$

$$\leq \frac{D}{k} \frac{m^\alpha (mH)^{\beta+1}}{M_\alpha M_{\beta+1}} |x^{\beta+1} \varphi^{(\alpha)}(x)| \leq \frac{C}{k}.$$

Therefore, for every $m > 0$ and $\varphi \in \mathcal{S}^{(M_p)}$,

$$\frac{m^{\alpha+\beta}}{M_\alpha M_\beta} |x^\beta \varphi^{(\alpha)}(x)| \to 0 \qquad (2.72)$$

as $|x| \to \infty$ uniformly in $\alpha, \beta \in \mathbb{N}_0$. From the definition of the space $\mathcal{S}^{(M_p)}$ it follows that, for every $m > 0$, the convergence to zero in (2.72) is uniform in $x \in \mathbb{R}$ as $\alpha + \beta \to \infty$.

Hence, for a given element φ of $\mathcal{S}^{(M_p)}$ and every $m > 0$ there are $\alpha_0, \beta_0 \in \mathbb{N}_0$ and $x_0 \in \mathbb{R}$ such that

$$\sup_{\alpha, \beta \in \mathbb{N}_0} \frac{m^{\alpha+\beta}}{M_\alpha M_\beta} \| x^\beta \varphi^{(\alpha)} \|_{L^\infty} = \frac{m^{\alpha_0+\beta_0}}{M_{\beta_0} M_{\alpha_0}} |x_0^{\beta_0} \varphi^{(\alpha_0)}(x_0)|$$

$$= \Big\| \sup_{\beta \in \mathbb{N}_0} \big[\sup_{\alpha \in \mathbb{N}_0} \frac{m^{\alpha+\beta}}{M_\alpha M_\beta} |x^\beta \varphi^{(\alpha)}| \big] \Big\|_{L^\infty}$$

$$= \Big\| \sup_{\alpha \in \mathbb{N}_0} \big[\sup_{\beta \in \mathbb{N}_0} \frac{m^{\alpha+\beta}}{M_\alpha M_\beta} |x^\beta \varphi^{(\alpha)}| \big] \Big\|_{L^\infty}$$

$$= \sup_{\alpha \in \mathbb{N}_0} \big[\big\| \sup_{\beta \in \mathbb{N}_0} \frac{m^{\alpha+\beta}}{M_\alpha M_\beta} |x^\beta \varphi^{(\alpha)}| \big\|_{L^\infty} \big]$$

$$= \sup_{\alpha \in \mathbb{N}_0} \Big(\frac{m^\alpha}{M_\alpha} \| \varphi^{(\alpha)} \exp [M(m| \cdot |)] \|_{L^\infty} \Big).$$

The proof of Theorem 2.5.2 is completed. $\square$

Remark 2.5.1. It is easy to verify that the proofs of the theorems of this section hold in the n-dimensional case. In particular, if $(M.2)$ holds, they can be presented in the same way.

Corollary 2.5.1. $\mathcal{S}^{\{M_p\}} = \underset{(a_p),(b_p) \in \mathcal{R}}{\text{proj lim}} \mathcal{S}^{(M_p)}_{(a_p),(b_p),\infty}.$

Using an analogous idea as in the proof of Theorem 1 in [50] (p. 29, Satz 1), one can prove the following assertion:

Theorem 2.5.3. *A set B is relatively compact in $\mathcal{S}_2^{(M_p),m}$ if and only if*

(i) *the set $B_\beta^\alpha := \{\langle x \rangle^\beta \varphi^{(\alpha)} : \varphi \in B\}$ is a relatively compact subset of L^2 for each $\alpha, \beta \in \mathbb{N}_0$ and*

(ii) *the series $\displaystyle\sum_{\alpha,\beta \in \mathbb{N}_0} \int_\mathbb{R} \left| \frac{m^{\alpha+\beta}}{M_\alpha M_\beta} \langle x \rangle^\beta \varphi^{(\alpha)}(x) \right|^2 dx$ converges uniformly for $\varphi \in B$.*

Proof. We give the proof in the case $n = 1$.

We are going to prove that B fulfills (i) by checking that the set B_β^α : $\alpha, \beta \in \mathbb{N}_0$, fulfills the assumptions of Kolmogorov's theorem (see [50]). It is obvious that for each $\alpha, \beta \in \mathbb{N}_0$ the set $B_\beta^\alpha = \{\langle x \rangle^\beta \varphi^{(\alpha)}, \varphi \in B\}$ is bounded in the space L^2. Applying the Cauchy-Schwarz inequality and the Fubini-Tonelli theorem we see that, for arbitrary $\varphi \in B$ and $\alpha, \beta \in \mathbb{N}_0$,

$$\int_\mathbb{R} \left| \langle x+h \rangle^\beta \varphi^{(\alpha)}(x+h) - \langle x \rangle^\beta \varphi^{(\alpha)}(x) \right|^2 dx$$

$$\leq \int_\mathbb{R} \left(\int_0^1 \left| \frac{d}{dt}(\langle x+th \rangle^\beta \varphi^{(\alpha)}(x+th)) \right| dt \right)^2 dx$$

$$\leq \int_\mathbb{R} \left(\int_0^1 \left| \frac{d}{dt}[\langle x+th \rangle^\beta \varphi^{(\alpha)}(x+th)] \right|^2 dt \right) dx$$

$$\leq \beta^2 h^2 \int_0^1 \left(\int_\mathbb{R} \left| \langle x+th \rangle^\beta \varphi^{(\alpha)}(x+th) \right|^2 dx \right) dt$$

$$+ h^2 \int_0^1 \left(\int_\mathbb{R} \left| \langle x+th \rangle^\beta \varphi^{(\alpha+1)}(x+th) \right|^2 dx \right) dt$$

$$= \beta^2 h^2 \left(\int_\mathbb{R} \left| \langle \xi \rangle^\beta \varphi^{(\alpha)}(\xi) \right|^2 d\xi \right) + h^2 \left(\int_\mathbb{R} \left| \langle \xi \rangle^\beta \varphi^{(\alpha+1)}(\xi) \right|^2 d\xi \right)$$

$$\leq h^2 \left(\beta^2 \frac{M_\alpha M_\beta}{\tilde{m}^{\alpha+\beta}} + \frac{M_{\alpha+1} M_\beta}{\tilde{m}^{\alpha+\beta+1}} \right).$$

Hence, the integral

$$\int_\mathbb{R} |\langle x+h \rangle^\beta \varphi^{(\alpha)}(x+h) - \langle x \rangle^\beta \varphi^{(\alpha)}(x)|^2 dx$$

converges to zero as $h \to 0$ uniformly for $\varphi \in B$.

For each $\varphi \in B$ and $k > 0$

$$\langle k \rangle^2 \int_{\mathbb{R} \setminus [-k,k]} |\langle x \rangle^\beta \varphi^{(\alpha)}(x)|^2 \, dx \le \int_{\mathbb{R} \setminus [-k,k]} |\langle x \rangle^{\beta+1} \varphi^{(\alpha)}(x)|^2 \, dx$$

$$\le \frac{M_\alpha M_{\beta+1}}{\tilde{m}^{\alpha+\beta+1}}.$$

Therefore

$$\int_{\mathbb{R} \setminus [-k,k]} |\langle x \rangle^\beta \varphi^{(\alpha)}(x)|^2 \, dx \le \langle k \rangle^{-2} \frac{M_\alpha M_{\beta+1}}{\tilde{m}^{\alpha+\beta+1}}, \qquad \varphi \in B.$$

According to the theorem of Kolmogorov, it follows that the set B_α^β, $\alpha, \beta \in \mathbb{N}_0$, is relatively compact in L^2.

Let us prove that B fulfills condition (ii). For each $\varepsilon > 0$ there exists $\mu \in \mathbb{N}_0$ such that $m^\alpha \le \varepsilon \tilde{m}^\alpha$ for all $\alpha \ge \mu$. Hence

$$\sum_{\substack{\alpha \ge \mu \\ \beta \in \mathbb{N}_0}} \int_{\mathbb{R}} \left| \frac{m^{\alpha+\beta}}{M_\alpha M_\beta} \langle x \rangle^\beta \varphi^{(\alpha)}(x) \right|^2 dx \le \varepsilon^2 \sum_{\substack{\alpha \ge \mu \\ \beta \in \mathbb{N}_0}} \int_{\mathbb{R}} \left| \frac{\tilde{m}^{\alpha+\beta}}{M_\alpha M_\beta} \langle x \rangle^\beta \varphi^{(\alpha)}(x) \right|^2 dx$$

$$\le \varepsilon^2$$

for each $\varphi \in B$ and the proof of assertion (ii) and the whole theorem is completed. $\square$

Now we are in a position to prove the following theorem:

Theorem 2.5.4.
    1. The spaces $\mathcal{S}^{(M_p)}$ and $\mathcal{S}^{\{M_p\}}$ are $(F\tilde{S})$ and (LS) spaces, respectively.
    2. If $(M.2')$ is satisfied, then

$$\mathcal{D}^* \hookrightarrow \mathcal{S}^* \hookrightarrow \mathcal{E}^*, \quad \mathcal{S}^* \hookrightarrow \mathcal{S};$$
$$\mathcal{E}'^* \hookrightarrow \mathcal{S}'^* \hookrightarrow \mathcal{D}'^*, \quad \mathcal{S}' \hookrightarrow \mathcal{S}'^*.$$

Proof. Again, we give the proof for the case $n = 1$. Recall that a locally convex topological vector space is a (FS) space (respectively, a (LS) space) if it is a projective limit (respectively, an inductive limit) of a countable, compact specter of spaces. If the mentioned specter is also nuclear, the space is called a (FN) space (respectively, a (LN) space); for more details see [50].

1. In order to prove the first part of the assertion it is enough to show that the inclusion mapping

$$i\colon \mathcal{S}_2^{(M_p),\tilde{m}} \to \mathcal{S}_2^{(M_p),m}, \qquad m < \tilde{m},$$

is compact. Since $\mathcal{S}_2^{(M_p),\tilde{m}}$ and $\mathcal{S}_2^{(M_p),m}$ are Banach spaces, it suffices to prove that the unit ball $B := \{\varphi\colon \sigma_{\tilde{m},2}(\varphi) \leq 1\}$ of the space $\mathcal{S}_2^{(M_p),\tilde{m}}$ is a relatively compact set in $\mathcal{S}_2^{(M_p),m}$, where $\sigma_{\tilde{m},2}(\varphi)$ is defined in (2.53). But this follows from Theorem 2.5.3 and the proof of assertion 1 is completed.

2. Since the proofs of the second assertion in the Beurling case $* = (M_p)$ and the Roumieu case $* = \{M_p\}$ are analogous, we will prove the assertion only in the first case. Let $\varphi \in \mathcal{D}^{(M_p)}$ and $\operatorname{supp}\varphi \subset [-k,k]$, $k > 1$. Condition $(M.3')$ implies that for each $m > 0$ there exists $C > 0$ such that

$$\sup_{\alpha,\beta \in \mathbb{N}_0} \frac{m^{\alpha+\beta}}{M_\alpha M_\beta}\|\langle x\rangle^\beta \varphi^{(\alpha)}\|_\infty = \sup_{\alpha,\beta \in \mathbb{N}_0} \frac{(mk)^\beta m^\alpha}{M_\beta M_\alpha}\|\varphi^{(\alpha)}\|_\infty$$

$$\leq C \sup_{\alpha \in \mathbb{N}_0} \frac{m^\alpha}{M_\alpha}\|\varphi^{(\alpha)}\|_\infty.$$

It follows that the inclusion mapping $i\colon \mathcal{D}^{(M_p)} \to \mathcal{S}^{(M_p)}$ is continuous.

The sequence (φ_j) with $\varphi_j(x) := \rho(x/j)\varphi(x)$, $j \in \mathbb{N}$, where ρ is a function of the class $\mathcal{D}^{(M_\alpha)}$ such that $\rho = 1$ in a neighborhood of 0, converges to φ in the space $\mathcal{S}^{(M_\alpha)}$, since $\dfrac{m^{\alpha+\beta}}{M_\alpha M_\beta}|x^\beta \varphi^{(\alpha)}(x)|$ converges uniformly in $\alpha,\beta \in \mathbb{N}_0$ as $|x|$ tends to infinity for arbitrary fixed $\varphi \in \mathcal{S}^{(M_p)}$ and $m > 0$. It follows that $\mathcal{D}^{(M_p)}$ is dense in $\mathcal{S}^{(M_p)}$ and the assertion of Theorem 2.5.4 is proved. $\square$

2.6 Tempered ultradistributions

A non-trivial example, in case $n = 1$, of an element of the space $\mathcal{S}'^*$ is

$$\langle f,\varphi\rangle = \int_{\mathbb{R}} f\varphi\, dx, \qquad \varphi \in \mathcal{S}^*,$$

where f is a locally integrable function in $\mathbb{R}$ of ultrapolynomial growth of the class $*$, that is

$$|f(x)| \leq P(x), \qquad x \in \mathbb{R},$$

where P is an ultrapolynomial of the class $*$. Note that if $(M.2')$ is satisfied, then the function f is of ultrapolynomial growth of the class (M_p)

(respectively, $\{M_p\}$) if and only if there exists an $m > 0$ and there exists a $C > 0$ (respectively, for every $m > 0$ there exists a $C > 0$) such that

$$|f(x)| \leq C \exp[M(m|x|)], \qquad x \in \mathbb{R}.$$

Let us now give the structure theorems for the space $\mathcal{S}'^*$.

Theorem 2.6.1. *Assume that condition $(M.2')$ holds, $r \in (1, \infty]$ and $f \in \mathcal{D}'^{(M_p)}$ (respectively, $f \in \mathcal{D}'^{\{M_p\}}$). Then*

1. $f \in \mathcal{S}'^{(M_p)}$ (respectively, $f \in \mathcal{S}'^{\{M_p\}}$) if and only if f is of the form

$$f = \sum_{\alpha, \beta \in \mathbb{N}_0} (\langle x \rangle^\beta F_{\alpha,\beta})^{(\alpha)}, \qquad \tag{2.73}$$

in the sense of convergence in $\mathcal{S}'^{(M_p)}$ (respectively, $\mathcal{S}'^{\{M_p\}}$), where $(F_{\alpha,\beta})_{\alpha,\beta \in \mathbb{N}_0}$ is a sequence of elements of L^r such that for some (respectively, each) $m > 0$ we have

$$\left(\sum_{\alpha, \beta \in \mathbb{N}_0} \int_{\mathbb{R}} \left| \frac{M_\alpha M_\beta}{m^{\alpha+\beta}} F_{\alpha,\beta}(x) \right|^r dx \right)^{1/r} < \infty, \qquad \tag{2.74}$$

in case $r \in (1, \infty)$, and

$$\sup_{\substack{\alpha, \beta \in \mathbb{N}_0 \\ x \in \mathbb{R}}} \left(\frac{M_\alpha M_\beta}{m^{\alpha+\beta}} |F_{\alpha,\beta}(x)| \right) < \infty, \qquad \tag{2.75}$$

in case $r = \infty$.

2. Let $(M.2)$ and $(M.3)$ be satisfied. Then $f \in \mathcal{S}'^*$ if and only if f is of the form

$$f = P(D)F, \qquad \tag{2.76}$$

where P is an ultradifferentiable operator of the class $*$ and F is a continuous function of $\mathbb{R}$ of ultrapolynomial growth of the class $*$.

Proof. First notice that the weak and the strong sequential convergence are equivalent in $\mathcal{S}'^*$.

1. In the Beurling case $* = (M_p)$, the proof of assertion 1 is quite analogous to the proof given in [115]. In the Roumieu case $* = \{M_p\}$, it follows easily that (2.73) determines an element of $\mathcal{S}'^{\{M_p\}}$. To prove the converse we will use the dual Mittag-Leffler lemma (see [82], Lemma 1.4) similarly as in the proof of [82], Proposition 8.6.

Denote

$$X_m := \mathcal{S}_q^{(M_p),m}, \qquad Y_m := \{(\varphi_{\alpha,\beta})_{\alpha,\beta \in \mathbb{N}_0} : \|(\varphi_{\alpha,\beta})\|_{Y_m} < \infty\},$$

where $q = r/(r-1)$ and

$$\|(\varphi_{\alpha,\beta})\|_{Y_m} := \sup_{\alpha,\beta\in\mathbb{N}_0} \frac{m^{\alpha+\beta}}{M_\alpha M_\beta} \|\varphi_{\alpha,\beta}\|_q.$$

The space Y_m is a reflexive Banach space. According to Alaoglu's theorem, a bounded set in Y_m is weakly compact in Y_m. Therefore the inclusion mapping $i : Y_{m'} \to Y_m$, $m' > m$ is weakly compact. We will identify X_m with a closed subspace of Y_m in which X_m is mapped by

$$\langle x\rangle^\beta D^\alpha : \quad X_m \ni \varphi \mapsto (\langle x\rangle^\beta \varphi^{(\alpha)})_{\alpha,\beta} \in Y_m.$$

Clearly (X_m) and (Y_m) are injective sequences of Banach spaces and if $m' > m$, then $X_{m'} \cap Y_m = X_m$. It follows that the quotient space $Z_m = Y_m/X_m$ (with the quotient topology) is also an injective weakly compact sequence of Banach spaces. It follows from the dual Mittag-Leffler lemma that

$$\{0\} \longleftarrow (\text{ind}\lim_{m\to 0} X_m)' \xleftarrow{\sum(-1)^\alpha D^\alpha \langle x\rangle^\beta} (\text{ind}\lim_{m\to 0} Y_m)'$$

is topologically exact (see [82]). The above fact together with the identities:

$$\text{proj}\lim_{m\to 0} X_m' = (\text{ind}\lim_{m\to 0} X_m)'$$

and

$$\text{proj}\lim_{m\to 0} Y_m' = (\text{ind}\lim_{m\to 0} Y_m)'$$

imply that the space $\lim\text{ind}_{m\to 0}X_m$ has the same strong dual as the closed subspace $\lim\text{ind}_{m\to 0}X_m$ of $\lim\text{ind}_{m\to 0}Y_m$. Since Y_m' is the Banach space of all $F = (F_{\alpha,\beta})$, $F_{\alpha,\beta} \in L^r$, with

$$\|f\|_{Y_m'} := \begin{cases} \left(\sum_{\alpha,\beta\in\mathbb{N}_0} \int_\mathbb{R} \left|\frac{M_\alpha M_\beta}{m^{\alpha+\beta}}F_{\alpha,\beta}(x)\right|^r dx\right)^{1/r} < \infty, & r\in(1,\infty),\\[3mm] \sup_{\substack{\alpha,\beta\in\mathbb{N}_0\\ x\in\mathbb{R}}} \left(\frac{M_\alpha M_\beta}{m^{\alpha+\beta}}|F_{\alpha,\beta}(x)|\right) < \infty, & r=\infty, \end{cases}$$

the assertion is proved. $\square$

In [75], the spaces $\mathcal{S}^*$ are characterized in terms of Hermite expansions. The following theorem is based on these expansions.

Theorem 2.6.2. *If condition (M.2) is satisfied, then* $\mathcal{S}^{(M_p)}$, $\mathcal{S}'^{\{M_p\}}$ *are (FN) spaces and* $\mathcal{S}^{\{M_p\}}$, $\mathcal{S}'^{(M_p)}$ *are (LN) spaces, respectively.*

2.7 Laplace transform

Suppose that conditions $(M.1)$, $(M.2)$ and $(M.3')$ are satisfied.

We will give the definition of the Laplace transform based on (1.7) in the case $n = 1$. For $n > 1$ the definitions can be extended easily.

Denote by $\mathcal{S}'^{*}_{+}(\mathbb{R})$ the subspace of $\mathcal{S}'^{*}$ consisting of elements supported by $[0, \infty)$. Let $g \in \mathcal{S}'^{*}_{+}$. For fixed $y > 0$, we define $g \exp\left[-2\pi y \cdot\right]$ as an element of $\mathcal{S}'^{*}$ by

$$\langle g \exp\left(-2\pi y \cdot\right), \varphi \rangle := \langle g, \varrho \exp\left(-2\pi y \cdot\right)\varphi \rangle, \quad \varphi \in \mathcal{S}^{*}(\mathbb{R}),$$

where ϱ is an element of $\mathcal{E}^{*}(\mathbb{R})$ such that, for some $\varepsilon > 0$, $\varrho(x) := 1$ if $x \in (-\varepsilon, \infty)$ and $\varrho(x) := 0$ if $x \in (-\infty, -2\varepsilon)$. It is easy to see that the definition does not depend on the choice of ϱ.

An example of such a function is $\varrho := f * \omega$, where f is a function such that $f(x) := 1$ for $x \geq -3\varepsilon/2$ and $f(x) := 0$ for $x < -3\varepsilon/2$, and ω is a function in $\mathcal{D}^{*}$ such that $\int \omega = 1$ and $\operatorname{supp}\omega \subset [-\varepsilon/2, \varepsilon/2]$; for the existence of such a function ω see [82], Theorem 4.2. Clearly, the function ϱ so defined belongs to $\mathcal{E}^{*}$.

As in the case of $\mathcal{S}'_{+}$ (see e.g. [151]), we define the Laplace transform of $g \in \mathcal{S}'^{*}_{+}$ by

$$(\mathcal{L}g)(\zeta) := \mathcal{F}[g \exp\left(-2\pi y\cdot\right)](x), \quad \zeta = x + iy \in \mathbb{C}_{+}.$$

Clearly, if $y > 0$ is fixed, $\mathcal{L}g$ is an element of $\mathcal{S}'^{*}$.

Let

$$G(\zeta) := \langle g, \eta \exp\left(2\pi i \zeta \cdot\right)\rangle, \quad \zeta = x + iy \in \mathbb{C}_{+},$$

where η is as chosen above. The function G is analytic on $\mathbb{C}_{+}$, and its definition does not depend on η.

Chapter 3

Boundedness

3.1 Boundedness in $\mathcal{D}'(*, L^s)$

Denote by $\mathcal{C}_0$ the space of continuous functions f on $\mathbb{R}^n$ vanishing at ∞, i.e., such that $\lim_{|x|\to\infty} f(x) = 0$, equipped with the norm $\|\cdot\|_{L^\infty}$. Its dual space, the space of measures, is denoted by $\mathcal{M}^1$ as in [60]. We denote the dual norm in $\mathcal{M}^1$ by $\|\cdot\|_{\mathcal{M}^1}$. Note that, under conditions $(M.1)$ and $(M.3')$, the space $\mathcal{D}^*$ is dense in $\mathcal{C}_0$.

Theorem 3.1.1. *If (M_p) satisfies conditions $(M.1)$ and $(M.3')$, then*

(i) *a set $B \subset \mathcal{D}'((M_p), L^t)$, $t \in (1, \infty]$, is bounded if and only if every $f \in B$ can be represented in the form*

$$f = \sum_{\alpha=0}^{\infty} \partial^\alpha f_\alpha, \qquad f_\alpha \in L^t, \qquad \alpha \in \mathbb{N}^n,$$

and, moreover, there exist $d > 0$ and $C > 0$, independent of $f \in B$, such that

$$\sum_{\alpha=0}^{\infty} d^\alpha M_\alpha \|f_\alpha\|_{L^t} < C; \tag{3.1}$$

(ii) *a set $B \subset \mathcal{D}'((M_p), L^1)$ is bounded if and only if the representation of f in (i) holds with $f_\alpha \in \mathcal{M}^1$ and the condition in (i) holds with the norm $\|f_\alpha\|_{\mathcal{M}^1}$;*

(iii) *an element f of $\mathcal{D}'^{\{M_p\}}$ belongs to $\mathcal{D}'(\{M_p\}, L^t)$ for $t \in [1, \infty]$ if and only if f is of the form*

$$f = \sum_{\alpha \in \mathbb{N}_0} \partial^\alpha f_\alpha,$$

where $f_\alpha \in L^t$ if $t \in (1, \infty]$ and $f_\alpha \in \mathcal{M}^1$ if $t = 1$ for $\alpha \in \mathbb{N}_0^n$, moreover, for every $d > 0$, we have

$$\sum_{\alpha \in \mathbb{N}_0} d^\alpha M_\alpha \|f_\alpha\|_{L^t} < \infty \qquad \text{in case} \quad t \in (1, \infty];$$

$$\sum_{\alpha \in \mathbb{N}_0} d^\alpha M_\alpha \|f_\alpha\|_{\mathcal{M}^1} < \infty \qquad in \ case \quad t = 1.$$

Proof. Clearly, the conditions given in (*i*) - (*iii*) are sufficient. We will prove that they are necessary.

(*i*) Notice that $\mathcal{D}((M_p), L^s)$ with $s = t/(t-1) \in [1, \infty)$ is barrelled , the set B is equicontinuous in $\mathcal{D}'((M_p), L^t)$ and, for some $d > 0$ and $C > 0$,

$$|\langle f, \varphi \rangle| \le C\|\varphi\|_{L^s, d}, \qquad f \in B, \quad \varphi \in \mathcal{D}((M_p), L^s).$$

Hence, by the Hahn-Banach theorem, elements of B can be extended to constitute an equicontinuous set B_1 on $\mathcal{D}_{L^s, d}^{(M_p)}$. Let $Y_{s,d}$ be the space of all sequences (φ_α) in L^s such that

$$\|(\varphi_\alpha)\|_{L^s, d} = \sup \left\{ \frac{\|\varphi_\alpha\|_{L^s}}{d^\alpha M_\alpha} : \ \alpha \in \mathbb{N}_0^n \right\} < \infty$$

equipped with this norm. Again by the Hahn-Banach theorem, elements of B_1 can be extended to constitute an equicontinuous set B_2 on $Y_{s,d}$. An equicontinuous set on $Y_{s,d}$ consists of all sequences (f_α) from L^t for which (3.1) holds and this implies assertion (*i*).

(*ii*) Let $X_{\infty, h}$ be the space of all smooth functions φ such that $\varphi^{(\alpha)} \in \mathcal{C}_0$ and $\|\varphi\|_{L^\infty, h} < \infty$ for every $\alpha \in \mathbb{N}_0^n$, equipped with the norm $\|\cdot\|_{L^\infty, h}$. We have

$$\dot{\mathcal{B}}^{(M_p)} = \underset{h \to 0}{\text{proj lim}} \ X_{\infty, h},$$

which implies that $\dot{\mathcal{B}}^{(M_p)}$ is barrelled. Thus, using the same reasoning as in (*i*), the proof of (*ii*) follows.

(*iii*) Let $Y_{s,h}$ with $s \in [1, \infty]$ and $h > 0$ be the space of all sequences $(\varphi_\alpha) = (\varphi_\alpha)_{\alpha \in \mathbb{N}_0^n}$ in L^s, for $s \in [1, \infty)$, and in $\mathcal{C}_0$, for $s = \infty$, such that

$$\|(\varphi_\alpha)\|_{L^s, h} = \sup \left\{ \frac{\|\varphi_\alpha\|_{L^s}}{h^\alpha M_\alpha} : \ \alpha \in \mathbb{N}_0^n \right\} < \infty,$$

equipped with the above norm.

For a given $h > 0$, let $X_{s,h} = \mathcal{D}((M_p), h, L^s)$, for $s \in [1, \infty)$, and $X_{\infty, h}$ be as in the proof of (*ii*). We identify $X_{s,h}$ with the corresponding subspace of $Y_{s,h}$, for $s \in [1, \infty]$ and $h > 0$ via the mapping $\varphi \to (\varphi^{(\alpha)})$. Notice that

$$\dot{\mathcal{B}}^{\{M_p\}} = \underset{h \to \infty}{\text{ind lim}} \ X_{s,h}.$$

According to this identification, we have

$$\mathcal{D}(\{M_p\}, L^s) \subset Y_s = \underset{h \to \infty}{\text{ind lim}} \ Y_{s,h}, \qquad s \in [1, \infty),$$

and

$$\dot{\mathcal{B}}^{\{M_p\}} \subset Y_\infty = \underset{h \to \infty}{\text{ind lim }} Y_{\infty,h}.$$

Since the inclusion mappings are continuous, every continuous linear functional on $\mathcal{D}(\{M_p\}, L^s)$ or on $\dot{\mathcal{B}}^{\{M_p\}}$ is continuous on this space, equipped with the induced topology from Y_s for $s \in [1, \infty]$. Thus the Hahn-Banach theorem implies the assertion (iii), because

$$(\underset{h \to \infty}{\text{ind lim }} Y_{s,h})' = \underset{h \to \infty}{\text{proj lim }} Y'_{s,h}, \qquad s \in [1, \infty],$$

in the set theoretical sense, which completes the proof of the last assertion of Theorem 3.1.1. $\square$

Remark 3.1.1. With the notation as in (iii) for $s \in (1, \infty)$, the sequence $(Y_{s,h})_{h \in \mathbb{N}^n}$ is weakly compact. This implies that the sequences $(X_{s,h})_{h \in \mathbb{N}^n}$ and $(Z_{s,h})_{h \in \mathbb{N}^n}$, where $Z_{s,h} := Y_{s,h}/X_{s,h}$, are weakly compact, as well. Thus the dual Mittag - Leffler lemma (see [82], Lemma 1.4) implies that the sequence

$$\{0\} \longleftarrow \underset{h \to \infty}{\text{proj lim }} X'_{s,h} \longleftarrow \underset{h \to \infty}{\text{proj lim }} Y'_{s,h}$$

is exact, where

$$\underset{h \to \infty}{\text{proj lim }} X'_{s,h} = X'_s = (\underset{h \to \infty}{\text{ind lim }} X_{s,h})'$$

and

$$\underset{h \to \infty}{\text{proj lim }} Y'_{s,h} = Y'_s = (\underset{h \to \infty}{\text{ind lim }} Y_{s,h})'$$

in the topological sense. This implies that $\mathcal{D}(\{M_p\}, L^s)$ and X_s, equipped with the induced topology from Y_s, have the same strong duals (see [82], Lemma 1.4, (iii)). We do not know whether the space X_s with the induced topology is quasi-barrelled and, consequently, we do not have a characterization of bounded sets in $\mathcal{D}'(\{M_p\}, L^t)$ for $t \in (1, \infty)$.

Denote by $\mathcal{D}(\{M_p\}, (a_p), L^s)$, with $(a_p) \in \mathcal{R}$ and $s \in [1, \infty]$, the space of all smooth functions φ such that

$$\|\varphi\|_{L^s,(a_p)} = \sup\{\frac{\|\partial^\alpha \varphi\|_{L^s}}{M_\alpha(\prod_{j=1}^\alpha a_j)} : \quad \alpha \in \mathbb{N}_0^n\} < \infty,$$

equipped with the above norm, and let

$$\tilde{\mathcal{D}}(\{M_p\}, L^s) = \underset{(a_p) \in \mathcal{R}}{\text{proj lim }} \mathcal{D}(\{M_p\}, (a_p), L^s).$$

For the completion of $\mathcal{D}^{\{M_p\}}$ in the space $\tilde{\mathcal{D}}(\{M_p\}, L^\infty)$ we use the symbol $\dot{\mathcal{B}}^{\{M_p\}}$. The corresponding dual spaces are denoted by $\tilde{\mathcal{D}}'(\{M_p\}, L^t)$ for $t = s/(s-1) \in (1, \infty]$ and by $\tilde{\mathcal{D}}'(\{M_p\}, L^1)$, respectively.

Theorem 3.1.2. *If (M_p) satisfies (M.1) and (M.3'), then*

(i) $\tilde{\mathcal{D}}(\{M_p\}, L^s) = \mathcal{D}(\{M_p\}, L^s)$ *for* $s \in [1, \infty)$ *in the set theoretical sense; the same is true for $\dot{\mathcal{B}}^{\{M_p\}}$ and $\tilde{\dot{\mathcal{B}}}^{\{M_p\}}$;*

(ii) the inclusion mappings $i_1 \colon \mathcal{D}(\{M_p\}, L^s) \to \tilde{\mathcal{D}}(\{M_p\}, L^s)$ *for* $s \in [1, \infty)$ *and* $i_2 \colon \dot{\mathcal{B}}^{\{M_p\}} \to \tilde{\dot{\mathcal{B}}}^{\{M_p\}}$ *are continuous.*

Proof. Note that
$$\tilde{\dot{\mathcal{B}}}^{\{M_p\}} = \underset{(a_p) \in \mathcal{R}}{\text{proj lim}} \; X_{\infty, (a_p)},$$
where $X_{\infty, (a_p)}$ is the space of all smooth functions φ such that $\varphi^{(\alpha)} \in \mathcal{C}_0$ for $\alpha \in \mathbb{N}_0^n$ and $\|\varphi\|_{L^\infty, (a_p)} < \infty$, equipped with this norm.

The set theoretical equalities in (i) follow from Lemma 3.4 in [82]. The continuity of the inclusion mappings i_1 and i_2 in (ii) follows from the inequality
$$\|\varphi\|_{L^s, (a_p)} \le C_{(a_p), h} \|\varphi\|_{L^s, h}, \qquad \varphi \in \mathcal{D}(\{M_p\}, h, L^s),$$
where $(a_p) \in \mathcal{R}$, $h > 0$ and $C_{(a_p), h} > 0$ is a suitable constant. The proof of Theorem 3.1.2 is thus completed. $\square$

From here to the end of this section we shall assume that conditions (M.1), (M.2) and (M.3) are satisfied.

The following assertion of Komatsu will be used. Note that the first part of this assertion is also proved in [43].

Lemma 3.1.1. (see [84]) *Let K be a compact neighborhood of zero, $r > 0$ and $(a_p) \in \mathcal{R}$. Then*

(i) there are a $u \in \mathcal{D}((M_p), r/2, K)$ and $\psi \in \mathcal{D}((M_p), K)$ such that
$$P_r(D)u = \delta + \psi, \tag{3.2}$$
where P_r is of the form (2.15).

(ii) there are a $u \in C^\infty$ and $\psi \in \mathcal{D}(\{M_p\}, K)$ such that
$$P_{(a_p)}(D)u = \delta + \psi, \qquad \text{supp}\, u \subset K \tag{3.3}$$
and
$$\sup_{x \in K} \frac{|u^{(\alpha)}(x)|}{A_\alpha M_\alpha} \to 0, \qquad as \qquad \alpha \to \infty, \tag{3.4}$$
where $P_{(a_p)}$ is of the form (2.17).

Theorem 3.1.3. *If $A \subset \mathcal{D}'^*$, then*

(i) *A is a bounded subset of $\mathcal{D}'((M_p), L^t)$ for $t \in [1, \infty]$ if and only if there are an $r > 0$ and bounded sets A_1 and A_2 in L^t such that every $f \in A$ is of the form*

$$f = P_r(D)F_1 + F_2, \qquad F_1 \in A_1, \quad F_2 \in A_2;$$

(ii) *A is an equicontinuos subset of $\tilde{\mathcal{D}}'(\{M_p\}, L^t)$ for $t \in [1, \infty]$ if and only if there are an $(a_p) \in \mathcal{R}$ and bounded sets A_1 and A_2 in L^t such that every $f \in A$ is of the form*

$$f = P_{(a_p)}(D)F_1 + F_2, \qquad F_1 \in A_1, \quad F_2 \in A_2. \tag{3.5}$$

Proof. Notice that we do not know whether the basic space is quasi-barrelled and therefore we assume in (ii) that A is equicontinuous. We shall prove only assertion (ii), because it is more complicated. Since $P_{(a_p)}$ maps continuously the spaces $\mathcal{D}(\{M_p\}, L^s)$, $s \in [1, \infty)$ and $\dot{\mathcal{B}}^{\{M_p\}}$ into themselves, (3.5) implies that A is bounded in $\mathcal{D}'(\{M_p\}, L^t)$.

We shall now prove the converse for $t = s/(s-1)$ with $s \geq 1$. For $t = 1$ (i.e. $s = \infty$) the proof is similar. Let Ω be a bounded open set in $\mathbb{R}^n$ containing zero, $K = \bar{\Omega}$ and $\varphi \in \mathcal{D}(\{M_p\}, K)$.

First we show that, for every $f \in A$, the functional T defined by $T(\varphi) = f * \varphi$ is a continuous linear functional on the space $\mathcal{D}^{\{M_p\}}$ endowed with the topology of L^s. Since A is equicontinuous, there are a constant $C > 0$ (which does not depend on $f \in A$) and an $(a_p) \in \mathcal{R}$ such that

$$\begin{aligned} | <f * \varphi, \psi> | = | <f, \tilde{\varphi} * \psi> | &\leq C \|\tilde{\varphi} * \psi\|_{L^s, (a_p)} \\ &\leq C \|\varphi\|_{K, (a_p)} \|\psi\|_{L^s} \leq C_1 \|\psi\|_{L^s}, \end{aligned} \tag{3.6}$$

for every $\psi \in \mathcal{D}^{\{M_p\}}$, where $\tilde{\varphi}(-x) = \varphi(x)$.

Since $\mathcal{D}^{\{M_p\}}$ is dense in L^s, it follows that $\{f * \varphi \colon f \in A\}$ is a set of (continuous) functions bounded in L^t. Moreover, (3.6) implies that

$$\sup\{\|f * \varphi\|_{L^t} \colon f \in A\} \leq C \|\varphi\|_{K, (a_p)}.$$

Consequently, if B is a bounded set of $\mathcal{D}(\{M_p\}, K)$ then

$$\sup\{\|f * \varphi\|_{L^t} \colon f \in A, \; \varphi \in B\} < \infty.$$

Next, we show that there is (another) $(a_p) \in \mathcal{R}$ such that $\{f * \theta \colon f \in A\}$ is a bounded set in L^t for every $\theta \in \mathcal{D}(\{M_p\}, (a_p), \Omega)$. Let B_1 be the unit ball in L^s and B be a bounded subset of $\mathcal{D}(\{M_p\}, K)$.

We have

$$| < f * \check{\psi}, \check{\varphi} > | = | < f * \varphi, \psi > | \leq \|f * \varphi\|_{L^t} \|\psi\|_{L^s} = \|f * \varphi\|_{L^t}$$

$$\leq D < \infty$$

for all $f \in A$, $\psi \in B_1 \cap \mathcal{D}^{\{M_p\}}$ and $\varphi \in B$, where D does not depend on φ and f. This implies that the set

$$\{f * \check{\psi}: \ f \in A, \ \psi \in B_1 \cap \mathcal{D}^{\{M_p\}}\}$$

is bounded in $\mathcal{D}'(\{M_p\}, K)$. Since $\mathcal{D}(\{M_p\}, K)$ is barrelled, this family is equicontinuous in $\mathcal{D}'(\{M_p\}, K)$. This implies that there exists a neighborhood $V_{(a_p)}(\varepsilon)$ of zero in $\mathcal{D}(\{M_p\}, K)$ of the form:

$$V_{(a_p)}(\varepsilon) = \{\theta \in \mathcal{D}(\{M_p\}, K): \ \|\theta\|_{K,(a_p)} \leq \varepsilon\}$$

with $\varepsilon > 0$, such that $\theta \in V_{(a_p)}(\varepsilon)$ implies that

$$| < f * \check{\psi}, \check{\theta} > | = | < f * \theta, \psi > | \leq 1,$$

for $f \in A$ and $\psi \in B_1 \cap \mathcal{D}^{\{M_p\}}$. The same is true for the closure of $V_{(a_p)}(\varepsilon)$ in $\mathcal{D}(\{M_p\}, (a_p), K)$.

Let now $\delta_k(t) = k\omega(kt)$ for $t \in \mathbb{R}$ and $k \in \mathbb{N}$, where $\omega \in \mathcal{D}^{\{M_p\}}$, $0 \leq \omega \leq 1$ and

$$\int_{\mathbb{R}} \omega(t) \, dt = 1,$$

and let $\delta_k(x) = \delta_k(x_1) \cdot \ldots \cdot \delta_k(x_n)$ for $x \in \mathbb{R}^n$ and $k \in \mathbb{N}$.

One can easily prove that, for every $\mu \in \mathcal{D}(\{M_p\}, (a_p), \Omega)$, the sequence $(\mu * \delta_k)$ of elements of $\mathcal{D}(\{M_p\}, K)$ converges to μ in the norm $\| \cdot \|_{K,(a_p)}$. For an arbitrary $\theta \in \mathcal{D}(\{M_p\}, (a_p), \Omega)$, there is an $N > 0$ such that $\|\theta/N\|_{K,(a_p)} < \varepsilon$ and there exists a sequence in $\overline{V_{(a_p)}(\varepsilon)}$ which converges to θ/N in the norm $\| \cdot \|_{K,(a_p)}$. This implies that, for every $\theta \in \mathcal{D}(\{M_p\}, (a_p), \Omega)$, there is a constant $C > 0$ such that

$$| < f * \check{\psi}, \check{\theta} > | = | < f * \theta, \psi > | \leq C,$$

for $f \in A$ and $\psi \in B_1 \cap \mathcal{D}^{\{M_p\}}$. Consequently,

$$| < f * \theta, \psi > | \leq C\|\psi\|_{L^s}$$

for $f \in A$ and $\psi \in \mathcal{D}^{\{M_p\}}$. This proves that $\{f * \theta: \ f \in A\}$ is a bounded set in L^t for every $\theta \in \mathcal{D}(\{M_p\}, (a_p), \Omega)$.

Lemma 3.1.1 implies that there exist functions $u \in \mathcal{D}(\{M_p\}, (a_p), \Omega)$ and $\psi \in \mathcal{D}(\{M_p\}, \Omega)$ such that

$$f = P_{(a_p)}(u * f) - \psi * f$$

for every $f \in A$. Since $\{u * f : \ f \in A\}$ and $\{\psi * f : \ f \in A\}$ are bounded sets in L^t, the proof is completed. $\square$

Let $r > 0$ (respectively, $(a_p) \in \mathcal{R}$) be given. There is a $\tilde{r} > 0$ (respectively, $(\tilde{a}_p) \in \mathcal{R}$) such that the function $P_r\varphi$ (respectively, $P_{(a_p)}\varphi$) is continuous for $\varphi \in \mathcal{D}((M_p), \tilde{r}/2, K)$ (respectively, $\varphi \in \mathcal{D}(\{M_p\}, (\tilde{a}_p), K)$).

Hence Theorems 3.1.2 and 3.1.3 imply the following corollary:

Corollary 3.1.1. *Suppose that* $f \in \mathcal{D}'^*$. *An ultradistribution* f *is an element of the space* $\mathcal{D}'((M_p), L^t)$ *(respectively,* $\mathcal{D}'(\{M_p\}, L^t)$*) for* $t \in [1, \infty]$ *if and only if for every compact set* K *there is an* $r > 0$ *(respectively,* $(a_p) \in \mathcal{R}$*) such that* $f * \varphi \in L^t$ *for every* $\varphi \in \mathcal{D}((M_p), r/2, K)$ *(respectively,* $\varphi \in \mathcal{D}(\{M_p\}, (a_p), K)$*).*

3.2 Boundedness in $\mathcal{S}'^*$

The following structural theorem is true for tempered ultradistributions.

Theorem 3.2.1. *Let* (M_p) *satisfy conditions* $(M.1)$ *and* $(M.3')$. *A set* $B \subset \mathcal{S}'^{(M_p)}$ *(respectively,* $B \subset \mathcal{S}'^{\{M_p\}}$*) is bounded if and only if every* $f \in B$ *can be represented in the form*

$$f = \sum_{\alpha,\beta \in \mathbb{N}_0^n} \partial^\alpha [< x >^\beta f_{\alpha,\beta}],$$

where $f_{\alpha,\beta} \in L^2$ *for* $\alpha, \beta \in \mathbb{N}_0^n$ *are functions with the following property: for some* $d > 0$ *(respectively, for every* $d > 0$*) there exists a* $D > 0$, *independent of* $f \in B$, *such that*

$$\sum_{\alpha,\beta \in \mathbb{N}_0^n} d^{\alpha+\beta} M_\alpha M_\beta \|f_{\alpha,\beta}\|_{L^2} < D.$$

Proof. We shall prove the assertion only in the more difficult Roumieu case $* = \{M_p\}$.

Note that the space $\mathcal{S}^{\{M_p\}}$ is barrelled and thus B is an equicontinuous subset of $\mathcal{S}'^{\{M_p\}}$.

Let W_h, $h > 0$, be the space of all sequences $(\varphi_{\alpha,\beta}) = (\varphi_{\alpha,\beta})_{\alpha,\beta \in \mathbb{N}_0^n}$ in L^2 such that

$$\|(\varphi_{\alpha,\beta})\|_{L^2,h} = \sup\{\frac{\|\varphi_{\alpha,\beta}\|_{L^2}}{h^{\alpha+\beta} M_\alpha M_\beta} : \ \alpha, \beta \in \mathbb{N}_0^n\} < \infty,$$

equipped with the above defined norm. We identify $\mathcal{S}_h^{(M_p)}$ with the corresponding subspace of W_h. Since the inductive sequences $(W_h)_{h\in\mathbb{N}}$ and $(\mathcal{S}_h^{(M_p)})_{h\in\mathbb{N}}$ are weakly compact and compact, respectively, it follows from Lemma 1.4, (iii) in [82] that the sequence

$$\{0\} \leftarrow \operatorname*{proj\,lim}_{h\,\to\,\infty}\ (\mathcal{S}_h^{(M_p)})' \leftarrow \operatorname*{proj\,lim}_{h\,\to\,\infty}\ W_h'$$

is exact, where

$$\operatorname*{proj\,lim}_{h\,\to\,\infty}\ (\mathcal{S}_h^{(M_p)})' = \mathcal{S}'^{\{M_p\}} = (\operatorname*{ind\,lim}_{h\,\to\,\infty}\ \mathcal{S}_h^{(M_p)})'$$

and

$$\operatorname*{proj\,lim}_{h\,\to\,\infty}\ W_h' = W' = (\operatorname*{ind\,lim}_{h\,\to\,\infty}\ W_h)'.$$

Since the space $\mathcal{S}^{\{M_p\}}$ is Montel, by Lemma 1.4, (v) in [82], it is a closed subspace of W; and, by the Hahn-Banach theorem, the equicontinuous set $B \subset \mathcal{S}'^{\{M_p\}}$ can be extended to the equicontinuous set $\tilde{B}$ in W'. Thus $\tilde{B}$ consists of all sequences $(f_{\alpha,\beta}) = (f_{\alpha,\beta})_{\alpha,\beta\in\mathbb{N}_0^n} \in L^2$ from L^2 such that for every $d \in \mathbb{N}$ there is a constant $C > 0$, independent of the elements of $\tilde{B}$, such that

$$\sum_{\alpha,\beta\in\mathbb{N}_0^n} d^{\alpha+\beta} M_\alpha M_\beta \|f_{\alpha,\beta}\|_{L^2} < C.$$

The mapping T given by

$$T:\ g \mapsto \sum_{\alpha,\beta\in\mathbb{N}_0^n} (-1)^\alpha \partial^\alpha [< x >^\beta g]$$

maps W' onto $\mathcal{S}'^{\{M_p\}}$ and

$$B \subset T(\tilde{B}),$$

which implies the assertion. $\square$

Theorem 3.2.2. *Assume that the sequence (M_p) satisfies conditions $(M.1)$, $(M.2)$ and $(M.3)$. Let $B \subset \mathcal{D}'^{(M_p)}$ (respectively, $B \subset \mathcal{D}'^{\{M_p\}}$). The set B is a bounded subset of $S'^{(M_p)}$ (respectively, B is a bounded subset of $S'^{\{M_p\}}$) if and only if every $f \in B$ is of the form*

$$f = P(D)F, \qquad F \in B_1, \tag{3.7}$$

where P is an operator of class (M_p) (respectively, of class $\{M_p\}$) and B_1 is the set of all continuous functions on $\mathbb{R}^n$ such that for some $k > 0$ and some $C > 0$ (respectively, for every $k > 0$ there is a $C > 0$)

$$|F(x)| \le C \exp[M(k|x|)], \qquad x \in \mathbb{R}^n. \tag{3.8}$$

for all $F \in B_1$.

Proof. We shall prove again the Roumieu case $* = \{M_p\}$, since the idea of the proof is similar in both cases and this case is more complicated.

Clearly, (3.8) implies that the set of all elements f of the form given by (3.7) is a bounded set in $S'^{\{M_p\}}$.

Let B be a bounded set in $S'^{\{M_p\}}$. For the Fourier transform $\hat{f}$ of the ultradistribution $f \in B$ there are $(a_j), (b_j) \in \mathcal{R}$ and a constant $A > 0$ (which do not depend on $f \in B$) such that

$$| < \hat{f}, \varphi > | < A \gamma_{(a_j),(b_j)}(\varphi), \qquad \varphi \in S^{\{M_p\}}. \tag{3.9}$$

To simplify the notation, for given sequences $(a_j), (b_j) \in \mathcal{R}$ and a given $\alpha = (\alpha_1, \ldots, \alpha_n) \in \mathbb{N}_0^n$, denote

$$A_\alpha = \prod_{j=1}^{\bar{\alpha}} a_j, \quad B_\alpha = \prod_{j=1}^{\bar{\alpha}} b_j, \quad S'_\alpha = \prod_{j=1}^{\bar{\alpha}} a_j/2,$$

where $\bar{\alpha} = \alpha_1 + \ldots + \alpha$. For some $D > 0$ and $c > 0$, we have

$$\sup_{\alpha \in \mathbb{N}_0^n} \frac{(1 + |x|^2)^{\alpha/2}}{M_\alpha A_\alpha} \le D \exp[N(c|x|)], \qquad x \in \mathbb{R}^n. \tag{3.10}$$

Let $(\tilde{a}_p)$ and ρ_0 correspond to $(a_p), c$ and C in (2.18), where C is given by (2.20) and c by (3.10). If $\varphi \in \mathcal{D}^{\{M_p\}}$, it follows from (2.18) - (2.20) that

$$\gamma_{(a_j),(b_j)}(\varphi/P_{(\tilde{a}_p)})$$

$$\le \sup_{\alpha,\beta \in \mathbb{N}_0^n} \frac{\|(1 + |x|^2)^{\alpha/2} \sum_{0 \le k \le \beta} \binom{\beta}{k} \partial^{\beta-k} \varphi \partial^k (1/P_{(\tilde{a}_p)})\|_{L^2}}{M_\alpha A_\alpha M_{\beta-k} B_{\beta-k} M_k B_k}$$

$$\le \sup_{\substack{k,\beta \in \mathbb{N}_0^n \\ k \le \beta}} \left\| \sup_{\alpha \in \mathbb{N}_0} \frac{(1 + |x|^2)^{\alpha/2}}{2^\beta M_\alpha A_\alpha} \sum_{0 \le k \le \beta} \binom{\beta}{k} \sup_{0 \le k \le \beta} \frac{|\partial^k (1/P_{(\tilde{a}_p)}) \partial^{\beta-k} \varphi|}{M_k S'_k M_{k-\beta} S'_{k-\beta}} \right\|_{L^2}$$

$$\le D \sup_{\substack{k,\beta \in \mathbb{N}_0^n \\ k \le \beta}} \sup_{0 \le k \le \beta} \frac{k! d^{-k}}{2^\beta M_k S'_k} \left\| e^{N(c|\cdot|) - N(|\cdot|)/C} \sum_{0 \le k \le \beta} \binom{\beta}{k} \frac{|\partial^{\beta-k} \varphi|}{M_{\beta-k} S'_{\beta-k}} \right\|_{L^2}$$

$$\le C_1 \sup_{\substack{k,\beta \in \mathbb{N}_0^n \\ k \le \beta}} 2^{-\beta} \sum_{0 \le k \le \beta} \binom{\beta}{k} \frac{\|\partial^{\beta-k} \varphi\|_{L^2}}{M_{\beta-k} S'_{\beta-k}} \le C_1 \|\varphi\|_{L^2,(b_j/2)}$$

for a certain constant $C_1 > 0$. Thus, (3.9) implies that

$$|\langle \hat{f}/P_{(\tilde{a}_p)}, \varphi \rangle| = |\langle \hat{f}, \varphi/P_{(\tilde{a}_p)} \rangle| \le C_1 \|\varphi\|_{L^2,(b_p/2)}$$

for a suitable constant $C_1 > 0$, all $f \in B$ and all $\varphi \in \mathcal{D}(\{M_p\}, L^2)$. This implies that the set $\{\hat{f}/P_{(\tilde{a}_p)} : f \in B\}$ is equicontinuous in $\tilde{\mathcal{D}}'(\{M_p\}, L^2)$. Hence, by Theorem 3.1.3, every $\hat{f}$ for $f \in B$ is of the form

$$\hat{f}(\xi) = P_{(\tilde{a}_p)}(\xi)[P_{(\tilde{r}_p)}(D)\tilde{F}_1(\xi) + \tilde{F}_2(\xi)], \qquad \tilde{F}_1 \in \tilde{B}_1, \quad \tilde{F}_2 \in \tilde{B}_2,$$

where $\tilde{B}_1$ and $\tilde{B}_2$ are bounded subsets of L^2. Using the inverse Fourier transform, we obtain

$$f(x) = P_{(\tilde{a}_p)}(D)[P_{(\tilde{r}_p)}(x)F_1(x) + F_2(x)], \qquad F_1 \in B_1, \quad F_2 \in B_2,$$

where B_1 and B_2 are bounded subsets of L^2. Put

$$F(x) = \int_0^{x_1} \cdots \int_0^{x_n} [P_{(\tilde{r}_p)}(t)F_1(t) + F_2(t)]\, dt_1 \ldots dt_n,$$

where $x = (x_1, \ldots, x_n) \in \mathbb{R}^n$ and $t = (t_1, \ldots, t_n) \in \mathbb{R}^n$, with $F_1 \in B_1$ and $F_2 \in B_2$ and

$$P(D) = P_{(\tilde{a}_p)}(D)\frac{\partial^n}{\partial x_1 \ldots \partial x_n}.$$

From (2.16) it follows that

$$|F(x)| \le C \exp[\tilde{N}(|x|)](1 + |x|^2)^n \int_0^x \frac{|F_1(t) + F_2(t)|}{(1 + |t|^2)^n}\, dt$$

$$\le C(\|F_1\|_{L^2} + \|F_2\|_{L^2}) \exp[\tilde{N}(|x|)]$$

for $x \in \mathbb{R}^n$. Now, for every $k > 0$ there is a $\rho_k > 0$ such that

$$\tilde{N}(|x|) \le M(k|x|),$$

whenever $|x| > \rho_k$ (see [82], 3.8); and the theorem is proved. $\square$

Chapter 4

Cauchy and Poisson Integrals

4.1 Cauchy and Poisson kernels as ultradifferentiable functions

The first authors who have studied the representations of the Schwartz distributions $\mathcal{D}'_{L^r}$ as boundary values of analytic functions were Tillmann [140] and Łuszczki and Zieleźny [96]. The one-dimensional case of functions analytic in half planes was studied in [96]. In [140], the n-dimensional case was analyzed for functions analytic in the 2^n tubes of the form $T^{C_u} := \mathbb{R}^n + iC_u$, where C_u is any of the 2^n n-rants in $\mathbb{R}^n$ defined in (1.11) by

$$C_u := \{y = (y_1, \ldots, y_n) \in \mathbb{R}^n : \ u_j y_j > 0, \quad j = 1, \ldots, n\} \qquad (4.1)$$

for an arbitrary $u \in \Theta$, where

$$\Theta := \{u = (u_1, \ldots, u_n) \in \mathbb{R}^n : \ u_j = \pm 1, \ j = 1, \ldots, n\}. \qquad (4.2)$$

The values of r considered in these papers were $1 < r < \infty$ and fundamental to the analysis was the property that the Cauchy kernel function corresponding to half planes or tubes $\mathbb{R}^n + iC_u$ is an element of $\mathcal{D}_{L^s}$, $1 < s < \infty$. In [16], Carmichael noticed that the Cauchy kernel for tubes $\mathbb{R}^n + iC_u$ is also an element of $\dot{\mathcal{B}} \cap \mathcal{D}_{L^\infty}$.

In this section we will prove that the general Cauchy kernel defined in Section 1.3 corresponding to a regular cone C is an element of $\mathcal{D}(*, L^s)$, $1 < s \le \infty$, where $*$ is both (M_p) and $\{M_p\}$, and hence is in $\mathcal{D}_{L^s}$, $1 < s \le \infty$. Additionally, we will prove that the Poisson kernel corresponding to the tube $\mathbb{R}^n + iC$ is in $\mathcal{D}(*, L^s)$, $1 \le s \le \infty$, and hence is in $\mathcal{D}_{L^s}$, $1 \le s \le \infty$.

In later sections we will use the Cauchy and Poisson kernels to construct the Cauchy and Poisson integrals of ultradistributions in the corresponding spaces $\mathcal{D}'(*, L^s)$, and we will prove some results about these integrals. We will conclude that an analysis similar to that of Tillmann and of Łuszczki and Zieleźny can be obtained in the more general tube setting of $\mathbb{R}^n + iC$ for all values of r, $1 < r < \infty$, for which their results were obtained in the special cases of $\mathbb{R}^n + iC_u$ and half planes, respectively.

Let C be a regular cone in $\mathbb{R}^n$. We shall consider now the Cauchy and Poisson kernels, corresponding to $\mathbb{R}^n + iC$, defined in Section 1.3.

Theorem 4.1.1. *Let the sequence (M_p) of positive numbers satisfy conditions $(M.1)$ and $(M.3')$. We have $K(z - \cdot) \in \mathcal{D}(*, L^s), 1 < s \leq \infty$, for $z \in T^C$, where the symbol $*$ means either (M_p) or $\{M_p\}$.*

Proof of Theorem 4.1.1 for dimension $n = 1$. First let C be the cone $C = (0, \infty)$ in $\mathbb{R}^1$. We have $C^* = [0, \infty)$ and $K(z - t) = 1/2\pi i(t - z)$ as usual for $z = x + iy \in \mathbb{R}^1 + i(0, \infty)$ and $t \in \mathbb{R}^1$. Let α be a nonnegative integer. We have

$$\frac{d^\alpha K(z - t)}{dt^\alpha} = (-1)^\alpha K^{(\alpha)}(z - t) = (2\pi i)^{-1}(-1)^\alpha \alpha!(t - z)^{-\alpha - 1}.$$

For $1 < s < \infty$, we have

$$\left\| K^{(\alpha)}(z - \cdot) \right\|_{L^s} = \left(\int_{-\infty}^\infty |\alpha!/2\pi i(t - z)^{\alpha + 1}|^s \, dt \right)^{1/s}$$

$$\leq (\alpha!/2\pi y^\alpha)(\int_{-\infty}^\infty ((t - x)^2 + y^2)^{-s/2} \, dt)^{1/s}$$

$$\leq K(s, x, y)(\alpha!/y^\alpha), \tag{4.3}$$

where $K(s, x, y)$ is a constant depending on s, x, and y; we recall that $y > 0$ here. Let $h > 0$ be arbitrary and apply definition (2.7). Since (M_p) satisfies $(M.1)$ and $(M.3')$, we have

$$\alpha! \prec M_\alpha,$$

by Lemma 2.1.2. Thus for $L := hy$ there is a constant $B > 0$ which is independent of α such that (2.6) holds, i.e.,

$$\alpha! \leq B(hy)^\alpha M_\alpha, \qquad \alpha \in \mathbb{N}_0. \tag{4.4}$$

Applying (4.4) to (4.3), we obtain

$$\left\| K^{(\alpha)}(z - \cdot) \right\|_{L^s} \leq BK(s, x, y)h^\alpha M_\alpha, \qquad \alpha \in \mathbb{N}_0, \tag{4.5}$$

for all $h > 0$ which proves that $K(z - \cdot) \in \mathcal{D}((M_p), L^s), 1 < s < \infty$, for $z \in \mathbb{R}^1 + i(0, \infty)$.

For $s = \infty$, and $z \in \mathbb{R}^1 + i(0, \infty)$, we have

$$\left| \frac{d^\alpha K(z - t)}{dt^\alpha} \right| \leq \frac{\alpha!}{2\pi}((t - x)^2 + y^2)^{-(\alpha+1)/2} \leq \frac{\alpha!}{2\pi y^{\alpha+1}}.$$

Again let $h > 0$ be arbitrary. Using (4.4) in the above inequality, we have

$$\left|\frac{d^\alpha K(z-t)}{dt^\alpha}\right| \le (B/2\pi y)h^\alpha M_\alpha, \qquad \alpha \in \mathbb{N}_0,$$

for all $h > 0$ which proves that $K(z - \cdot) \in \mathcal{D}((M_p), L^\infty)$ and, combining our results, we have $K(z - \cdot) \in \mathcal{D}((M_p), L^s), 1 < s \le \infty$, for $z \in \mathbb{R}^1 + iC$ with $C = (0, \infty) \subset \mathbb{R}^1$.

If $C = (-\infty, 0) \subset \mathbb{R}^1$ then $C^* = (-\infty, 0]$; and $K(z - \cdot) \in \mathcal{D}((M_p), L^s), 1 < s \le \infty$, as a function of $t \in \mathbb{R}^1$ for $z = x + iy \in \mathbb{R}^1 + i(-\infty, 0)$ is proved like that for $C = (0, \infty)$ with $|y|$ in place of y in the proof. Since $\mathcal{D}((M_p), L^s) \subset \mathcal{D}(\{M_p\}, L^s)$ we thus have $K(z - \cdot) \in \mathcal{D}(\{M_p\}, L^s)$ also for both cases $C = (0, \infty)$ and $C = (-\infty, 0)$. This completes the proof of Theorem 4.1.1 for dimension $n = 1$. $\square$

Let $u = (u_1, u_2, \ldots, u_n)$ such that $u_j = \pm 1, j = 1, \ldots, n$. Each n-rant $C_u, u \in \Theta$, is a regular cone in $\mathbb{R}^n$ and $C_u^* = \bar{C}_u$. The Cauchy kernel corresponding to the tube $T^{C_u} = \mathbb{R}^n + iC_u$ takes the form

$$K(z - t) = (2\pi i)^{-n} \prod_{j=1}^n \frac{\operatorname{sgn} y_j}{t_j - z_j}, \qquad z = x + iy \in T^{C_u}, \quad t \in \mathbb{R}^n, \quad (4.6)$$

where

$$\operatorname{sgn} y_j = \begin{cases} 1, & y_j > 0, \\ -1, & y_j < 0, \end{cases}$$

for $j = 1, \ldots, n$. Thus for the tubes $\mathbb{R}^n + iC_u$ in $\mathbb{C}^n$, it is clear from the form of $K(z-t)$ that a proof like that in the one dimensional case will yield the desired result of Theorem 4.1.1. This case for the tubes $\mathbb{R}^n + iC_u$ also follows as a special case of the general proof of Theorem 4.1.1 for dimension $n \ge 2$ given below.

Before giving the proof of Theorem 4.1.1 for dimension $n \ge 2$ we first adopt some notation and prove a needed lemma.

Let $a > 0$ be arbitrary but fixed. Let C be a regular cone in $\mathbb{R}^n$ and C^* be its dual cone. Let $x \in \mathbb{R}^n$ and $t \in \mathbb{R}^n$ be arbitrary but fixed. Put

$$C_a^* := \{\eta \in C^*: |\eta| \le a\}; \tag{4.7}$$
$$C_{a,x,t}^* := \{\zeta: \zeta = \eta + i|\eta|(x - t)\eta \in C_a^*\}; \tag{4.8}$$
$$J_{a,x,t}^* := \{\zeta: \zeta = \eta + iau(x - t), \eta \in C^*, |\eta| = a, 0 \le u \le 1.\}. \tag{4.9}$$

The differential form properties used in the following lemma can be found in [1] and [49]. This lemma will be used in the proof of Theorem

4.1.1 for dimension $n \geq 2$ given below; in particular it will be used to obtain equation (4.19) below.

In the sequel of this chapter, it will be convenient to use the notation introduced for exponents in formulas (1.3) and (1.4) of Chapter 1, namely

$$E_z(\zeta) := \exp\left[2\pi i \langle z, \zeta \rangle\right], \qquad z, \zeta \in \mathbb{C}^n, \tag{4.10}$$

and

$$e_y(t) := \exp\left[-2\pi \langle y, t \rangle\right], \qquad y \in \mathbb{R}^n, \ t \in \mathbb{R}^n. \tag{4.11}$$

Lemma 4.1.1. *For every $\alpha \in \mathbb{N}_0^n$, we have*

$$\lim_{a \to +\infty} \int_{J_{a,x,t}^*} \zeta^\alpha E_{z-t}(\zeta) \, d\zeta = 0$$

for $z = x + iy \in T^C$ and $t \in \mathbb{R}^n$, where the symbol E_{z-t} is defined in (4.10).

Proof. The form $d\eta := d\eta_1 \ldots d\eta_n$ with $\eta = (\eta_1, \ldots, \eta_n) \in \mathbb{R}^n$ is an n-form and the set $\{\eta \in \mathbb{R}^n : |\eta| = a\}$ has dimension less than n for every $a > 0$. Thus

$$d\eta_1 \ldots d\eta_n = 0,$$

whenever $|\eta| = a$. Since $du\,du = du \wedge du = 0$, we have

$$d\zeta = d\zeta_1 \ldots d\zeta_n = ia \sum_{j=1}^n (x_j - t_j) d_j,$$

where

$$d_j := d\eta_1 \ldots d\eta_{j-1} \, du \, d\eta_{j+1} \ldots d\eta_n.$$

Letting $E(C^*, a) := \{\eta \in C^* : |\eta| = a\}$, we obtain

$$I_a = \int_{J_{\alpha,x,t}^*} \zeta^a E_{z-t}(\zeta) \, d\zeta = ia \int_{E(C^*,a)} \int_0^1 (\zeta_\eta(x-t))^\alpha E_{z-t}(\zeta_\eta(x-t))\langle x-t, d \rangle, \tag{4.12}$$

where the symbol $E_z(\zeta)$ for $z, \zeta \in \mathbb{C}^n$ is defined in (4.10), $\zeta_\eta(x) := \eta + iaux$ and $\langle x, d \rangle := \sum_{j=1}^n x_j d_j$ for $x = (x_1, \ldots, x_n) \in \mathbb{R}^n$. From the last term in (4.12), we have $I_a = 0$ for $x = t$ and for every $a > 0$. Thus the remainder of the proof proceeds under the assumption that $x \neq t$. For an arbitrary $a > 0$, it follows from (4.12) that

$$|I_a| \leq a \int_{E(C^*,a)} \int_0^1 e_{a|x-t|^2}(u) e_y(\eta) \prod_{j=1}^n (|\eta_j| + au|x_j - t_j|)^{\alpha_j} \langle |x-t|, d \rangle, \tag{4.13}$$

where $\langle |x-t|, d \rangle := \sum_{j=1}^{n} |x_j - t_j| d_j$ and the symbol e_y is defined in (4.11).

Let now $y \in C$ be fixed. By Lemma 1.2.2, there exists a $\delta = \delta_y > 0$ depending on y such that

$$< y, \eta > \geq \delta |y| |\eta|, \qquad \eta \in C^*. \tag{4.14}$$

Using in (4.13) inequality (4.14) and the estimate

$$\prod_{j=1}^{n} (|\eta_j| + au|x_j - t_j|)^{\alpha_j} \leq (|\eta| + au|x - t|)^{\alpha},$$

we have

$$|I_a| \leq a \int_{E(C^*,a)} \int_0^1 (|\eta| + au|x - t|)^{\alpha} e_{a|x-t|^2}(u) e_{\delta|y|}(|\eta|) \sum_{j=1}^{n} |x_j - t_j| d_j$$

$$= a^{\alpha+1} e_{\delta a}(|y|) \|x - t\| \int_0^1 (1 + u|x - t|)^{\alpha} e_{a|x-t|^2}(u) \, du \cdot I(C^*, a)$$

$$\leq a^{\alpha+1} S(a) e_{\delta a}(|y|) \|x - t\| \int_0^1 (1 + u|x - t|)^{\alpha} e_{a|x-t|^2}(u) \, du, \tag{4.15}$$

where

$$I(C^*, a) := \int_{E(C^*,a)} 1 \, d\eta_1 \dots d\eta_{j-1} \, d\eta_{j+1} \dots d\eta_n$$

and $S(a) := (2\pi^{n/2} a^{n-1}/\Gamma(n/2))$ is the surface area of the sphere of radius $a > 0$ in $\mathbb{R}^n$. Since

$$(1 + u|x - t|)^{\alpha} \leq (1 + |x - t|)^{\alpha}, \qquad 0 \leq u \leq 1,$$

we get from (4.15), for $x \neq t$, the inequality

$$|I_a| \leq a^{\alpha+1} S(a)(1 + |x - t|)^{\alpha} \|x - t\| \exp\left[-2\pi\delta a|y|\right]$$
$$\cdot (1 - \exp\left[-2\pi a|x - t|^2\right])(2\pi a|x - t|^2)^{-1}. \tag{4.16}$$

For $x \neq t$, the right side of (4.16) approaches 0 as $a \to +\infty$. It suffices to combine this with the previously noted fact that $I_a = 0$ for every $a > 0$ with $x = t$ to complete the proof of Lemma 4.1.1. $\square$

Proof of Theorem 4.1.1 for dimension $n \geq 2$.

Let α be an arbitrary n-tuple of nonnegative integers. For fixed $t \in \mathbb{R}^n$ and $z = x + iy \in T^C$, the function $\Psi_{z,t}$ given by

$$\Psi_{z,t}(\zeta) := \zeta^{\alpha} \exp\left[2\pi i < z - t, \zeta >\right]$$

is an entire analytic function of ζ in $\mathbb{C}^n$. Thus by the discussion made in [147] (Section IV.22.6, p. 198), the form

$$\Psi_{z,t}(\zeta)\, d\zeta_1\, d\zeta_2\, \ldots\, d\zeta_n$$

is a closed differential form; that is

$$d(\zeta^\alpha \exp\left[2\pi i < z - t, \zeta >\right] d\zeta_1\, d\zeta_2\, \ldots\, d\zeta_n) = 0$$

and

$$\int_{E_{a,x,t}^*} \zeta^\alpha \exp\left[2\pi i < z - t, \zeta >\right] d\zeta_1\, d\zeta_2\, \ldots\, d\zeta_n = 0 \qquad (4.17)$$

for each $a > 0$, $x \in \mathbb{R}^n$, and $t \in \mathbb{R}^n$, where

$$E_{a,x,t}^* := C_a^* \cup \overline{C_{a,x,t}^*} \cup J_{a,x,t}^* \qquad (4.18)$$

with $\overline{C_{a,x,t}^*}$ denoting $C_{a,x,t}^*$, as defined in (4.8), suitably oriented. Now recall the definition of the Cauchy kernel function in (1.15), the sets defined in (4.7)-(4.9), and Lemma 4.1.1. Using these we let $a \to +\infty$ in (4.17) and obtain

$$K^{(\alpha)}(z - t) = \int_{C^*} \eta^\alpha E_{z-t}(\eta)\, d\eta = \int_{C_{x,t}^*} \zeta^\alpha E_{z-t}(\zeta)\, d\zeta, \qquad (4.19)$$

where the symbol E_{z-t} is defined in (4.10) and

$$C_{x,t}^* := \{\zeta\colon\ \zeta = \eta + i|\eta|(x - t),\ \ \eta \in C^*\}. \qquad (4.20)$$

Between $d\eta = d\eta_1 \ldots d\eta_n$ in the first integral and $d\zeta = d\zeta_1 \ldots d\zeta_n$ in the second integral in (4.19), we have the following relationship:

$$\begin{aligned}
d\zeta_j &= d\eta_j + i(x_j - t_j)\left(\frac{\partial|\eta|}{\partial\eta_1}d\eta_1 + \ldots + \frac{\partial|\eta|}{\partial\eta_n}d\eta_n\right) \\
&= d\eta_j + i(x_j - t_j)\left(\frac{\eta_1}{|\eta|}d\eta_1 + \ldots + \frac{\eta_n}{|\eta|}d\eta_n\right).
\end{aligned} \qquad (4.21)$$

Using the following differential properties:

$$d\eta_j d\eta_k = d\eta_j \wedge d\eta_k = -d\eta_k \wedge d\eta_j = -d\eta_k d\eta_j, \qquad j \neq k,$$
$$d\eta_j d\eta_j = d\eta_j \wedge d\eta_j = 0$$

(see e.g. [1] or [49]), we obtain, exactly as in [60] (p. 359, the last line), the relation

$$d\zeta = \left(1 + i\frac{< x - t, \eta >}{|\eta|}\right) d\eta \qquad (4.22)$$

from (4.21). Thus from (4.19) and (4.22) we get

$$K^{(\alpha)}(z - t) = \int_{C^*} (k_{x-t}(\eta))^\alpha E_{z-t}(k_{x-t}(\eta)) l_{x-t}(\eta) \, d\eta, \qquad (4.23)$$

where

$$k_{x-t}(\eta) := \eta + i|\eta|(x - t); \qquad l_{x-t}(\eta) := 1 + i\frac{<x - t, \eta>}{|\eta|}.$$

Now

$$|(k_{x-t}(\eta))^\alpha| = \left| \prod_{j=1}^{n} (\eta_j + i|\eta|(x_j - t_j))^{\alpha_j} \right| \le |\eta|^\alpha (1 + |x - t|)^\alpha \qquad (4.24)$$

and

$$|l_{x-t}(\eta)| \le 1 + |\langle x - t, \eta \rangle|/|\eta| \le 1 + |x - t||\eta|/|\eta| = 1 + |x - t|. \qquad (4.25)$$

Using (4.14), (4.24) and (4.25) in (4.23), we have

$$|K^{(\alpha)}(z - t)| \le (1 + |x - t|)^{\alpha+1} \int_{C^*} |\eta|^\alpha e_{\delta|y| + |x-t|^2}(|\eta|) \, d\eta, \qquad (4.26)$$

where the symbol e_y is defined in (4.11). Letting pr(C^*) denote the projection of C^*, which is the intersection of C^* with the unit sphere in $\mathbb{R}^n$, we change variables in the integral in (4.26) by letting $\psi \in \mathrm{pr}(C^*)$ and $0 < r < \infty$ and obtain

$$|D_t^\alpha K(z - t)| \le (1 + |x - t|)^{\alpha+1} \int_{\mathrm{pr}(C^*)} \int_0^\infty e_{\delta|y| + |x-t|^2}(r) r^{\alpha+n-1} \, dr \, d\psi$$

$$\le S_{C^*}(1 + |x - t|)^{\alpha+1} \int_0^\infty e_{\delta|y| + |x-t|^2}(r) r^{\alpha+n-1} \, dr, \qquad (4.27)$$

where S_{C^*} is the surface area of pr(C^*). It follows (see [33], p.93, (4.6) or [34] p. 60, (3.5)) that

$$\sup_{r \ge 0} r^\alpha \exp\left[-\pi r(\delta|y| + |x - t|^2)\right] \le B_{\alpha, \delta, y}(|x - t|), \qquad (4.28)$$

where

$$B_{\alpha, \delta, y}(u) := \begin{cases} 1, & \alpha = 0, \\ (\pi(\delta|y| + u^2)/\bar{\alpha})^{-\alpha}, & \alpha \ne 0, \end{cases}$$

and $\bar{\alpha} := \alpha_1 + \ldots + \alpha_n$. Using (4.28) and integrating by parts, we obtain from (4.27) the estimate

$$|D_t^\alpha K(z - t)| \le S_{C^*}\Gamma(n)(1 + |x - t|)^{\alpha+1}\pi^{-n-\alpha}$$

$$\cdot (\delta|y| + |x - t|^2)^{-n-\alpha}\bar{\alpha}^\alpha \qquad (4.29)$$

with the convention that $0^0 := 1$. Thus we have, for $1 < s < \infty$,

$$\int_{\mathbb{R}^n} |D_t^\alpha K(z - t)|^s \, dt \leq \left(\frac{S_{C^*} \Gamma(n)}{\pi^n}\right)^s \pi^{-s\alpha} \bar{\alpha}^{s\alpha}$$
$$\cdot \int_{\mathbb{R}^n} \frac{(1 + |x - t|)^{s(\alpha+1)}}{(\delta|y| + |x - t|^2)^{s(\alpha+n)}} \, dt. \qquad (4.30)$$

Here $z + iy \in T^C$ is arbitrary but fixed, $\delta = \delta_y > 0$ is fixed, and $n \geq 2$. Using (4.30), straightforward estimation shows

$$\int_{\mathbb{R}^n} |D_t^\alpha K(z - t)|^s \, dt \leq (S_{C^*} \Gamma(n)/\pi^n)^s \pi^{-s\alpha} \bar{\alpha}^{s\alpha} M_\alpha(s, \delta, y, n), \qquad (4.31)$$

where

$$M_\alpha(s, \delta, y, n) := \begin{cases} M_1(s, \delta, y, n)(\frac{2}{\delta|y|})^{s\alpha} + M_1'(s, n), & \text{if } \delta|y| \geq 1, \\ M_2(s, \delta, y, n)(\frac{2}{\delta|y|})^{s\alpha}, & \text{if } \delta|y| < 1. \end{cases}$$

Here $M_1(s, \delta, y, n), M_1'(s, n)$, and $M_2(s, \delta, y, n)$ are positive constants which depend on the parameters listed.

Using (4.31), we get

$$\|K^{(\alpha)}(z - \cdot)\|_{L^s} \leq (S_{C^*} \Gamma(n)/\pi^n) \pi^{-\alpha} \bar{\alpha}^\alpha N_\alpha(s, \delta, y, n), \qquad (4.32)$$

where

$$N_\alpha(s, \delta, y, n) := \begin{cases} N_1(s, \delta, y, n)(\frac{2}{\delta|y|})^\alpha, & \text{if } \delta|y| \leq 2, \\ N_2(s, \delta, y, n), & \text{if } \delta|y| > 2, \end{cases}$$

for $1 < s < \infty$. Since the sequence (M_p) satisfies $(M.1)$ and $(M.3')$ we have $\bar{\alpha}^\alpha \prec M_\alpha$, by Lemma 2.1.2.

Let $h > 0$ be arbitrary. For the fixed $y = \text{Im } z \in C$ and $\delta = \delta_y > 0$ put $L = (h\pi\delta|y|/2)$ if $\delta|y| \leq 2$ or $L = h\pi$ if $\delta|y| > 2$. From (2.6) and (4.32) there is a constant $B > 0$ which is independent of α such that

$$\|K^{(\alpha)}(z - \cdot)\|_{L^s} \leq (S_{C^*} \Gamma(n)/\pi^n) R(s, \delta, y, n) B h^\alpha M_\alpha \qquad (4.33)$$

for all $h > 0$, where $R(s, \delta, y, n)$ is a constant depending on the stated parameters. Now (4.33) proves that $K(z - \cdot) \in \mathcal{D}((M_p), L^s)$, $1 < s < \infty$, for $z \in T^C$ and $n \geq 2$.

For the case $s = \infty$ and $n \geq 2$ we return to (4.29) where $z = x + iy \in T^C$ and $\delta = \delta_y > 0$ are fixed. We have

$$\frac{(1 + |x - t|)^{\alpha+1}}{(\delta|y| + |x - t|^2)^{\alpha+n}} \leq \max\left\{\frac{2^{\alpha+1}}{(\delta|y|)^{\alpha+n}}, 2^{\alpha+1}\right\} \qquad (4.34)$$

for all $t \in \mathbb{R}^n$ and for fixed $y \in C$ and $\delta = \delta_y > 0$. Let $h > 0$ be arbitrary and put $L = (h\pi\delta|y|)/2$ if $\delta|y| < 1$ or $L = h\pi/2$ if $\delta|y| \geq 1$. From (4.29), (4.34) and the fact that $\bar{a}^\alpha \prec M_\alpha$ here, we obtain

$$|D_t^\alpha K(z-t)| \leq (S_{C^*}\Gamma(n)/\pi^n)\max\{2/(\delta|y|)^n, 2\}Bh^\alpha M_\alpha, \qquad (4.35)$$

where the constant $B > 0$ is independent of α, which proves that $K(z - \cdot) \in \mathcal{D}((M_p), L^\infty)$ for $z \in T^C$ and $n \geq 2$.

Thus we have $K(z - \cdot) \in \mathcal{D}((M_p), L^s) \subset \mathcal{D}(\{M_p\}, L^s)$ for $1 < s \leq \infty$, for $z \in T^C$ and for $n \geq 2$. The proof is complete. $\square$

We recall that $K(z-\cdot)$ cannot be in $\mathcal{D}(*, L^1)$ or $\mathcal{D}_{L^1}$ since $K(z-\cdot)$ is not in L^1 for $z \in T^C$. Theorem 4.1.1 additionally proves that $K(z - \cdot) \in \mathcal{D}_{L^s}$ for $1 < s \leq \infty$ and for $z \in T^C$, since $\mathcal{D}(*, L^s) \subset \mathcal{D}_{L^s}$ for $1 < s \leq \infty$.

For a regular cone C we now consider the Poisson kernel $Q(z;t)$, $z \in T^C = \mathbb{R}^n + iC$, $t \in \mathbb{R}^n$, defined in equation (1.17).

Theorem 4.1.2. *Let the sequence of positive real numbers* (M_p), $p = 0, 1, 2, \ldots$, *satisfy* (M.1) *and* (M.3′). *We have* $Q(z; \cdot) \in \mathcal{D}(*, L^s)$, $1 \leq s \leq \infty$ *for* $z \in T^C$ *where* $*$ *is either* (M_p) *or* $\{M_p\}$.

Proof. From (1.15) and Lemma 1.2.1 we first note that $K(2iy) > 0$, $y \in C$, in (1.17), and Lemma 1.3.5 holds since C is a regular cone by assumption. Now let $s = 1$. Let $z = x + iy \in T^C$ be arbitrary but fixed, and let α be an arbitrary n-tuple of nonnegative integers. By the generalized Leibniz rule

$$D_t^\alpha Q(z;t) = \frac{1}{K(2iy)} \sum_{\beta+\gamma=\alpha} \frac{\alpha!}{\beta!\gamma!} D_t^\beta K(z-t) D_t^\gamma \overline{K(z-t)}, \qquad (4.36)$$

and $Q(z;t)$ is infinitely differentiable as a function of $t \in \mathbb{R}^n$. Using (4.36) and Hölder's inequality we have

$$\int_{\mathbb{R}^n} |D_t^\alpha Q(z;t)|\, dt$$

$$\leq \frac{1}{K(2iy)} \sum_{\beta+\gamma=\alpha} \frac{\alpha!}{\beta!\gamma!} \int_{\mathbb{R}^n} |D_t^\beta K(z-t) D_t^\gamma \overline{K(z-t)}|\, dt$$

$$\leq \frac{1}{K(2iy)} \sum_{\beta+\gamma=\alpha} \frac{\alpha!}{\beta!\gamma!} \|K^{(\beta)}(z-\cdot)\|_{L^2} \|\overline{K}^{(\gamma)}(z-\cdot)\|_{L^2} \qquad (4.37)$$

for every n-tuple α of nonnegative integers. By the proof of Theorem 4.1.1, $\overline{K(z-\cdot)} \in \mathcal{D}(*, L^s)$, $1 < s \leq \infty$, for $z \in T^C$. Let $h > 0$ be arbitrary.

Thus by Theorem 4.1.1 there exist constants N and N' which are both independent of h and α such that

$$\|K^{(\beta)}(z-\cdot)\|_{L^s} \leq Nh^\beta M_\beta, \qquad \beta \in \mathbb{N}_0^n, \tag{4.38}$$

and

$$\|\overline{K}^{(\gamma)}(z-\cdot)\|_{L^2} \leq N'h^\gamma M_\gamma, \qquad \gamma \in \mathbb{N}_0^n. \tag{4.39}$$

Combining (4.37), (4.38) and (4.39) and using (2.3), we have

$$\int_{R^n} |D_t^\alpha Q(z;t)|\, dt \leq \frac{1}{K(2iy)} \sum_{\beta+\gamma=\alpha} \frac{\alpha!}{\beta!\gamma!} NN'h^{\beta+\gamma} M_\beta M_\gamma$$

$$\leq \frac{NN'M_0}{K(2iy)} \Big(\sum_{\beta+\gamma=\alpha} \frac{\alpha!}{\beta!\gamma!} \Big) h^\alpha M_\alpha \tag{4.40}$$

for each $\alpha \in \mathbb{N}_0^n$. Thus

$$Q(z;\cdot) \in \mathcal{D}((M_p), L^1) \subset \mathcal{D}(\{M_p\}, L^1).$$

Now let $1 < s \leq \infty$ and let $h > 0$ be arbitrary. Let α be an arbitrary n-tuple of nonnegative integers. By Theorem 4.1.1 (see (4.33) and (4.35)) there exist constants N and N' which are independent of h and of $\alpha \in \mathbb{N}_0^n$, such that

$$|D_t^\gamma \overline{K(z-t)}| \leq Nh^\gamma M_\gamma, \qquad \gamma \in \mathbb{N}_0^n, \tag{4.41}$$

and

$$\|D^\beta K(z-\cdot)\|_{L^s} \leq N'h^\beta M_\beta, \qquad \beta \in \mathbb{N}_0^n, \tag{4.42}$$

where $\beta + \gamma = \alpha$. Using (4.36), (4.41), (4.42), and (2.3), we proceed as in (4.40) to obtain

$$\|D^\alpha Q(z;\cdot)\|_{L^s} \leq \frac{1}{K(2iy)} \sum_{\beta+\gamma=\alpha} \frac{\alpha!}{\beta!\gamma!} \|K^{(\beta)}(z-\cdot)\overline{K}^{(\gamma)}(z-\cdot)\|_{L^s}$$

$$\leq \frac{1}{K(2iy)} \sum_{\beta+\gamma=\alpha} \frac{\alpha!}{\beta!\gamma!} (Nh^\gamma M_\gamma)\|K^{(\beta)}(z-\cdot)\|_{L^s}$$

$$\leq \frac{1}{K(2iy)} \sum_{\beta+\gamma=\alpha} \frac{\alpha!}{\beta!\gamma!} (Nh^\gamma M_\gamma)(N'h^\beta M_\beta)$$

$$\leq \frac{NN'M_0}{K(2iy)} \Big(\sum_{\beta+\gamma=\alpha} \frac{\alpha!}{\beta!\gamma!} \Big) h^\alpha M_\alpha \tag{4.43}$$

for each $\alpha \in \mathbb{N}_0^n$. Thus $Q(z;\cdot) \in \mathcal{D}((M_p), L^s) \subset \mathcal{D}(\{M_p\}, L^s)$ for $1 < s \leq \infty$. The proof is complete. $\square$

For the n-rant C_u, defined in the introduction to this section, the Cauchy kernel takes the form

$$K(z - t) = (2\pi i)^{-n} \prod_{j=1}^{n} \frac{\operatorname{sgn} y_j}{t_j - z_j}$$

for $z = x + iy \in T^{C_u}$ and $t \in \mathbb{R}^n$, where

$$\operatorname{sgn} y_j := \begin{cases} 1, & y_j > 0, \\ -1, & y_j < 0, \end{cases}$$

for $j = 1, \ldots, n$. From (1.17), the Poisson kernel takes the form

$$Q(z; t) = (\pi)^{-n} \prod_{j=1}^{n} \frac{(\operatorname{sgn} y_j) y_j}{(t_j - x_j)^2 + y_j^2}$$

for $z = x + iy \in T^{C_u}$ and $t \in \mathbb{R}^n$, where $\operatorname{sgn} y_j$ is as above. These forms of $K(z - t)$ and $Q(z; t)$ are special cases to which Theorems 4.1.1 and 4.1.2, respectively, are applicable.

4.2 Cauchy integral of ultradistributions

Let C be a regular cone in $\mathbb{R}^n$, and let the sequence of positive real numbers (M_p) satisfy $(M.1)$ and $(M.3')$. Let $U \in \mathcal{D}'(*, L^s)$, $1 < s \leq \infty$, where we recall that $*$ means either (M_p) or $\{M_p\}$. Because of Theorem 4.1.1 we can form

$$C(U; z) = \langle U, K(z - \cdot) \rangle, \qquad z \in T^C, \tag{4.44}$$

which is the Cauchy integral of U. In this section we show that this Cauchy integral is an analytic function in T^C, has both pointwise and norm growth properties, and has boundary value properties.

Theorem 4.2.1. *Let C be a regular cone in $\mathbb{R}^n$ and let the sequence (M_p) satisfy $(M.1)$ and $(M.3')$. Let $U \in \mathcal{D}'(*, L^s)$, $1 < s < \infty$. The Cauchy integral $C(U; z)$ is analytic in T^C.*

Proof. We first give the proof for $\mathcal{D}'((M_p), L^s)$. From Theorem 2.3.2 we have

$$C(U; z) = \sum_{\alpha \in \mathbb{N}_0} (-1)^\alpha \int_{\mathbb{R}^n} g_\alpha(t) D_t^\alpha K(z - t) \, dt, \qquad z \in T^C, \tag{4.45}$$

where the $g_\alpha \in L^r$, $1/r + 1/s = 1$, such that (2.47) holds for some $k > 0$. Let Q be a compact subset of T^C. From the proof of Theorem 4.1.1 (recall

(4.5) and (4.33)) there is a constant $B(s, Q, n)$ depending on s, the compact set Q, and the dimension n such that

$$\|K^{(\alpha)}(z - \cdot)\|_{L^s} \leq B(s, Q, n) h^\alpha M_\alpha \qquad (4.46)$$

for all $h > 0$, where $z \in Q \subset T^C$ and $\alpha \in \mathbb{N}_0^n$. For each g_α in (4.45) we see, from (4.46) and Hölder's inequality, that

$$\int_{\mathbb{R}^n} |g_\alpha(t) D_t^\alpha K(z - t)| \, dt \leq \|g_\alpha\|_{L^r} \|K^{(\alpha)}(z - \cdot)\|_{L^s}$$

$$\leq B(s, Q, n) h^\alpha M_\alpha \|g_\alpha\|_{L^r} \qquad (4.47)$$

for all $h > 0$, where $z \in Q \subset T^C$. Recall that (2.47) is equivalent to (2.49) on the functions g_α in (4.45). Using (4.47) and (2.49) and choosing the h in (4.47) to be $k/2$ for the k in (2.49) we conclude that there exists a constant $D > 0$ such that

$$\int_{\mathbb{R}^n} |g_\alpha(t) D_t^\alpha K(z - t)| \, dt \leq B(s, Q, n) D (1/2)^\alpha$$

and, consequently,

$$\sum_{\alpha \in \mathbb{N}_0} \int_{\mathbb{R}^n} |g_\alpha(t) D_t^\alpha K(z - t)| \, dt \leq B(s, Q, n) D \sum_{\alpha \in \mathbb{N}_0} (1/2)^\alpha < \infty. \qquad (4.48)$$

The bound on the right of (4.48) is independent of $z \in Q \subset T^C$. The resulting uniform and absolute convergence of the series in (4.45) for $z \in Q$, where Q is an arbitrary compact subset of T^C, proves that $C(U; z)$ is analytic in T^C.

Using Theorem 2.3.1, the result that $C(U; z)$ is analytic in T^C for $U \in \mathcal{D}'(\{M_p\}, L^s)$, $1 < s < \infty$, can similarly be proved. This completes the proof of Theorem 4.2.1. $\square$

We obtain a pointwise growth estimate for the Cauchy integral after proving a needed lemma.

Lemma 4.2.1. *Let C be a regular cone in $\mathbb{R}^n$. For every n-tuple α of nonnegative integers,*

$$h_{y,\alpha} \in L^r \qquad (4.49)$$

for all $r, 1 \leq r \leq \infty$, and for $y \in C$, where

$$h_{y,\alpha}(t) := t^\alpha I_{C^*}(t) e_y(t), \qquad t \in \mathbb{R}^n,$$

with the symbol e_y defined in (4.11), and I_{C^*} is the characteristic function of the dual cone C^* of C.

Proof. Let $y \in C$. By Lemma 1.2.2 there is a $\delta = \delta_y > 0$ such that (1.12) holds for $t \in C^*$. Let α be an arbitrary n-tuple of nonnegative integers. For $r = \infty$ we apply (1.12) and, using the notation from (4.11), obtain

$$|t^\alpha I_{C^*}(t) e_y(t)| \leq \sup_{t \in C^*} |t|^\alpha e^{-2\pi\delta|y||t|} \leq \sup_{\rho \geq 0, u \in \mathrm{pr}(C^*)} |\rho u|^\alpha e_{\delta\rho|y|}(|u|)$$

$$= \sup_{\rho \geq 0} \rho^\alpha e^{-2\pi\delta\rho|y|} \qquad (4.50)$$

for all $t \in \mathbb{R}^n$ and for $y \in C$. From the above calculations we conclude that

$$\sup_{\rho \geq 0} \rho^\alpha e^{-2\pi\delta\rho|y|} \leq \begin{cases} 1, & \alpha = 0, \\ \left(\dfrac{\bar{\alpha}}{2\pi\delta|y|}\right)^\alpha, & \alpha \neq 0. \end{cases} \qquad (4.51)$$

From (4.50) and (4.51) we conclude (4.49) for $r = \infty$, where $y \in C$.

Now, let $1 \leq r < \infty$. From (1.12) and a calculation as in (1.20) we have

$$\int_{\mathbb{R}^n} |t^\alpha I_{C^*}(t) e_y(t)|^r \, dt \leq \int_{\mathbb{R}^n} I_{C^*}(t) |t|^{r\alpha} e_{r\delta|y|}(|t|) \, dt$$

$$\leq \Omega_n \int_0^\infty w^{r\alpha+n-1} e_{r\delta|y|}(w) \, dw = \Omega_n \Gamma(r\alpha + n)(2\pi r\delta|y|)^{-r\alpha-n}$$

$$(4.52)$$

for $y \in C$, where Ω_n is the surface area of the unit sphere in $\mathbb{R}^n$ and the change of variable for $u = 2\pi r\delta w|y|$ was used to obtain the gamma function Γ. Now (4.52) proves (4.49) for $1 \leq r < \infty$. $\square$

Theorem 4.2.2. *Let the cone C and the sequence (M_p) satisfy the hypotheses of Theorem 4.2.1. If $U \in \mathcal{D}'((M_p), L^s)$, $2 \leq s < \infty$, for each compact subcone $C' \subset\subset C$ there are constants $A = A(n, C', s) > 0$ and $T = T(C') > 0$ such that*

$$|C(U; z)| \leq A|y|^{-n/r} \exp[M^*(T/|y|)], \qquad z = x + iy \in \mathbb{R}^n + iC', \quad (4.53)$$

where n is the dimension, $1/r + 1/s = 1$, and M^* is the function defined in (2.9). If $U \in \mathcal{D}'(\{M_p\}, L^s)$, $2 \leq s < \infty$, for each compact subcone $C' \subset\subset C$ and arbitrary constant $T > 0$, which is independent of $C' \subset\subset C$, there is a constant $A = A(n, C', s) > 0$ such that (4.53) holds.

Proof. Let $U \in \mathcal{D}'((M_p), L^s)$, $2 \leq s < \infty$, and let C' be an arbitrary compact subcone of C. From (4.45) we have

$$|C(U; z)| \leq \sum_{\alpha \in \mathbb{N}_0} \|K^{(\alpha)}(z - \cdot)\|_{L^s} \|g_\alpha\|_{L^r}, \qquad z \in T^C, \quad (4.54)$$

where $g_\alpha \in L^r$, $1/r + 1/s = 1$, such that (2.47) holds for some $k > 0$. For $2 \leq s < \infty$ we have $1 < r \leq 2$, $1/r + 1/s = 1$. For these s and subsequent r we apply Lemma 4.2.1 and so we can write, applying the notation from (4.10),

$$D_t^\alpha K(z - t) = \mathcal{F}^{-1}[I_{C^*} x^\alpha E_z](t), \qquad z \in T^C, \ t \in \mathbb{R}^n, \qquad (4.55)$$

where the inverse Fourier transform can be interpreted in both the L^1 and L^r sense. By the Plancherel theory of Fourier analysis and the analysis of (4.50), (4.51), and (4.52), it follows from (4.55) that, for $z = x + iy \in \mathbb{R}^n + iC'$ and $C' \subset\subset C$,

$$\|K^{(\alpha)}(z - \cdot)\|_{L^s} \leq \left(\int_{C^*} |\eta|^{r\alpha} e^{-2\pi r \langle y, \eta \rangle} \, d\eta \right)^{1/r}$$

$$\leq (\Omega_n)^{1/r} \left(\int_0^\infty w^{r\alpha + n - 1} \exp\left[-2\pi \delta r w |y| \right] dw \right)^{1/r}$$

$$\leq (\Omega_n)^{1/r} (\sup_{w \geq 0} (w^{r\alpha} e^{-\pi \delta r w |y|}))^{1/r} \left(\int_0^\infty u^{n-1} e^{-\pi \delta r w |y|} \, dw \right)^{1/r}$$

$$= (\Omega_n (n - 1)!)^{1/r} (\pi \delta r |y|)^{-n/r} \sup_{w \geq 0} (w^\alpha e^{-\pi \delta w |y|})$$

$$\leq \begin{cases} (\Omega_n (n - 1)!)^{1/r} (\pi \delta r |y|)^{-n/r}, & \alpha = 0, \\ (\Omega_n (n - 1)!)^{1/r} (\pi \delta r |y|)^{-n/r} \left(\dfrac{\bar{\alpha}}{\pi \delta |y|} \right)^\alpha, & \alpha \in \mathbb{N}^n. \end{cases} \qquad (4.56)$$

Using (4.56) in (4.54) we have

$$|C(U; z)| \leq \tilde{A}(n, C', s) |y|^{-n/r} \sum_{\alpha \in \mathbb{N}_0} \left(\frac{\bar{\alpha}}{\pi \delta |y|} \right)^\alpha \|g_\alpha\|_{L^r} \qquad (4.57)$$

for $z = x + iy \in T^{C'} = \mathbb{R}^n + iC'$, where

$$\tilde{A}(n, C', s) = (\Omega_n (n - 1)!)^{1/r} (\pi \delta r)^{-n/r}$$

with $1/r + 1/s = 1$, $2 \leq s < \infty$. Using (2.49), the fact that

$$\bar{\alpha}^\alpha \leq e^\alpha \bar{\alpha}!, \qquad \bar{\alpha} = 1, 2, 3, \dots, \qquad (4.58)$$

from the proof of Stirling's formula, and our convention that $\bar{\alpha}^\alpha = 1$ if $\alpha = 0$, and putting $T = T(C') = (2e/k\pi\delta)$ for the k in (2.49), we continue (4.57) as

$$|C(U; z)| \leq B\tilde{A}(n, C', s) |y|^{-n/r} \sum_{\alpha \in \mathbb{N}_0} \left(\frac{1}{2} \right)^\alpha \left(\frac{T}{|y|} \right)^\alpha \frac{\bar{\alpha}!}{M_\alpha} \qquad (4.59)$$

for some $B > 0$. Recalling the definition of the associated function $M^*(\rho)$ in (2.9), we have for each $\bar{\alpha} = 1, 2, 3, \ldots$ that

$$\left(\frac{T}{|y|}\right)^\alpha \frac{\bar{\alpha}!}{M_\alpha} = \frac{1}{M_0}\left(\left(\frac{T}{|y|}\right)^\alpha \bar{\alpha}! \frac{M_0}{M_\alpha}\right) = \frac{1}{M_0} \exp\left[\log\left(\left(\frac{T}{|y|}\right)^\alpha \bar{\alpha}! \frac{M_0}{M_\alpha}\right)\right]$$

$$\leq \frac{1}{M_0} \exp\left[M^*(T/|y|)\right]. \tag{4.60}$$

The constant $T = T(C') = (2e/k\pi\delta)$ depends on $C' \subset\subset C$, because δ depends on C' but not on $y \in C'$. Using (4.60) in (4.59) and putting

$$A = A(n, C', s) = (B/M_0)\tilde{A}(n, C', s) \sum_{\alpha \in \mathbb{N}_0} (1/2)^\alpha$$

the desired estimate (4.53) follows in case $U \in \mathcal{D}'((M_p), L^s)$, $2 \leq s < \infty$.

For $U \in \mathcal{D}'(\{M_p\}, L^s)$, $2 \leq s < \infty$, we use Theorem 2.3.1 and the analysis of (4.54) and (4.56) to obtain the conclusion (4.53) for this case similarly as we did for the case $U \in \mathcal{D}'((M_p), L^s)$ above. For the case of $\mathcal{D}'(\{M_p\}, L^s)$ the constant $T > 0$ in (4.53) is arbitrary and independent of $C' \subset\subset C$, which is the consequence of the fact that relation (2.39) in Theorem 2.3.1 is true for arbitrary $k > 0$ and does not depend on $C' \subset\subset C$. The proof of Theorem 4.2.2 is completed. $\square$

We now obtain a norm growth estimate for the Cauchy integral of ultradistributions.

Theorem 4.2.3. *Let the cone C and the sequence (M_p) satisfy the hypotheses of Theorem 4.2.1.*

If $U \in \mathcal{D}'((M_p), L^s)$, $2 \leq s < \infty$, then for each compact subcone $C' \subset\subset C$ there exists a constant $T = T(C') > 0$ depending on C' such that, in case $s = 2$,

$$\|C(U; \cdot + iy)\|_{L^s} = \left(\int_{\mathbb{R}^n} |C(U; x + iy)|^s \, dx\right)^{1/s}$$

$$\leq K(U) \exp\left[M^*(T/|y|)\right] \tag{4.61}$$

for $y \in C' \subset\subset C$ and, in case $2 < s < \infty$,

$$\|C(U; \cdot + iy)\|_{L^s} \leq K(U, C', s, r, n)|y|^{-n(s-r)/rs} \exp\left[M^*(T/|y|)\right] \tag{4.62}$$

for $y \in C' \subset\subset C$ with $1/r + 1/s = 1$, where $K(U)$ in (4.61), i.e., in case $s = 2$, is a constant depending on U, and $K(U, C', s, r, n)$ in (4.62), i.e. in case $2 < s < \infty$, is a constant depending on U, C', s, r and n.

If $U \in \mathcal{D}'(\{M_p\}, L^s)$, $2 \leq s < \infty$, then for each compact subcone $C' \subset\subset C$ and an arbitrary constant $T > 0$, which may or may not depend on $C' \subset\subset C$, there is a constant $K(U)$ if $s = 2$ or a constant $K(U, C', s, r, n)$ if $2 < s < \infty$ such that (4.61) and (4.62) hold, respectively.

Proof. For $U \in \mathcal{D}'((M_p), L^s)$, $2 \leq s < \infty$, we use Theorem 2.3.2 and Fubini's theorem to obtain

$$C(U; z) = \sum_{\alpha \in \mathbb{N}_0} \langle g^{(\alpha)}, K(z - \cdot) \rangle$$

$$= \sum_{\alpha \in \mathbb{N}_0} (-1)^\alpha \langle g_\alpha, K^{(\alpha)}(z - \cdot) \rangle = \sum_{\alpha \in \mathbb{N}_0} h_\alpha(z), \qquad (4.63)$$

where $g_\alpha \in L^r$ for all $\alpha \in \mathbb{N}_0$ (with $1/r + 1/s = 1$), I_{C^*} is the characteristic function of C^* and

$$h_\alpha(z) := \int_{\mathbb{R}^n} \eta^\alpha e^{2\pi i \langle z, \eta \rangle} \mathcal{F}^{-1}[g_\alpha](\eta) I_{C^*}(\eta)\, d\eta = \int_{C^*} \eta^\alpha (E_z \mathcal{F}^{-1}[g_\alpha])(\eta)\, d\eta$$

$$(4.64)$$

with the notation (4.10) used. Since $g_\alpha \in L^r$ ($1 < r \leq 2$), we have $\mathcal{F}^{-1}[g_\alpha] \in L^s$ if $2 \leq s < \infty$ (since $1/r + 1/s = 1$) for each α. If $s = 2$, then $r = 2$ and, by Lemma 4.2.1, the integrand in (4.64) is in $L^1 \cap L^2$. In the case $s = 2$, we use Parseval's equality to get

$$\|h_\alpha(\cdot + iy)\|_{L^2} = \|\mathcal{F}[\eta^\alpha e_y \mathcal{F}^{-1}[g_\alpha] I_{C^*}]\|_{L^2} = \|\eta^\alpha e_y \mathcal{F}^{-1}[g_\alpha] I_{C^*}\|_{L^2}$$

$$\leq \|\eta^\alpha e_y I_{C^*}\|_{L^\infty} \|\mathcal{F}^{-1}[g_\alpha]\|_{L^2} \qquad (4.65)$$

with the symbol e_y meant as in (4.11). Now let $s > 2$ and note that by Hölder's inequality

$$\int_{\mathbb{R}^n} \left| \eta^\alpha e_y(\eta) \mathcal{F}^{-1}[g_\alpha](\eta) I_{C^*}(\eta) \right|^r d\eta$$

$$\leq (\|\mathcal{F}^{-1}[g_\alpha]\|_{L^s})^r \|\eta^{\alpha r}(e_y)^r I_{C^*}\|_{L^{s/(s-r)}} < \infty, \qquad (4.66)$$

since $\mathcal{F}^{-1}[g_\alpha] \in L^s$ and because of Lemma 4.2.1. Thus, by (4.66) and Lemma 4.2.1, each integrand in the last sum in (4.63) is in $L^1 \cap L^r$ for the case $s > 2$, $1/r + 1/s = 1$. By the Parseval inequality and by (4.66), we have, in the case $s > 2$,

$$\|h_\alpha(\cdot + iy)\|_{L^s} = \|\mathcal{F}[\eta^\alpha e_y \mathcal{F}^{-1}[g_\alpha] I_{C^*}]\|_{L^s} \leq \|\eta^\alpha e_y \mathcal{F}^{-1}[g_\alpha] I_{C^*}\|_{L^r}$$

$$\leq \|\mathcal{F}^{-1}[g_\alpha]\|_{L^s} \|\eta^\alpha e_y I_{C^*}\|_{L^{rs/(s-r)}} < \infty. \qquad (4.67)$$

Due to Lemma 1.2.2, for a given compact subcone $C' \subset\subset C$ there is a $\delta = \delta(C') > 0$ such that (1.12) holds for all $y \in C'$ and all $t \in C^*$. Hence, using (1.12) and the estimates (4.50) and (4.51), we have

$$\|\eta^\alpha e_y I_{C^*}\|_{L^\infty} \leq \sup_{\eta \in C^*} |\eta|^\alpha e^{-2\pi\delta|y||\eta|} \leq \begin{cases} 1, & \alpha = 0, \\[2mm] \left(\dfrac{\bar{\alpha}}{2\pi\delta|y|} \right)^\alpha, & \alpha \neq 0, \end{cases} \qquad (4.68)$$

for $y \in C' \subset\subset C$ and $\delta = \delta(C') > 0$.

For the case $s = 2$ we use (4.63), (4.65), (4.68), and the Parseval equality to obtain, for $y \in C' \subset C$,

$$\|C(U; z)\|_{L^2} \leq \sum_{\alpha \in \mathbb{N}_0} \|h_\alpha(\cdot + iy)\|_{L^2} \leq \sum_{\alpha \in \mathbb{N}_0} \|\eta^\alpha e_y I_{C^*}\|_{L^\infty} \|\mathcal{F}^{-1}[g_\alpha]\|_{L^2}$$

$$\leq \sum_{\alpha \in \mathbb{N}_0} \left(\frac{\bar{\alpha}}{2\pi\delta|y|}\right)^\alpha \|g_\alpha\|_{L^2}, \tag{4.69}$$

where the convention $\bar{\alpha}^\alpha = 1$ if $\alpha = 0$ is used. For $s > 2$ and $C' \subset C$ let us write $\delta = \delta(C') > 0$ in (1.12) as $\delta = \delta_1 + \delta_2$, where $\delta_1 = \delta_1(C') > 0$ and $\delta_2 = \delta_2(C') > 0$.

Now using (4.63), (4.67), Parseval's inequality, (1.12) with $\delta = \delta_1 + \delta_2$, (4.68) and applying the notation of (4.10) and (4.11), we have for $y \in C' \subset C$

$$\|C(U; \cdot + iy)\|_{L^s} \leq \sum_{\alpha \in \mathbb{N}_0} \left\| \mathcal{F}[I_{C^*} \eta^\alpha e_y \mathcal{F}^{-1}[g_\alpha]] \right\|_{L^s}$$

$$\leq \sum_{\alpha \in \mathbb{N}_0} \|\mathcal{F}^{-1}[g_\alpha]\|_{L^s} \|I_{C^*} \eta^\alpha e_y\|_{L^{rs/(s-r)}}$$

$$\leq \sum_{\alpha \in \mathbb{N}_0} \|g_\alpha\|_{L^r} \|I_{C^*} \eta^\alpha e_{\delta_1|y|}(|\cdot|)\|_{L^\infty} \|I_{C^*} e_{\delta_2|y|}(|\cdot|)\|_{L^{rs/(s-r)}}.$$

Denoting, as in (4.52), by Ω_n the surface area of the unit sphere in $\mathbb{R}^n$, we have

$$\|C(U; \cdot + iy)\|_{L^s}$$

$$\leq \Omega_n \left(\int_0^\infty w^{n-1} e_{\delta_2|y|}(wrs/(s-r)) \, dw \right)^{\frac{s-r}{rs}} \sum_{\alpha \in \mathbb{N}_0} \|g_\alpha\|_{L^r} \left(\frac{\bar{\alpha}}{2\pi\delta_1|y|}\right)^\alpha$$

$$= \Omega_n \left(\frac{(n-1)!(s-r)^n}{(2\pi\delta_2|y|rs)^n} \right)^{(s-r)/rs} \sum_{\alpha \in \mathbb{N}_0} \|g_\alpha\|_{L^r} \left(\frac{\bar{\alpha}}{2\pi\delta_1|y|}\right)^\alpha. \tag{4.70}$$

For the fixed $k > 0$ in the converse part of Theorem 2.3.2 we know that (2.49) holds; that is

$$\sup_\alpha (k^\alpha M_\alpha \|g_\alpha\|_{L^r}) < \infty, \tag{4.71}$$

where $2 \leq s < \infty$ and $1/r + 1/s = 1$.

For the case $s = 2$ and the fixed $k > 0$ we return to (4.69) to obtain

$$\|C(U; \cdot + iy)\|_{L^2} \le \sup_\alpha(k^\alpha M_\alpha \|g_\alpha\|_{L^2}) \sum_{\alpha \in \mathbb{N}_0} \left(\frac{\bar{\alpha}}{2k\pi\delta|y|}\right)^\alpha \frac{1}{M_\alpha}. \quad (4.72)$$

Recall (4.58) and our convention that $\bar{\alpha}^\alpha = 1$ if $\alpha = 0$. Using this in (4.72) and putting $T := e/k\pi\delta$, we obtain from (4.72) the estimate

$$\|C(U; \cdot + iy)\|_{L^2} \le \sup_\alpha(k^\alpha M_\alpha \|g_\alpha\|_{L^2}) \sum_{\alpha \in \mathbb{N}_0} \left(\frac{1}{2}\right)^\alpha \left(\frac{T}{|y|}\right)^\alpha \frac{\bar{\alpha}!}{M_\alpha}$$

$$\le \frac{2}{M_0} \sup_\alpha(k^\alpha M_\alpha \|g_\alpha\|_{L^2}) \exp\left[M^*(T/|y|)\right] \quad (4.73)$$

for $y \in C' \subset C$, where the calculation from (4.60) was used. This proves (4.61) in the case $U \in \mathcal{D}'((M_p), L^2)$, where

$$K(U) := \frac{2}{M_0} \sup_\alpha(k^\alpha M_\alpha \|g_\alpha\|_{L^2}).$$

For $s > 2$ we return to (4.70) and proceed using (4.71) and (4.58) to obtain (4.62) for the case $U \in \mathcal{D}'((M_p), L^s)$, $2 < s < \infty$, similarly as we did for the case $U \in \mathcal{D}'((M_p), L^2)$, where

$$K(U, C', s, r, n) := \frac{2\Omega_n}{M_0} \left(\frac{(n-1)!(s-r)^n}{(2\pi\delta_2 rs)^n}\right)^{\frac{s-r}{rs}} \sup_\alpha(k^\alpha M_\alpha \|g_\alpha\|_{L^r})$$

and $T := e/k\pi\delta_1$.

Under the assumption that $U \in \mathcal{D}'(\{M_p\}, L^s)$ with $2 \le s < \infty$, the conclusions in (4.61)-(4.62) are obtained due to the characterization given in Theorem 2.3.1 and an analysis similar to that in the case $U \in \mathcal{D}'((M_p), L^s)$, given previously in this proof. For $U \in \mathcal{D}'(\{M_p\}, L^s)$, $T > 0$ in (4.61)-(4.62) is an arbitrary constant, because it depends on arbitrary $k > 0$ in (2.39), and T may or may not depend on C' depending on the choice of the arbitrary $k > 0$. A complete proof of inequalities (4.61)-(4.62) for $U \in \mathcal{D}'(\{M_p\}, L^s)$, $2 \le s < \infty$, is given in [35]. The proof of Theorem 4.2.3 is complete. $\square$

We desire to extend the growth results of Theorems 4.2.2 and 4.2.3 to the case $1 < s < 2$. The proofs of Theorems 4.2.2 and 4.2.3 given above depend considerably on properties of the Fourier transform for elements in L^s, $2 \le s < \infty$, properties which are not available in general for the cases $1 < s < 2$. A detailed analysis of integrals, as in the proof of Theorem 4.1.1, may yield these growth results for $1 < s < 2$. We consider this in future research.

Throughout the remainder of this section the sequence (M_p) will satisfy $(M.1)$, $(M.2)$ and $(M.3')$.

We now proceed to investigate the boundary value properties of the analytic function $C(U; z)$, $z \in T^C$, for $U \in \mathcal{D}'(*, L^s)$. We first define a convolution which corresponds to that given in the definition in [82], p. 71. Let $U \in \mathcal{D}'(*, L^s)$ and $\varphi \in \mathcal{D}(*, L^s)$, $1 < s < \infty$. Then the convolution of U with φ is given by

$$(U * \varphi)(x) := \langle U, \varphi(x - \cdot) \rangle, \qquad x \in \mathbb{R}^n. \tag{4.74}$$

Lemma 4.2.2. *Let $U \in \mathcal{D}'(*, L^s)$ and $\varphi \in \mathcal{D}(*, L^s)$, where $1 < s < \infty$. We have $(U * \varphi) \in \mathcal{D}(*, L^\infty)$.*

Proof. Let $U \in \mathcal{D}'((M_p), L^s)$ and $\varphi \in \mathcal{D}((M_p), L^s)$. By (4.74) and Theorem 2.3.2, we have

$$(U * \varphi)(x) = \sum_{\alpha \in \mathbb{N}_0} (-1)^\alpha \int_{\mathbb{R}^n} g_\alpha(t) \varphi^{(\alpha)}(x - t) \, dt. \tag{4.75}$$

Let β be an arbitrary n-tuple of nonnegative integers. The βth derivative of the sum on the right of (4.75) can be taken under the summation and the integral sign. Hence $U * \varphi \in C^\infty(\mathbb{R}^n)$. With the aid of the chain rule and the definition of $\mathcal{D}((M_p), L^s)$ we find a constant $N > 0$ such that

$$|(U * \varphi)^{(\beta)}(x)| = \left| \sum_{\alpha \in \mathbb{N}_0} (-1)^{2\alpha} \int_{\mathbb{R}^n} g_\alpha(t) \varphi^{(\alpha+\beta)}(x - t) \, dt \right|$$

$$\leq \sum_{\alpha \in \mathbb{N}_0} \|g_\alpha\|_{L^r} \|\varphi^{(\alpha+\beta)}\|_{L^s} \leq N \sum_{\alpha \in \mathbb{N}_0} h^{\alpha+\beta} M_{\alpha+\beta} \|g_\alpha\|_{L^r} \tag{4.76}$$

for every $h > 0$. Due to condition $(M.2)$, there exist positive constants A and H such that

$$M_{\alpha+\beta} \leq AH^{\alpha+\beta} M_\alpha M_\beta.$$

Taking this into account in (4.76), we obtain

$$|(U * \varphi)^{(\beta)}(x)| \leq AN \sum_{\alpha \in \mathbb{N}_0} h^{\alpha+\beta} H^{\alpha+\beta} M_\alpha M_\beta \|g_\alpha\|_{L^r}$$

and, consequently,

$$\frac{|(U * \varphi)^{(\beta)}(x)|}{N(hH)^\beta M_\beta} \leq A \sum_{\alpha \in \mathbb{N}_0} (jH)^\alpha M_\alpha \|g_\alpha\|_{L^r}, \tag{4.77}$$

where $h > 0$ and $j > 0$ are arbitrary and A and H are positive constants. Note that we have renamed h to be j on the right side of (4.77). Recalling that the g_α satisfy (2.47) or (2.49) for some $k > 0$, we can appropriately choose the arbitrary $j > 0$ on the right of (4.77) to show, as in the proof of Theorem 4.2.1, that the series on the right side of (4.77) converges. Since $h > 0$ on the left of (4.77) is arbitrary, (4.77) proves the desired growth for $(U * \varphi)$ to be in $\mathcal{D}((M_p), L^\infty)$. The proof is therefore complete for the Beurling case.

The proof for the Roumieu $* = \{M_p\}$ case is obtained by a similar analysis with the use of (4.74) and Theorem 2.3.1. $\square$

The following two results lead us directly to the calculation of the boundary value of the Cauchy integral $C(U; z)$, $z \in T^C$.

Theorem 4.2.4. *Let C be a regular cone in $\mathbb{R}^n$. Suppose that $U \in \mathcal{D}'(*, L^s)$ with $1 < s < \infty$ and $\varphi \in \mathcal{D}(*, L^1)$. For a fixed $y = \operatorname{Im} z \in C$ we have*

$$\langle C(U; \cdot + iy), \varphi \rangle = \langle U, \psi_y \rangle, \tag{4.78}$$

where

$$\psi_y(t) := \langle K(\cdot + iy - t), \varphi \rangle, \qquad t \in \mathbb{R}^n.$$

Proof. By Theorem 4.1.1, we have $K(z - \cdot) \in \mathcal{D}(*, L^s)$ for all $z \in T^C$ and s, $1 < s \le \infty$. Put

$$K_y(x) := K(x + iy) = \int_{C^*} \exp\left[2\pi i \langle x + iy, \eta \rangle\right] d\eta, \qquad y \in C.$$

The proof of Theorem 4.1.1 can be adapted to show that $K_y \in \mathcal{D}(*, L^s)$ for all $y \in C$ and s, $1 < s \le \infty$. Since, by definition (4.74),

$$(U * K_y)(x) = \langle U, K(x - \cdot + iy) \rangle, \qquad z \in T^C,$$

it follows from Lemma 4.2.2 that

$$U * K_y = C(U; \cdot + iy) \in \mathcal{D}(*, L^\infty)$$

for $y \in C$ and $1 < s < \infty$ and thus

$$\langle C(U; \cdot + iy), \varphi \rangle = \langle U * K_y, \varphi \rangle$$

is well defined for $\varphi \in \mathcal{D}(*, L^1)$ with $y \in C$. Using the representation of $U \in \mathcal{D}'(*, L^s)$ given in either Theorem 2.3.1 or Theorem 2.3.2 and Fubini's theorem, we have

$$\langle C(U; \cdot + iy), \varphi \rangle = \langle U * K_y, \varphi \rangle$$

$$= \sum_{\alpha \in \mathbb{N}_0} (-1)^\alpha \int_{\mathbb{R}^n} \left[\int_{\mathbb{R}^n} g_\alpha(t) D_t^\alpha K(z - t) \, dt \right] \varphi(x) \, dx$$

$$= \sum_{\alpha \in \mathbb{N}_0} (-1)^\alpha \int_{\mathbb{R}^n} g_\alpha(t) \left[\int_{\mathbb{R}^n} D_t^\alpha K(z - t) \varphi(x) \, dx \right] dt$$

$$= \sum_{\alpha \in \mathbb{N}_0} (-1)^\alpha \langle g_\alpha(t), D^\alpha \psi_y \rangle = \langle U, \psi_y \rangle$$

for $y = \text{Im } z \in C$. The proof is complete. $\square$

We now show that the term $\langle K(\cdot + iy - t), \varphi \rangle$, $t \in \mathbb{R}^n$, converges in the topology of $\mathcal{D}(*, L^s)$, $2 \leq s < \infty$, to a certain limit function as $y \to 0$, $y \in C$. The choice of the space to which φ belongs depends on whether $*$ is (M_p) or $\{M_p\}$. For $\mathcal{D}((M_p), L^s)$ we have the following result.

Theorem 4.2.5. *Let C be a regular cone and let I_{C^*} be the characteristic function of the dual cone C^* of C. Let $\varphi \in \mathcal{D}((M_p), \mathbb{R}^n)$. We have*

$$\lim_{y \to 0, y \in C} \langle K(\cdot + iy - t), \varphi \rangle = \mathcal{F}^{-1}[I_{C^*}\widehat{\varphi}](t) \qquad (4.79)$$

in $\mathcal{D}((M_p), L^s)$, whenever $2 \leq s < \infty$.

Proof. Recall the definition of convergence in $\mathcal{D}((M_p), L^s)$ from Section 2.3. If $\varphi \in \mathcal{D}^{(M_p)}$, then $\widehat{\varphi}$ is an element of the Schwartz space $\mathcal{S}$ and hence $I_{C^*}\widehat{\varphi} \in L^r$ for all r, $1 \leq r \leq \infty$. Thus $\mathcal{F}^{-1}[I_{C^*}\widehat{\varphi}]$ can be interpreted as both the L^1 and L^r inverse Fourier transform, $1 < r \leq 2$. For every n-tuple α of nonnegative integers we have

$$D_t^\alpha \mathcal{F}^{-1}[I_{C^*}\widehat{\varphi}](t) = \mathcal{F}^{-1}[I_{C^*}\eta^\alpha \widehat{\varphi}](t). \qquad (4.80)$$

Since $\widehat{\varphi}(\eta) \in \mathcal{S}$, we have $I_{C^*}\eta^\alpha \widehat{\varphi} \in L^r$ for all α and for all r, $1 \leq r \leq \infty$. The Parseval theory for the Fourier transform now yields that the left side of (4.80) is an element of L^s, $1/r + 1/s = 1$, $1 \leq r \leq 2$, for all s and all α. That the left side of (4.80) satisfies the required boundedness condition for elements in $\mathcal{D}((M_p), L^s)$ for each α will follow from techniques that we present later in this proof. The preceding points yield that the right side of (4.79) is an element of $\mathcal{D}((M_p), L^s)$, $2 \leq s < \infty$. Similarly, details in the remaining analysis in this proof will imply that for each $y \in C$,

$$\langle K(\cdot + iy - t), \varphi \rangle \in \mathcal{D}((M_p), L^s), \qquad 2 \leq s < \infty,$$

as a function of $t \in \mathbb{R}^n$.

We now prove the desired convergence in L^s in order to obtain (4.79). Let $z = x + iy \in T^C$. By a change of order of integration

$$\langle K(\cdot + iy - t), \varphi \rangle \int_{\mathbb{R}^n} \left[\int_{C^*} e^{2\pi i \langle x + iy - t, \eta \rangle} d\eta \right] \varphi(x) \, dx$$

$$= \int_{\mathbb{R}^n} I_{C^*}(\eta) e^{-2\pi i \langle t, \eta \rangle} e^{-2\pi \langle y, \eta \rangle} \left[\int_{\mathbb{R}^n} \varphi(x) e^{2\pi i \langle x, \eta \rangle} \, dx \right] d\eta$$

$$= \int_{\mathbb{R}^n} I_{C^*}(\eta) \widehat{\varphi}(\eta) e^{-2\pi \langle y, \eta \rangle} e^{-2\pi i \langle t, \eta \rangle} \, d\eta. \qquad (4.81)$$

For every n-tuple α of nonnegative integers we have

$$\|D^\alpha \Phi_y(\cdot)\|_{L^s} = \|\mathcal{F}^{-1}[I_{C^*}\eta^\alpha\widehat{\varphi}(e_y - 1)]\|_{L^s}, \qquad (4.82)$$

where

$$\Phi_y(t) := \langle K(\cdot + iy - t), \varphi\rangle - \mathcal{F}^{-1}[I_{C^*}\widehat{\varphi}](t)$$

and e_y is defined in (4.11).

Since $\langle y, \eta\rangle \geq 0$ for $y \in C$ and $\eta \in C^*$, we have

$$|I_{C^*}(\eta)\eta^\alpha\widehat{\varphi}(\eta)(e^{-2\pi\langle y,\eta\rangle} - 1)| \leq 2|\eta^\alpha\widehat{\varphi}(\eta)|, \qquad y \in C. \qquad (4.83)$$

Hence, since $\widehat{\varphi}(\eta) \in \mathcal{S}$, we conclude that

$$(I_{C^*}(\eta)\eta^\alpha\widehat{\varphi}(\eta)(exp(-2\pi\langle y,\eta\rangle) - 1)) \in L^r$$

for all r, $1 \leq r \leq \infty$. Thus if $1 < r \leq 2$ and $1/r + 1/s = 1$, we have, by the Parseval inequality,

$$\|\mathcal{F}^{-1}[I_{C^*}\eta^\alpha\widehat{\varphi}(e_y - 1)]\|_{L^s} \leq \|I_{C^*}\eta^\alpha\widehat{\varphi}(e_y - 1)\|_{L^r}. \qquad (4.84)$$

By (4.83) and the Lebesgue dominated convergence theorem, we have

$$\lim_{y\to 0, y\in C}\int_{\mathbb{R}^n}|I_{C^*}(\eta)\eta^\alpha\widehat{\varphi}(\eta)(e^{-2\pi\langle y,\eta\rangle} - 1)|^r\, d\eta = 0 \qquad (4.85)$$

for $1 < r \leq 2$. Combining (4.82), (4.84), and (4.85), we have

$$\lim_{y\to 0, y\in C}\|D^\alpha \Phi_y\|_{L^s} = 0 \qquad (4.86)$$

for all n-tuples α of nonnegative integers and for $2 \leq s < \infty$ as desired.

To complete the proof it remains to show that there is a constant $N > 0$, which is independent of α and $y \in C$, such that for all $h > 0$

$$\|D^\alpha \Phi_y\|_{L^s} \leq Nh^\alpha M_\alpha \qquad (4.87)$$

for each α. As in (4.82), (4.83) and (4.84) we have, for $y \in C$,

$$\| D_t^\alpha\langle K(\cdot + iy - t), \varphi\rangle\|_{L^s}$$
$$= \|\mathcal{F}^{-1}[I_{C^*}\eta^\alpha\widehat{\varphi}e_y\|_{L^s} \leq \|I_{C^*}\eta^\alpha\widehat{\varphi}e_y\|_{L^r} \leq \|\eta^\alpha\widehat{\varphi}\|_{L^r} \qquad (4.88)$$

for $1 < r \leq 2$ and $1/r + 1/s = 1$. Recall the definition of $\mathcal{D}^{(M_p)}$ in Section 2.3 and let $\varphi \in \mathcal{D}((M_p), \mathbb{R}^n)$. Using known Fourier transform properties, we have

$$\int_{\mathbb{R}^n}|\eta^\alpha\widehat{\varphi}(\eta)|^r\, d\eta = \int_{\mathbb{R}^n}\left|\eta^\alpha\Big(1 + \sum_{j=1}^n\eta_j^{2n}\Big)\widehat{\varphi}(\eta)\Big/\Big(1 + \sum_{j=1}^n\eta_j^{2n}\Big)\right|^r\, d\eta$$

$$= \int_{\mathbb{R}^n}\left|\mathcal{F}\Big[\Big(\Big(1 + \sum_{j=1}^n D_j^{2n}\Big)(-1)^\alpha D^\alpha\Big)\varphi\Big](\eta)\Big/\Big(1 + \sum_{j=1}^n\eta_j^{2n}\Big)\right|^r\, d\eta.$$

Putting

$$R := \int_{\mathbb{R}^n} \left| 1 + \sum_{j=1}^{n} \eta_j^{2n} \right|^{-r} d\eta, \quad R' := \left(\int_{\text{supp } \varphi} 1 \, dt \right)^r,$$

we have

$$\int_{\mathbb{R}^n} |\eta^\alpha \widehat{\varphi}(\eta)|^r \, d\eta \leq R \sup_{\eta \in \mathbb{R}^n} \left| \mathcal{F}\left[\left(\left(1 + \sum_{j=1}^{n} D_j^{2n} \right) (-1)^\alpha D^\alpha \right) \varphi \right](\eta) \right|^r$$

$$\leq R \left(\int_{\mathbb{R}^n} \left| \left(\left(1 + \sum_{j=1}^{n} D_j^{2n} \right) (-1)^\alpha D^\alpha \right) \varphi(t) \right| dt \right)^r$$

$$\leq R \left(\int_{\mathbb{R}^n} \left(|D^\alpha \varphi(t)| + |D_1^{2n} D^\alpha \varphi(t)| + \ldots + |D_n^{2n} D^\alpha \varphi(t)| \right) dt \right)^r$$

$$\leq RR'(Bk^\alpha M_\alpha + Bk^{\alpha+2n} M_{\alpha+2n} + \ldots + Bk^{\alpha+2n} M_{\alpha+2n})^r$$

$$\leq RR'(Bn)^r (k^\alpha M_\alpha + k^{\alpha+2n} M_{\alpha+2n})^r$$

for all $k > 0$. By property $(M.2)$ of the sequence (M_p), there are constants $A > 0$ and $H > 0$ such that

$$M_{\alpha+2n} \leq AH^{\alpha+2n} M_\alpha M_{2n}.$$

Putting $H' := \max\{1, H\}$, we continue the preceding estimate as

$$\int_{\mathbb{R}^n} |\eta^\alpha \widehat{\varphi}(\eta)|^r \, d\eta \leq RR'(Bn)^r (k^\alpha M_\alpha + Ak^{\alpha+2n} H^{\alpha+2n} M_\alpha M_{2n})^r$$

$$\leq RR'(Bn)^r ((kH')^\alpha M_\alpha + A(kH')^{2n} (kH')^\alpha M_\alpha M_{2n})^r$$

$$\leq RR'(Bn)^r (1 + (A(kH')^{2n} M_{2n}))^r ((kH')^\alpha M_\alpha)^r. \tag{4.89}$$

Thus, by (4.88) and (4.89), we have

$$\|D^\alpha \Psi_y\|_{L^s} \leq (RR')^{1/r} Bn(1 + A(kH)^{2n} M_{2n}) h^\alpha M_\alpha, \tag{4.90}$$

for all $h = (kH') > 0$, where

$$\Psi_y(t) := \langle K(\cdot + iy - t), \varphi \rangle.$$

Recall that $k > 0$ is arbitrary and that $H' > 0$ is fixed. By the same analysis as in (4.90), we get

$$\|D^\alpha \mathcal{F}^{-1}[I_{C^*} \widehat{\varphi}]\|_{L^s} = \|\mathcal{F}^{-1}[I_{C^*} \eta^\alpha \widehat{\varphi}]\|_{L^s} \leq \|\eta^\alpha \widehat{\varphi}\|_{L^r}$$

$$\leq (RR')^{1/r} Bn(1 + A(kH')^{2n} M_{2n}) h^\alpha M_\alpha \tag{4.91}$$

for all $h > 0$. Combining (4.90) and (4.91) and using the Minkowski inequality, we have (4.87) for each α, where N can be chosen to be

$$N = (RR')^{1/r} Bn(1 + A(kH')^{2n} M_{2n})$$

and $k > 0$ can be chosen arbitrarily. The proof of Theorem 4.2.5 is complete.
□

Let us now consider the space $S_\infty(\{N_p\}, \{M_p\}) \subset \mathcal{S}^{(M_p)}$ of Roumieu [127] p. 70, where the sequences (M_p) and (N_p) satisfy conditions $(M.1)$, $(M.2)$ and $(M.3')$. Under these conditions, the sequences (M_p) and (N_p) satisfy the conditions of Gel'fand and Shilov in [51], (9), p. 245; thus the space $S_\infty(\{N_p\}, \{M_p\}) = S_{(N_p)}^{(M_p)}$ has the Fourier transform property

$$\mathcal{F}(S_{(M_p)}^{(N_p)}) = S_{(M_p)}^{(N_p)} \subset \mathcal{S}^{(M_p)} \tag{4.92}$$

in the notation of Gel'fand and Shilov (see [51], (11), p. 254). Because of the Fourier transform theory for the spaces $S_\infty(\{N_p\}, \{M_p\})$, we may take $\varphi \in S_\infty(\{N_p\}, \{M_p\})$ in the following result which corresponds to Theorem 4.2.5.

Theorem 4.2.6. *Let C be a regular cone and let $I_{C^*}(\eta)$ be the characteristic function of the dual cone C^* of C. Let $\varphi \in S_\infty(\{N_p\}, \{M_p\})$ (or $\varphi \in \mathcal{D}(\{M_p\}, \mathbb{R}^n)$), where both sequences (M_p) and (N_p) satisfy conditions $(M.1)$, $(M.2)$ and $(M.3')$. We have*

$$\lim_{y \to 0, y \in C} \langle K(\cdot + iy - t), \varphi \rangle = \mathcal{F}^{-1}[I_{C^*} \widehat{\varphi}](t) \tag{4.93}$$

in $\mathcal{D}(\{M_p\}, L^s)$, $2 \le s < \infty$.

Proof. The proof of Theorem 4.2.6 is very similar to that of Theorem 4.2.5. The difference is in the technique to estimate the term on the right side of (4.88) when we take $\varphi \in S_\infty(\{N_p\}, \{M_p\})$. We have

$$\int_{\mathbb{R}^n} |\eta^\alpha \widehat{\varphi}(\eta)|^r \, d\eta = \int_{|\eta| \le 1} |\eta^\alpha \widehat{\varphi}(\eta)|^r \, d\eta + \int_{|\eta| > 1} |\eta^{\alpha + \overline{2}} \widehat{\varphi}(\eta)|^r \eta^{-\overline{2}r} \, d\eta,$$

where $\overline{2}$ is the n-tuple $(2, 2, \ldots, 2)$. Now the boundedness condition for convergence in $\mathcal{D}(\{M_p\}, L^s)$ with $2 \le s < \infty$ follows by using (4.92), the defining growth of $S_\infty(\{N_p\}, \{M_p\})$ and $(M.2)$. Of course, (4.93) in Theorem 4.2.6 also holds for $\varphi \in \mathcal{D}(\{M_p\}, \mathbb{R}^n)$ by similar reasoning as in the proof of Theorem 4.2.5. □

Now let $\varphi \in \mathcal{D}(*, \mathbb{R}^n)$. Since $\mathcal{D}(*, \mathbb{R}^n) \subset \mathcal{D}(*, L^1)$, we have (4.78) for $U \in \mathcal{D}'(*, L^s)$, $1 < s < \infty$ and $\varphi \in \mathcal{D}(*, \mathbb{R}^n)$. This fact combined with Theorem 4.2.4, Theorem 4.2.5 and the continuity of U prove the following result.

Corollary 4.2.1. *Let C be a regular cone in $\mathbb{R}^n$. Let $U \in \mathcal{D}'(*, L^s)$, $2 \leq s < \infty$. Let $\varphi \in \mathcal{D}(*, \mathbb{R}^n)$. We have*

$$\lim_{y \to 0, y \in C} \langle C(U; \cdot + iy), \varphi \rangle = \langle U, \mathcal{F}^{-1}[I_{C^*}\widehat{\varphi}] \rangle. \qquad (4.94)$$

Using Theorems 4.2.1, 4.2.2 and Corollary 4.2.1, we obtain an analytic decomposition theorem ,for elements in $\mathcal{D}'(*, L^s)$, $2 \leq s < \infty$.

Theorem 4.2.7. *Let m be a positive integer and let C_j, $j = 1, \ldots, m$, be regular cones such that*

$$\mathbb{R}^n \setminus \cup_{j=1}^m C_j^* \qquad and \qquad C_j^* \cap C_k^*, \qquad j, k = 1, \ldots, m, \quad j \neq k, \qquad (4.95)$$

are sets of Lebesgue measure zero. Let $U \in \mathcal{D}'(*, L^s)$, $2 \leq s < \infty$, and $\varphi \in \mathcal{D}(*, \mathbb{R}^n)$. There exist functions f_j which are analytic in $\mathbb{R}^n + iC_j$, $j = 1, \ldots, m$, such that

$$\langle U, \varphi \rangle = \sum_{j=1}^m \lim_{y \to 0, y \in C_j} \langle f_j(\cdot + iy), \varphi \rangle. \qquad (4.96)$$

If $U \in \mathcal{D}'((M_p), L^s)$, then for each $j = 1, \ldots, m$ and for each compact subcone $C_j' \subset C_j$ there are constants $A_j = A_j(n, C_j', s) > 0$ and $T_j = T_j(C_j') > 0$ such that

$$|f_j(x+iy)| \leq A_j |y|^{-n/r} \exp\left[M^*(T_j/|y|)\right], \qquad z = x+iy \in \mathbb{R}^n + iC_j', \quad (4.97)$$

where n is the dimension, $1/r + 1/s = 1$ and M^* is the function defined in (2.9).

If $U \in \mathcal{D}'(\{M_p\}, L^s)$, then for each $j = 1, \ldots, m$, each compact subcone $C_j' \subset C_j$ and arbitrary constant $T_j > 0$, which is independent of $C_j' \subset C_j$, there is a constant $A_j = A_j(n, C_j', s) > 0$ such that (4.97) holds.

Proof. For each $j = 1, \ldots, m$ put

$$f_j(z) := \langle U, \int_{C_j^*} \exp\left[2\pi i \langle z - \cdot, \eta \rangle\right] d\eta \rangle, \qquad z = x + iy \in \mathbb{R}^n + iC_j.$$

By Theorems 4.2.1 and 4.2.2 each f_j is an analytic function of $z \in \mathbb{R}^n + iC_j$ and satisfies the relevant version of (4.97) for $*$ being (M_p) or $\{M_p\}$. To prove (4.96) first note that

$$\lim_{y \to 0, y \in C_j} \langle f_j(\cdot + iy), \varphi \rangle = \langle U, \mathcal{F}^{-1}[I_{C_j^*}\widehat{\varphi}] \rangle, \qquad j = 1, \ldots, m, \qquad (4.98)$$

by Corollary 4.2.1. Since $\varphi \in \mathcal{D}(*, \mathbb{R}^n) \subset \mathcal{S}^{(M_p)}$, we have $\widehat{\varphi} \in \mathcal{S}^{(M_p)}$ and $\mathcal{F}^{-1}[\widehat{\varphi}] = \varphi$. We now use the linearity of U, the assumptions (4.95) on the dual cones C_j^*, $j = 1, \ldots, m$, and (4.98) to obtain

$$\sum_{j=1}^{m} \lim_{y \to 0, y \in C_j} \langle f_j(\cdot + iy), \varphi \rangle = \sum_{j=1}^{m} \langle U, \mathcal{F}^{-1}[I_{C_j^*} \widehat{\varphi}] \rangle$$

$$= \langle U, \sum_{j=1}^{m} \mathcal{F}^{-1}[I_{C_j^*} \widehat{\varphi}] \rangle = \langle U, \mathcal{F}^{-1}[\widehat{\varphi}] \rangle = \langle U, \varphi \rangle.$$

The proof is complete. □

The 2^n n-rants in $\mathbb{R}^n$ are an example of a finite number of regular cones for which (4.95) holds. Let $u = (u_1, \ldots, u_n)$ be any of the 2^n n-tuples whose components are -1 or 1. The set $C_u = \{y \in \mathbb{R}^n : u_j y_j > 0, \; j = 1, \ldots, n\}$ is an n-rant in $\mathbb{R}^n$ and is a regular cone with the property that $C_u^* = \overline{C}_u$. Thus Theorem 4.2.7 holds, in particular, for $m = 2^n$ with each C_j being an n-rant C_u in $\mathbb{R}^n$.

Of course, we can also state a L^s norm growth estimate on each function f_j of the variable $z \in \mathbb{R}^n + iC_j'$, $C_j' \subset\subset C_j$, $j = 1, \ldots, m$, because of Theorem 4.2.3.

We desire to obtain Theorems 4.2.5, 4.2.6, 4.2.7, and Corollary 4.2.1 for $1 < s < 2$ as well. We consider this in future research.

4.3　Poisson integral of ultradistributions

As in Section 4.2 we let C be a regular cone in $\mathbb{R}^n$ and let the sequence of positive real numbers (M_p), $p = 0, 1, 2, \ldots$, satisfy the conditions $(M.1)$ and $(M.3')$.

Let $U \in \mathcal{D}'(*, L^s)$, $1 \le s \le \infty$, where $*$ represents either (M_p) or $\{M_p\}$. Because of Theorem 4.1.2, we can form

$$P(U; z) := \langle U, Q(z; \cdot) \rangle, \qquad z \in T^C, \tag{4.99}$$

where $Q(z; t)$, $z \in T^C$, $t \in \mathbb{R}^n$, is the Poisson kernel corresponding to the tube T^C. The expression $P(U; z)$ defined in (4.99) is the Poisson integral of the ultradistribution U with respect to the tube T^C. In contrast to the Cauchy integral, the Poisson integral $P(U; z)$ is not an analytic function of $z \in T^C$, in general. If the cone C is a half-line $(0, \infty)$ or $(-\infty, 0)$ in $\mathbb{R}^1$, then $P(U; z)$ is a harmonic function; if C is an n-rant in $\mathbb{R}^n$, i.e., $C = C_u$ for some $u \in \Theta$, then $P(U; z)$ is an n-harmonic function.

Throughout the remainder of this section the sequence (M_p) satisfies the conditions $(M.1)$, $(M.2)$ and $(M.3')$.

We will prove that the Poisson integral $P(U; z)$ defined in (4.99) obtains U as its boundary value as $y \to 0$, $y \in C$, for an arbitrary regular cone C; thus the Poisson integral $P(U; z)$ has boundary values in analogy with Poisson integrals of L^p functions in tubes. To obtain the desired boundary value result we need two preliminary theorems.

Theorem 4.3.1. *Let C be a regular cone in $\mathbb{R}^n$. Let $U \in \mathcal{D}'(*, L^s)$, $1 < s < \infty$. Let $\varphi \in \mathcal{D}(*, L^1)$. We have*

$$\langle P(U; \cdot + iy), \varphi \rangle = \langle U, \psi_y \rangle, \qquad y \in C, \qquad (4.100)$$

where

$$\psi_y(t) := \langle Q(\cdot + iy; t), \varphi \rangle.$$

Proof. Put

$$K_y(x) := K(x + iy) = \int_{C^*} \exp\left[2\pi i \langle x + iy, \eta \rangle\right] d\eta, \qquad y \in C,$$

as in the proof of Theorem 4.2.4; and put

$$Q_y(x) := \frac{K(x + iy)\overline{K(x + iy)}}{K(2iy)}, \qquad x \in \mathbb{R}^n, \quad y \in C. \qquad (4.101)$$

As in Lemma 1.3.5, we have $Q_y(x) \geq 0$ for $x \in \mathbb{R}^n$, $y \in C$, and $Q(z; t) \geq 0$ for $z = x + iy \in T^C$, $t \in \mathbb{R}^n$. Further, we have

$$\int_{\mathbb{R}^n} Q_y(x)\, dx = 1, \qquad y \in C,$$

and, if $\delta > 0$,

$$\lim_{y \to 0, y \in C} \int_{|x| \geq \delta} Q_y(x)\, dx = 0.$$

The proof of Theorem 4.1.2 can be adapted to show that $Q_y \in \mathcal{D}(*, L^s)$, $1 \leq s \leq \infty$ for $y \in C$. By Lemma 4.2.4,

$$(U * Q_y)(x) = \langle U, Q_y(x - \cdot) \rangle = \langle U, Q(x + iy; \cdot) \rangle = P(U; x + iy)$$

is an element of $\mathcal{D}(*, L^\infty)$; hence $\langle P(U; \cdot + iy), \varphi \rangle$ is a well defined function of $y \in C$ for $\varphi \in \mathcal{D}(*, L^1)$. The proof of (4.100) follows as in the proof of Theorem 4.2.4, due to the representation of $U \in \mathcal{D}'(*, L^s)$ given in Theorems 2.3.1 and 2.3.2, by Fubini's theorem. $\square$

Theorem 4.3.2. *Let C be a regular cone in $\mathbb{R}^n$ and $\varphi \in \mathcal{D}(*, L^s)$, $1 \leq s < \infty$. We have*

$$\lim_{y \to 0, y \in C} \langle Q(\cdot + iy; t), \varphi \rangle = \varphi(t) \qquad (4.102)$$

in $\mathcal{D}(*, L^s)$.

Proof. First we consider the Beurling case where $*$ denotes (M_p). Let $\varphi \in \mathcal{D}((M_p), L^s)$. Differentiation under the integral sign shows that the function Q^y given by $Q^y(t) := \langle Q(\cdot + iy; t), \varphi \rangle$ is an infinitely differentiable function of the variable $t \in \mathbb{R}^n$ for every $y \in C$. Boundedness techniques which will be used later in this proof show that every derivative of the function Q^y is an element of L^s for each $y \in C$ and satisfies the defining growth (2.33) of elements in $\mathcal{D}((M_p), L^s)$ (see (4.107) below). Thus $Q^y \in \mathcal{D}((M_p), L^s)$ for each $y \in C$. For every n-tuple α of nonnegative integers, we make a change of variable and obtain

$$\left\| D^\alpha_. \int_{\mathbb{R}^n} Q(x + iy \, ; \cdot) \varphi(x) \, dx - D^\alpha \varphi \right\|_{L^s}$$

$$= \left\| D^\alpha_. \int_{\mathbb{R}^n} \varphi(x + \cdot) Q_y(x) \, dx - D^\alpha \varphi \right\|_{L^s}$$

$$= \left\| \int_{\mathbb{R}^n} \psi(x + \cdot) Q_y(x) \, dx - \psi \right\|_{L^s}, \qquad (4.103)$$

where $\psi(t) = (D^\alpha \varphi)(t)$ and $Q_y(x)$ is given in (4.101). Using the approximate identity properties of the Poisson kernel as given in the proof of Theorem 4.3.1, the proof of [27], Lemma 7, p. 213, shows that the right side of (4.103) approaches zero as $y \to 0$, $y \in C$. Thus

$$\lim_{y \to 0, y \in C} D^\alpha_t \langle Q(\cdot + iy; t), \varphi \rangle = (D^\alpha \varphi)(t) \qquad (4.104)$$

in L^s for every n-tuple α of nonnegative integers.

To complete the proof of the result for $*$ being (M_p), the boundedness condition (2.38) for convergence in $\mathcal{D}((M_p), L^s)$ remains to be shown. Proceeding as in (4.103) we use a change of variable and the chain rule to obtain

$$\left\| D^\alpha_. \int_{\mathbb{R}^n} Q(x + iy; \cdot) \varphi(x) \, dx \right\|_{L^s} = \left\| \int_{\mathbb{R}^n} D^\alpha_. \varphi(x + \cdot) Q_y(x) \, dx \right\|_{L^s}$$

$$= \left\| \int_{\mathbb{R}^n} D^\alpha_u \varphi(u) Q(u + iy; \cdot) \, du \right\|_{L^s}. \qquad (4.105)$$

By Jensen's inequality [48], 2.4.19, p. 91, Fubini's theorem and the approximate identity properties of the Poisson kernel stated in the proof of Theorem 4.3.1 we have

$$\int_{\mathbb{R}^n} \left| \int_{\mathbb{R}^n} D_u^\alpha \varphi(u) Q(u + iy; t) \, du \right|^s dt \leq \int_{\mathbb{R}^n} \int_{\mathbb{R}^n} |D_u^\alpha \varphi(u)|^s Q(u + iy; t) \, du \, dt$$

$$= \int_{\mathbb{R}^n} |D_u^\alpha \varphi(u)|^s \int_{\mathbb{R}^n} Q(u + iy; t) \, dt \, du = \int_{\mathbb{R}^n} |D_u^\alpha \varphi(u)|^s \, du. \tag{4.106}$$

Combining (4.105), (4.106), and the fact that $\varphi \in \mathcal{D}((M_p), L^s)$ we have the existence of a constant $N > 0$ from (2.33) such that

$$\left\| D^\alpha \int_{\mathbb{R}^n} Q(x + iy; \cdot) \varphi(x) \, dx \right\|_{L^s} \leq \|D^\alpha \varphi\|_{L^s} \leq N h^\alpha M_\alpha \tag{4.107}$$

for all $h > 0$. By (4.107) and the Minkowski inequality we have

$$\left\| D^\alpha \int_{\mathbb{R}^n} Q(x + iy; \cdot) \varphi(x) \, dx - D^\alpha \varphi \right\|_{L^s} \leq 2 N h^\alpha M_\alpha \tag{4.108}$$

for all $h > 0$ and all $y \in C$. Now we combine formulae (4.104) and (4.108), which hold for all α, to prove (4.102) in $\mathcal{D}((M_p), L^s)$ for $1 \leq s < \infty$.

The proof of (4.102) for $\mathcal{D}(\{M_p\}, L^s)$ is similar. The proof is completed. □

We can now obtain the desired boundary value property of the Poisson integral $P(U; z)$.

Theorem 4.3.3. *Let C be a regular cone in $\mathbb{R}^n$. Let $U \in \mathcal{D}'(*, L^s)$, $1 < s < \infty$, and let $\varphi \in \mathcal{D}(*, \mathbb{R}^n)$. We have*

$$\lim_{y \to 0, y \in C} \langle P(U; \cdot + iy), \varphi \rangle = \langle U, \varphi \rangle. \tag{4.109}$$

Proof. First recall that $\mathcal{D}(*, \mathbb{R}^n) \subset \mathcal{D}(*, L^s)$ for all s, $1 \leq s \leq \infty$. Formula (4.109) now follows by combining Theorems 4.3.1, 4.3.2 and the continuity of U. □

It is interesting that the boundary value of the Cauchy integral $C(U; z)$ as obtained in equation (4.94) of Corollary 4.2.1 depends on the cone C, whereas the boundary value of the Poisson integral $P(U; z)$ in (4.109) is always U independently of the cone C.

Let us note that in the Roumieu case, with $* = \{M_p\}$, Theorem 4.3.3 can be slightly generalized, as was done in Theorem 4.2.6, by taking φ

in the space of Roumieu $\mathcal{S}_{\infty}(\{N_p\}, \{M_p\}) = \mathcal{S}_{(M_p)}^{(N_q)}$ (see [127], p. 70, and Gel'fand and Shilov [51], p. 245). Of course, both sequences (M_p) and (N_p) are taken to satisfy the conditions $(M.1)$, $(M.2)$ and $(M.3')$. We write $\mathcal{S}_{\infty}(\{M_p\}, \{N_p\})$ here, instead of $\mathcal{S}_{\infty}(\{N_p\}, \{M_p\})$ as in Theorem 4.2.6 because the Fourier transform is not considered in the proof here; recall (4.92).

Chapter 5

Boundary Values of Analytic Functions

Analytic functions in tubes which satisfy certain growth conditions involving the (M_p) sequences are shown in this chapter to obtain ultradistributional boundary values. As a basis for the boundary value results we define and study generalizations of the Hardy spaces H^r corresponding to tubes in $\mathbb{C}^n$. For analytic functions considered in the chapter, representations are obtained in terms of the Fourier-Laplace and Cauchy integrals.

5.1 Generalizations of H^r functions in tubes

Let B be a proper open subset of $\mathbb{R}^n$. The set of analytic functions f in $T^B = \mathbb{R}^n + iB$ of the variable $z = x + iy, x \in \mathbb{R}^n, y \in B$, which satisfy the estimate

$$\|f(\cdot + iy)\|_{L^r} \leq M, \qquad y \in B,$$

where the constant M is independent of $y \in B$, is called the Hardy space $H^r(T^B)$, $r > 0$. Stein and Weiss in [137] have obtained representation and boundary value results for the Hardy spaces; see [137] for additional references concerning H^r functions. Generalizations of the spaces $H^r(T^B)$ have been considered and analyzed by several authors including Vladimirov [149], Carmichael and Hayashi [28], and Carmichael [17]-[26].

As in [137], let B denote a proper open subset of $\mathbb{R}^n$, the base of the tube T^B. Let $d(y)$ denote the distance from $y \in B$ to the complement of B in $\mathbb{R}^n$. The space $S_A^r(T^B)$, where $0 < r < \infty$ and $A \geq 0$, is the set of all analytic functions f of the variable $z = x + iy$ in $T^B = \mathbb{R}^n + iB$, which satisfy the inequality

$$\|f(\cdot + iy)\|_{L^r} \leq M(1 + (d(y))^{-m})^q \exp\left[2\pi A|y|\right], \qquad y \in B,$$

for some constants $m \geq 0$, $q \geq 0$ and $M > 0$ which can depend on f, r, and A, but not on $y \in B$. If $B = C$ is a cone, $d(y)$ is interpreted to be the distance from $y \in C$ to the boundary of C. The spaces $S_A^r(T^B)$ were defined and studied by Carmichael in [17]-[23]. For various values of r and various bases B of the tube T^B, Carmichael has obtained Cauchy, Poisson, and Fourier-Laplace integral representations of the $S_A^r(T^B)$ functions and boundary value results. The H^r functions, the functions of Vladimirov [150], and the functions of Carmichael and Hayashi [28] are all special cases of the $S_A^r(T^B)$ functions.

As an example let us consider the cone $(0, \infty)$ in $\mathbb{R}^1$ and the corresponding tube $T^{(0,\infty)}$ which is the upper half plane in $\mathbb{C}^1$. The function g defined by

$$g(z) := \frac{\exp\left[-2\pi i z\right]}{(i + z)z}, \qquad z \in T^{(0,\infty)}$$

is an element of the space $S_1^2(T^{(0,\infty)})$ but is not in the Hardy space $H^2(T^{(0,\infty)})$.

In this section we wish to consider other generalizations of H^r functions, generalizations associated with sequences (M_p). They are introduced by the norm growth (4.61)-(4.62) obtained on the Cauchy integral of ultradistributions in $\mathcal{D}'(*, L^s)$.

The results of this section will be useful in obtaining the ultradistributional boundary value results in the next section.

Again let B be a proper open subset of $\mathbb{R}^n$ and let $d(y)$ denote the distance from $y \in B$ to the complement of B in $\mathbb{R}^n$. Now let f satisfy the inequality

$$\|f(\cdot + iy)\|_{L^r} \leq K(1 + (d(y))^{-m})^q \exp\left[M^*(T/|y|)\right], \qquad y \in B, \qquad (5.1)$$

where $K > 0$, $T > 0$, $m \geq 0$, and $q \geq 0$ are all independent of $y \in B$ and M^* is the associated function of the sequence (M_p) defined in (2.9). The space of analytic functions in T^B which satisfy (5.1) will be denoted by $H_{(M_p)}^r(T^B)$. The spaces $H_{(M_p)}^r(T^B)$ for $0 < r < \infty$ are a generalization of the Hardy spaces.

If $B = C$, where C is an open connected cone in $\mathbb{R}^n$, we obtain from the formula

$$d(y) = \inf_{t \in \mathrm{pr}(C^*)} \langle t, y \rangle, \qquad y \in C,$$

given in [151], p. 159, the estimate $d(y) \leq |y|$, $y \in C$. From this inequality for $B = C$ we see that the term $(1 + (d(y))^{-m})^q$ in (5.1) is a generalizing factor of the growth in which the standard term $|y|$ is replaced by $d(y)$. Further, the right side of (5.1) allows for divergence to ∞ as y approaches an arbitrary point on the boundary of C and not only as y approaches 0 as would be the case if $|y|$ were in place of $d(y)$ in (5.1) .

In this section we shall obtain necessary and sufficient conditions for elements of $H^r_{(M_p)}(T^B)$ spaces, where B is a proper open connected subset of $\mathbb{R}^n$, to be representable by Fourier-Laplace integrals for certain values of r. We also obtain a Cauchy integral representation. As indicated previously, these results will lead us to boundary value and related results in the following section.

Throughout this section we assume that the sequence (M_p) satisfies conditions $(M.1)$ and $(M.3')$.

As previously, also in the present chapter we will use the simplified notation for exponents introduced in Chapter 1 in formulas (1.3) and (1.4), namely

$$E_z(\zeta) := \exp[2\pi i \langle z, \zeta \rangle], \qquad z, \zeta \in \mathbb{C}^n, \tag{5.2}$$

and

$$e_y(t) := \exp[-2\pi \langle y, t \rangle], \qquad y \in \mathbb{R}^n, \ t \in \mathbb{R}^n. \tag{5.3}$$

We begin by proving some lemmas.

Lemma 5.1.1. *Let B be a proper open connected subset of $\mathbb{R}^n$. Suppose that $1 \leq s < \infty$ and g is a measurable function on $\mathbb{R}^n$ which satisfies the estimate*

$$\|e_y g\|_{L^s} \leq K(1 + (d(y))^{-m})^q \exp[M^*(T/|y|)], \qquad y \in B, \tag{5.4}$$

where e_y is defined in (5.3), M^* is the associated function of the sequence (M_p) defined in (2.9), and $K > 0$, $T > 0$, $m \geq 0$, $q \geq 0$ are constants independent of $y \in B$. The function F given by

$$F(z) := \int_{\mathbb{R}^n} g(t) e^{2\pi i \langle z, t \rangle} \, dt = \mathcal{F}[e_y g](x), \qquad z = x + iy \in T^B, \tag{5.5}$$

is analytic in the tube T^B.

Proof. The proof is an extension of that in [28], Theorem 2.1. Let $y_0 \in B$ be arbitrary. Choose an open neighborhood $N(y_0)$ of y_0 such that $\overline{N(y_0)} \subset B$. There exists a $\delta > 0$ such that $\{y : |y - y_0| = \delta\} \subset N(y_0)$. Decompose $\mathbb{R}^n$ into a finite union of non-overlapping cones $C_1, C_2, \ldots, C_k$, each having vertex at the origin, such that $\langle u, v \rangle \geq (2^{1/2}/2)|u||v|$, whenever u and v are two points in a certain C_j for $j = 1, \ldots, k$.

For each $j = 1, \ldots, k$ choose a fixed y_j such that $y_0 - y_j \in C_j$ and $|y - y_0| = \delta$. Then, taking $\varepsilon := 2^{1/2}\pi s \delta > 0$, we have

$$-2\pi s \langle y_j - y_0, t \rangle \geq \varepsilon |t| = 2\pi s (2^{1/2}/2)|y_0 - y_j||t|, \qquad (5.6)$$

whenever $1 \leq s < \infty$ and $t \in C_j, j = 1, \ldots, k$. For each $j = 1, \ldots, k$, inequalities (5.6) and (5.4) yield

$$\int_{C_j} |g(t)|^s \exp\left[-2\pi s \langle y_0, t \rangle\right] \exp\left(\varepsilon|t|\right) dt \leq \int_{\mathbb{R}^n} |g(t)|^s \exp\left[-2\pi s \langle y_j, t \rangle\right] dt$$

$$\leq K^s (1 + (d(y_j))^{-m})^{sq} \exp\left[sM^*(T/|y_j|)\right], \qquad (5.7)$$

since

$$y_j \in \{y : |y - y_0| = \delta\} \subset N(y_0) \subset B, \qquad j = 1, \ldots, k.$$

Now, (5.7) yields

$$\int_{\mathbb{R}^n} |g(t)|^s \exp\left[-2\pi s \langle y_0, t \rangle\right] \exp\left(\varepsilon|t|\right) dt$$

$$\leq K^s \sum_{j=1}^{k} (1 + (d(y_j))^{-m})^{sq} \exp\left[sM^*(T/|y_j|)\right]. \qquad (5.8)$$

Hence, since $\varepsilon|t|/2 \leq \varepsilon|t|$ for $t \in \mathbb{R}^n$, we have, for $s = 1$,

$$\int_{\mathbb{R}^n} |g(t)| \exp\left[-2\pi \langle y_0, t \rangle\right] \exp\left(\varepsilon|t|/2\right) dt$$

$$\leq K \sum_{j=1}^{k} (1 + (d(y_j))^{-m})^{q} \exp\left[M^*(T/|y_j|)\right]. \qquad (5.9)$$

For $1 < s < \infty$, the identity

$$\exp\left(\varepsilon|t|/2s\right) = \exp\left(\varepsilon|t|/s\right) \exp\left(-\varepsilon|t|/2s\right),$$

Hölder's inequality and (5.8) imply

$$\int_{\mathbb{R}^n} |g(t)| \exp\left[-2\pi \langle y_0, t \rangle\right] \exp\left(\varepsilon|t|/2s\right) dt$$

$$\leq A \left(\int_{\mathbb{R}^n} |g(t)|^s \exp\left[-2\pi s \langle y_0, t \rangle\right] \exp\left(\varepsilon|t|\right) dt \right)^{1/s}$$

$$\leq AK \left(\sum_{j=1}^{k} (1 + (d(y_j))^{-m})^{sq} \exp\left[sM^*(T/|y_j|)\right] \right)^{1/s} \qquad (5.10)$$

where $1/s + 1/r = 1$ and

$$A := \left(\int_{\mathbb{R}^n} \exp\left[-\varepsilon r|t|/2s\right] dt \right)^{1/r}.$$

If $|y - y_0| < \varepsilon/4\pi s$, $y = \operatorname{Im} z$ and $1 \le s < \infty$, we have

$$
\begin{aligned}
\left| g(t) e^{2\pi i \langle z, t \rangle} \right| &= |g(t)| \exp\left[-2\pi \langle y - y_0, t \rangle\right] \exp\left[-2\pi \langle y_0, t \rangle\right] \\
&\le |g(t)| \exp\left(2\pi |y - y_0||t|\right) \exp\left[-2\pi \langle y_0, t \rangle\right] \\
&\le |g(t)| \exp\left(\varepsilon |t|/2s\right) \exp\left[-2\pi \langle y_0, t \rangle\right] \qquad (5.11)
\end{aligned}
$$

for all $t \in \mathbb{R}^n$. Estimates (5.9) and (5.10) now show that the right side of (5.11) is an L^1 function, independent of y such that $|y - y_0| < \varepsilon/4\pi s$, whenever $1 \le s < \infty$. Since $y_0 \in B$ is arbitrary, we conclude from (5.11) that F defined by (5.5) is an analytic function of $z \in T^B$. Estimate (5.11) also proves that $e_y g \in L^1$ for $y \in B$ whenever $1 \le s < \infty$. The proof is complete. $\square$

Lemma 5.1.2. *Let C be an open connected cone in $\mathbb{R}^n$ and $1 \le s < \infty$. Let g be a measurable function on $\mathbb{R}^n$ such that (5.4) holds for all $y \in C$. We have $\operatorname{supp} g \subseteq C^*$ almost everywhere.*

Proof. Assume that $g(t) \ne 0$ on a set of positive measure in

$$\mathbb{R}^n \setminus C^* = \{t : \langle y, t \rangle < 0 \quad \text{for some } y \in C\}.$$

Then there is a point $t_0 \in \mathbb{R}^n \setminus C^*$ such that $g(t) \ne 0$ on a set of positive measure in the neighborhood $N(t_0, \eta) = \{t : |t - t_0| < \eta\}$ of t_0 for arbitrary $\eta > 0$. Since $t_0 \in \mathbb{R}^n \setminus C^*$, there is a point $y_0 \in \operatorname{pr}(C) \subset C$ such that $\langle t_0, y_0 \rangle < 0$, where $\operatorname{pr}(C)$, as previously noted, denotes the projection of C which is $\{y \in C : |y| = 1\}$. Using the continuity of the function $t \mapsto \langle t, y_0 \rangle$, we can find a fixed $\sigma > 0$ and a fixed neighborhood $N(t_0, \eta')$ of t_0 such that

$$\langle t, y_0 \rangle < -\sigma < 0, \qquad t \in N(t_0, \eta').$$

Choose η above to be η'. For arbitrary $\lambda > 0$ we thus have

$$-\langle \lambda y_0, t \rangle = -\lambda \langle y_0, t \rangle > \lambda \sigma > 0, \qquad t \in N(t_0, \eta'), \quad \lambda > 0. \qquad (5.12)$$

Since $y_0 \in \mathrm{pr}(C) \subset C$ and C is a cone, we have $\lambda y_0 \in C$ for all $\lambda > 0$. Using (5.12) and (5.4) with $y := \lambda y_0$, we have, for all $\lambda > 0$,

$$
e^{2s\pi\lambda\sigma} \int_{N(t_0,\eta')} |g(t)|^s \, dt \leq \int_{N(t_0,\eta')} |g(t)|^s \exp\left[-2s\pi\langle\lambda y_0, t\rangle\right] dt
$$

$$
\leq \int_{\mathbb{R}^n} |g(t)|^s \exp\left[-2\pi s\langle\lambda y_0, t\rangle\right] dt
$$

$$
\leq K^s (1 + (d(\lambda y_0))^{-m})^{qs} \exp\left[sM^*(T/|\lambda y_0|)\right]
$$

$$
= K^s (1 + \lambda^{-m}(d(y_0))^{-m})^{qs} \exp\left[sM^*(T/\lambda)\right], \qquad (5.13)
$$

where M^* is the associated function of the sequence (M_p) defined in (2.9), since $d(\lambda y_0) = \lambda d(y_0)$ and $y_0 \in \mathrm{pr}(C)$ implies $|y_0| = 1$. The integral on the left of (5.13) is finite. We let $\lambda \to \infty$ in (5.13); thus for the fixed $T > 0$, which is independent of $y \in C$, we can consider $\lambda > 2T$. For the sequence (M_p) which defines M^* we have

$$
(T/\lambda)^p p! (M_0/M_p) < (1/2)^p p! (M_0/M_p), \qquad \lambda > 2T,
$$

for $p = 0, 1, 2, \ldots$. Consequently,

$$
M^*(T/\lambda) < M^*(1/2) < \infty, \qquad \lambda > 2T, \qquad (5.14)
$$

which follows from the definition of M^* and the fact that $(p! M_0/M_p) \to 0$ as $p \to \infty$ (see [82], p. 74), since (M_p) satisfies $(M.1)$ and $(M.3')$. By (5.13) and (5.14),

$$
e^{2\pi s\lambda\sigma} \int_{N(t_0,\eta')} |g(t)|^s \, dt \leq K^s [1 + (\lambda d(y_0))^{-m}]^{qs} \exp\left[sM^*(1/2)\right]
$$

and thus

$$
e^{2\pi s\lambda\sigma} [1 + (\lambda d(y_0))^{-m}]^{-qs} \int_{N(t_0,\eta')} |g(t)|^s \, dt \leq K^s \exp\left[sM^*(1/2)\right] \quad (5.15)
$$

for every $\lambda > 2T$. Since all constants in (5.15) are independent of λ, we let $\lambda \to \infty$. We conclude that g must be zero almost everywhere in $N(t_0,\eta')$, since $\exp(2\pi s\lambda\sigma) \to \infty$ for every $\sigma > 0$ and $[1 + (\lambda d(y_0))^{-m}]^{-qs} \to 1$ as $\lambda \to \infty$.

But this contradicts the fact that $g(t) \neq 0$ on a set of positive measure in $N(t_0,\eta')$. Thus g must be zero almost everywhere in $\mathbb{R}^n \setminus C^*$ and $\mathrm{supp}\, g \subseteq C^*$ almost everywhere, since the dual cone C^* of C is a closed set in $\mathbb{R}^n$. The proof is complete. $\square$

The next two lemmas concern the spaces $H^r_{(M_p)}(T^B)$, $0 < r < \infty$, defined in the paragraph containing (5.1).

In the following lemma B^c denotes the complement of B in $\mathbb{R}^n$, and

$$d(B', B^c) = \inf\{|y_1 - y_2| : y_1 \in B', \quad y_2 \notin B\}$$

is the distance from $B' \subset B$ to B^c.

Lemma 5.1.3. *Let B denote an open connected subset of $\mathbb{R}^n$ which does not contain 0. Suppose that $0 < r < \infty$ and $f \in H^r_{(M_p)}(T^B)$. Let B' be a subset of B such that $d(B', B^c) \geq 2\delta > 0$ for some $\delta > 0$. There is a positive constant K' depending on the $\delta > 0$, on the dimension n, and on f, but not on $z = x + iy \in T^B$, such that*

$$|f(x + iy)| \leq K' \exp\left[M^*(T/(|y| - \delta))\right], \qquad x + iy \in T^{B'}, \qquad (5.16)$$

where M^* is the associated function of the sequence (M_p) defined in (2.9).

Proof. Let $z_0 = x_0 + iy_0$ be an arbitrary point in $T^{B'}$. Put

$$R_\delta := \{z \in \mathbb{C}^n : |z - z_0| < \delta\}, \qquad N(y_0, \delta) := \{y \in \mathbb{R}^n : |y - y_0| < \delta\}.$$

Then $R_\delta \subset \mathbb{R}^n + iN(y_0, \delta) \subset T^B$ and

$$\left(\int_{R_\delta} |f(x + iy)|^r \, dx \, dy\right)^{1/r} \leq \left(\int_{N(y_0,\delta)} \int_{\mathbb{R}^n} |f(x + iy)|^r \, dx \, dy\right)^{1/r}$$

$$\leq K\left(\int_{N(y_0,\delta)} (1 + (d(y))^{-m})^{rq} \exp\left[rM^*(T/|y|)\right] dy\right)^{1/r}, \qquad (5.17)$$

by (5.1). We have

$$\exp\left[rM^*(T/|y|)\right] \leq \exp\left[rM^*(T/(|y_0| - \delta))\right], \qquad y \in N(y_0, \delta). \qquad (5.18)$$

Since $d(B', B^c) \geq 2\delta > 0$, then $y \in N(y_0, \delta)$ implies $d(y) \geq \delta$ and thus

$$(1 + (d(y))^{-m})^{rq} \leq (1 + \delta^{-m})^{rq}, \qquad y \in N(y_0, \delta). \qquad (5.19)$$

Combining (5.17), (5.18), and (5.19) we get

$$\left(\int_{R_\delta} |f(x + iy)|^r \, dx \, dy\right)^{1/r} \leq KN(1 + \delta^{-m})^q \exp\left[M^*(T/(|y_0| - \delta))\right], \quad (5.20)$$

where $N := (\text{measure}(N(y_0, \delta))^{1/r})$ is a number that is actually independent of every given $y_0 \in B'$, because N has the same value for each $y_0 \in B'$. Hence N depends only on δ and the dimension n and not on $y_0 \in B'$.

Since f is an analytic function of the variable $z = x + iy \in T^B$, the function $|f|^r$, $0 < r < \infty$, is a subharmonic function of the $2n$ variables $x_1, \ldots, x_n, y_1, \ldots, y_n$, where $z = x + iy \in T^B$ (see [137], p. 79). Thus [137], Chapter 2, Section 4 yields

$$|f(z_0)|^r \leq (\Omega_{2n} \delta^{2n})^{-1} \int_{R_\delta} |f(x + iy)|^r \, dx \, dy, \qquad (5.21)$$

where Ω_{2n} is the volume of the unit sphere in $\mathbb{R}^{2n}$. The desired estimate (5.16) follows now by combining (5.20) and (5.21) and observing that $z_0 = x_0 + iy_0 \in T^{B'}$ was arbitrary. $\square$

Lemma 5.1.4. *Let B denote an open connected subset of $\mathbb{R}^n$ which does not contain 0. Suppose that $1 < r \leq 2$ and $f \in H^r_{(M_p)}(T^B)$. For arbitrary y and y' in B, we have*

$$e^{2\pi \langle y, t \rangle} h_y(t) = e^{2\pi \langle y', t \rangle} h_{y'}(t) \qquad (5.22)$$

for almost every $t \in \mathbb{R}^n$, where

$$h_y(t) := \mathcal{F}^{-1}[f_y](t), \qquad y \in B, \ t \in \mathbb{R}^n,$$

is the L^s inverse Fourier transform, with $1/r + 1/s = 1$, of the function f_y of the variable $x \in \mathbb{R}^n$ defined by $f_y(x) := f(x + iy)$ for arbitrary fixed $y \in B$.

The proof of Lemma 5.1.4 follows from the growth (5.16) by analysis similar to that in [137], pp. 99-101, with Lemma 5.1.3 taking the place of Lemma 2.12 in [137], p. 99. Further, in our proof of Lemma 5.1.4 the Parseval inequality in the Fourier transform theory for $1 < r \leq 2$ takes the place of the use of the Parseval equality in the case $r = 2$ presented in [137], top of p. 101. We leave the details of the proof of Lemma 5.1.4 to the interested reader.

We can now give Fourier-Laplace and, in certain instances, Cauchy integral representations of the $H^r_{(M_p)}(T^B)$ functions for certain values of r.

Theorem 5.1.1. *Let B denote an open connected subset of $\mathbb{R}^n$ which does not contain $0 \in \mathbb{R}^n$, and let $f \in H^r_{(M_p)}(T^B)$ with $1 < r \leq 2$. There exists a measurable function g of the variable $t \in \mathbb{R}^n$ such that (5.4) holds for $y \in B$, where s satisfies the equality $1/r + 1/s = 1$ and the constants $K > 0$, $T > 0$, $m \geq 0$, $q \geq 0$ are independent of $y \in B$; and*

$$f(z) = \int_{\mathbb{R}^n} g(t) e^{2\pi i \langle z, t \rangle} \, dt = \mathcal{F}[e_y g](x), \qquad z = x + iy \in T^B. \qquad (5.23)$$

Proof. Put

$$g(t) := e^{2\pi\langle y,t\rangle} h_y(t), \qquad y \in B, \ t \in \mathbb{R}^n, \tag{5.24}$$

where $h_y := \mathcal{F}^{-1}[f_y]$ is the L^s inverse Fourier transform of $f_y := f(\cdot + iy)$ for $y \in B$, with s satisfying the equality $1/r + 1/s = 1$. By the Plancherel-Fourier transform theory, h_y is an element of L^s, since $f(\cdot + iy) \in L^r$ for $y \in B$, according to (5.1). By Lemma 5.1.4, g is independent of $y \in B$. From (5.24) it follows that

$$(e_y g)(t) = e^{-2\pi\langle y,t\rangle} g(t) = \mathcal{F}^{-1}[f_y](t), \qquad y \in B, t \in \mathbb{R}^n. \tag{5.25}$$

Since, as previously noted, $f_y \in L^r$ with $1 < r \le 2$ for $y \in B$, we have $e_y g \in L^s$ for $y \in B$, with $1/r + 1/s = 1$, by the Plancherel-Fourier transform theory. Moreover,

$$\|e_y g\|_{L^s} \le \|f(\cdot + iy)\|_{L^r} \le K(1 + (d(y))^{-m})^q \exp\left[M^*(T/|y|)\right]$$

for $y \in B$, where M^* is the associated function of the sequence (M_p) defined in (2.9), by the Parseval inequality and (5.1). Thus the growth (5.4) is obtained. From (5.25), by the Plancherel-Fourier transform theory, we also have

$$f(x + iy) = \mathcal{F}[e_y g](x), \qquad z = x + iy \in T^B, \tag{5.26}$$

with the Fourier transform $\mathcal{F}[e_y g]$ being in L^r. But by (5.4) and Lemma 5.1.1 the integral on the right side of (5.23), which is the L^1 transform of $e_y g$ for $y \in B$, is an analytic function of the variable $z \in T^B$; recall from the proof of Lemma 5.1.1 that $e_y g \in L^1$ for $y \in B$, since (5.4) is satisfied. From the Plancherel-Fourier transform theory we know that the L^1 and L^r Fourier transforms, with $1 < r \le 2$, of the same function are equal when both of them exists. Hence the desired equality (5.23) follows from (5.26). The proof is complete. $\square$

Corollary 5.1.1. *Let C be an open connected cone in $\mathbb{R}^n$ and assume that $f \in H^r_{(M_p)}(T^C)$ with $1 < r \le 2$. There exists a measurable function g of the variable $t \in \mathbb{R}^n$ such that (5.4) holds with s satisfying the equality $1/r + 1/s = 1$ for $y \in C$; the inclusion $\operatorname{supp} g \subseteq C^*$ holds almost everywhere; and equation (5.23) holds for $z \in T^C$.*

Proof. The proof is obtained by combining Theorem 5.1.1 and Lemma 5.1.2. Here the constants K, T, m, and q are as in (5.4). $\square$

In Theorem 5.1.1 and Corollary 5.1.1 we do not know whether $g \in L^s$. However, if g is an element of L^s and if it is the inverse Fourier transform of some function $G \in L^r$ with $1 < r \le 2$, then an additional representation of f from Corollary 5.1.1 is shown in Corollary 5.1.2 below in terms of the Cauchy integral.

Corollary 5.1.2. *Let C be a regular cone in $\mathbb{R}^n$. Let $f \in H^r_{(M_p)}(T^C)$ with $1 < r \le 2$. Suppose that the function g of Corollary 5.1.1 is the inverse Fourier transform of a function $G \in L^r$, $1 < r \le 2$. We have $g \in L^s$, where s satisfies the equality $1/r + 1/s = 1$; the inclusion supp $g \subseteq C^*$ holds almost everywhere; the equality (5.23) holds for $z \in T^C$; and*

$$f(z) = \int_{\mathbb{R}^n} G(t) K(z - t) \, dt, \qquad z \in T^C. \tag{5.27}$$

Proof. By assumption, the function g of Corollary 5.1.1 satisfies the equality

$$g(u) = \mathcal{F}^{-1}[G](u), \qquad u \in \mathbb{R}^n,$$

for some $G \in L^r$, $1 < r \le 2$. By the Plancherel theory, we have $g \in L^s$, where s satisfies the equality $1/r + 1/s = 1$. Note that the Cauchy integral in (5.27) is well defined because of the properties of the Cauchy kernel $K(z - t)$ formulated in Theorem 4.1.1. Using the definition of $K(z - t)$ for $t \in \mathbb{R}^n$ and $z \in T^C$, Fubini's theorem, and the representation (5.23), we obtain

$$\int_{\mathbb{R}^n} G(t) K(z - t) \, dt = \lim_{k \to \infty} \int_{|t| \le k} G(t) \int_{C^*} E_{z-t}(u) \, du \, dt$$

$$= \lim_{k \to \infty} \int_{C^*} E_z(u) \int_{|t| \le k} G(t) e_{it}(u) \, dt \, du = \int_{C^*} g(u) E_z(u) \, du = f(z)$$

$$\tag{5.28}$$

for $z \in T^C$, where the symbols E_z and e_y are defined in (5.2) and (5.3). The Cauchy integral representation (5.27) is thus obtained. $\square$

As a dual theorem to Theorem 5.1.1 we have the following result.

Theorem 5.1.2. *Let B be a proper open connected subset of $\mathbb{R}^n$ and let $1 < r \le 2$. Suppose that g is a measurable function on $\mathbb{R}^n$ which satisfies the estimate*

$$\|e_y g\|_{L^r} \le K(1 + (d(y))^{-m})^q \exp[M^*(T/|y|)], \qquad y \in B, \tag{5.29}$$

where e_y is defined in (5.3), M^* is the associated function of the sequence (M_p) defined in (2.9), and $K > 0$, $T > 0$, $m \geq 0$, $q \geq 0$ are constants which are independent of $y \in B$. Then the function f given by

$$f(z) := \int_{\mathbb{R}^n} g(t)e^{2\pi i \langle z,t \rangle} \, dt = \mathcal{F}[e_y g](x), \qquad z = x + iy \in T^B, \qquad (5.30)$$

is analytic in T^B and satisfies the inequality

$$\|f(\cdot + iy)\|_{L^s} \leq K(1 + (d(y))^{-m})^q \exp[M^*(T/|y|)], \qquad y \in B; \qquad (5.31)$$

and $f \in H^s_{(M_p)}(T^B)$, for s satisfying the equality $1/r + 1/s = 1$.

Proof. By the proof of Lemma 5.1.1, the function f is analytic in T^B and $e_y g \in L^1$ for $y \in B$. By (5.29), we have $e_y g \in L^r$, $1 < r \leq 2$, for $y \in B$. Thus the right side of (5.30) can be viewed as both the L^1 and the L^r Fourier transform of $e_y g$ for $y \in B$. By the Parseval inequality,

$$\|f(\cdot + iy)\|_{L^s} = \|\mathcal{F}[e_y g]\|_{L^s} \leq \|e_y g\|_{L^r} \qquad (5.32)$$

for $y \in B$ and s satisfying the equality $1/r + 1/s = 1$. The estimate in (5.31) now follows from (5.32) and (5.29). $\qquad \blacksquare$

In case B is an open connected subset of $\mathbb{R}^n$ which does not contain $0 \in \mathbb{R}^n$ and $r = 2$, Theorem 5.1.2 is a converse result to Theorem 5.1.1.

5.2 Boundary values in $\mathcal{D}'((M_p), L^s)$ for analytic functions in tubes

In this section we will consider analytic functions of the type $H^r_{(M_p)}(T^C)$, defined in Section 5.1, and obtain boundary values of the functions in $\mathcal{D}'((M_p), L^s)$. Before doing this, we will precisely state conditions that we need on the sequence $(M_p) = (M_p)_{p \in \mathbb{N}_0}$ in order to obtain the boundary value results. We also need to define new associated functions corresponding to these sequences. We will prove a number of lemmas which form a basis for our boundary value results.

If the sequence (M_p) satisfies condition $(M.1)$ of Section 2.1, we have

$$\frac{M_p}{M_{p-1}} \leq \frac{M_{p+1}}{M_p}, \qquad p \in \mathbb{N}. \qquad (5.33)$$

Put

$$m_p := \frac{M_p}{M_{p-1}}, \qquad p \in \mathbb{N}, \qquad (5.34)$$

(see [82], (3.10), p. 50). Note that (5.33) implies that the sequence $(m_p)_{p \in \mathbb{N}}$ is nondecreasing if $(M_p)_{p \in \mathbb{N}_0}$ satisfies $(M.1)$. Further, put

$$m_p^* := \frac{m_p}{p}, \qquad p \in \mathbb{N}. \tag{5.35}$$

To obtain several results in this section we will assume as, for example, Petzsche has done for his analysis in [110], p. 394, that the sequence (m_p^*) is nondecreasing. An example of a sequence (M_p) for which the sequence (m_p^*) is nondecreasing is $M_p := (p!)^s$ for $p \in \mathbb{N}_0$ and $s > 1$. If the sequence (m_p^*) is nondecreasing, we immediately obtain that the sequence (m_p) defined in (5.34) is a strictly increasing sequence; this follows directly from the definition (5.35).

Put

$$m(\lambda) := \text{the number of } m_p \leq \lambda, \tag{5.36}$$

(see [82], p. 50) and note that $m(\lambda)$ is finite for all $\lambda > 0$ if $m_p \to \infty$ as $p \to \infty$. For $(M_p)_{p \in \mathbb{N}_0}$ satisfying $(M.1)$, the sequence (m_p) is nondecreasing. Hence

$$m(\lambda) = \sup_p \{p : m_p \leq \lambda\} \tag{5.37}$$

and m is a nondecreasing function of λ. Now put

$$M_p^* := \frac{M_p}{p!}, \qquad p \in \mathbb{N}, \tag{5.38}$$

and note from (5.35) that

$$m_p^* = \frac{M_p^*}{M_{p-1}^*}, \qquad p \in \mathbb{N}. \tag{5.39}$$

Lemma 5.2.1. *The sequence (m_p^*) is nondecreasing if and only if the sequence (M_p^*) satisfies $(M.1)$.*

Proof. Suppose that the sequence (m_p^*) is nondecreasing. From (5.39) we have

$$\frac{M_{p+1}}{(p+1)\,M_p} \geq \frac{M_p}{p\,M_{p-1}}, \qquad p \in \mathbb{N}.$$

Hence

$$M_{p-1}\,M_{p+1} \geq (M_p)^2 \frac{(p+1)!}{p!} \frac{(p-1)!}{p!}$$

and

$$\frac{M_{p-1}}{(p-1)!} \frac{M_{p+1}}{(p+1)!} \geq \left(\frac{M_p}{p!}\right)^2$$

for all $p \in \mathbb{N}$. From definition (5.38) we thus have obtained

$$M_{p-1}^* M_{p+1}^* \geq (M_p^*)^2, \qquad p \in \mathbb{N},$$

which is condition $(M.1)$ of Section 2.1.

For the converse, let us assume that the sequence (M_p^*) satisfies $(M.1)$. It follows from (5.38) that

$$\frac{(M_p)^2}{(p!)^2} \leq \frac{M_{p-1}}{(p-1)!} \frac{M_{p+1}}{(p+1)!}$$

and

$$(M_p)^2 \leq M_{p-1} M_{p+1} \frac{p!}{(p-1)!} \frac{p!}{(p+1)!} = M_{p-1} M_{p+1} \frac{p}{p+1}$$

for $p \in \mathbb{N}$. This implies

$$\frac{M_p}{p M_{p-1}} \leq \frac{M_{p+1}}{(p+1) M_p}, \qquad p \in \mathbb{N},$$

or, by (5.34) and (5.35),

$$m_p^* \leq m_{p+1}^*. \qquad p \in \mathbb{N}.$$

Thus the sequence (m_p^*) is nondecreasing, and the proof is complete. $\square$

We define

$$m^*(\lambda) := \text{the number of } m_p^* \leq \lambda. \qquad (5.40)$$

If the sequence (m_p^*) is nondecreasing, we have

$$m^*(\lambda) = \sup_p \{p : m_p^* \leq \lambda\}. \qquad (5.41)$$

Recall the associated functions M and M^* defined in (2.8) and (2.9), respectively. If the sequence (M_p) satisfies $(M.1)$, we have

$$M(\rho) = \int\limits_0^\rho \frac{m(\lambda)}{\lambda} \, d\lambda, \qquad 0 < \rho < \infty, \qquad (5.42)$$

which is shown in [82], (3.11), p. 50. Since, by Lemma 5.2.1, the sequence (M_p^*) satisfies condition $(M.1)$ whenever the sequence (m_p^*) is nondecreasing, we have, similarly to (5.42),

$$M^*(\rho) = \int\limits_0^\rho \frac{m^*(\lambda)}{\lambda} \, d\lambda, \qquad 0 < \rho < \infty, \qquad (5.43)$$

if the sequence (m_p^*) is nondecreasing. In (5.42) and (5.43), the functions m and m^* are defined in (5.36) and (5.40), respectively.

Using the above information on sequences (M_p), we now prove four needed lemmas.

Lemma 5.2.2. *Assume that the sequence (M_p) satisfies condition (M.1) and the sequence (m_p^*) is nondecreasing. We have*

$$m^*(t/2m(t)) \le m(t), \qquad t \ge m_1. \tag{5.44}$$

Proof. Recall that the functions m and m^* are given by (5.37) and (5.41), respectively, under the assumptions of the lemma. Fix arbitrarily $t \ge m_1 = M_1/M_0$ and denote $p_0 := m(t)$ for this fixed t. We have

$$m_{p_0} \le t, \qquad m_{p_0+1} > t,$$

and

$$\frac{m_{p_0+1}}{p_0} > \frac{t}{m(t)}. \tag{5.45}$$

For $t \ge m_1$, we have $p_0 \ge 1$ and $2p_0 \ge p_0 + 1$. Hence

$$\frac{2m_{p_0+1}}{p_0+1} \ge \frac{p_0+1}{p_0} \cdot \frac{m_{p_0+1}}{p_0+1} = \frac{m_{p_0+1}}{p_0} > \frac{t}{m(t)},$$

which implies

$$m_{p_0+1}^* = \frac{m_{p_0+1}}{p_0+1} > \frac{t}{2m(t)}$$

and

$$m^*(t/(2m(t))) = \sup_p \{p : m_p^* \le t/(2m(t))\} < p_0 + 1.$$

Inequality (5.44) follows immediately from the above inequality for $t \ge m_1$. □

We know that the sequence (m_p) is strictly increasing if the sequence (m_p^*) is nondecreasing. In fact, we can say more. Recalling the definition (5.35) of the sequence (m_p^*), we see that if (m_p^*) is nondecreasing, then $m_{p+1}^* \ge m_p^*$ for each $p \in \mathbb{N}$ which implies $m_{p+1}/m_p \ge (p+1)/p$ for $p \in \mathbb{N}$. Thus, for each $p = 2, 3, 4, \ldots,$

$$\frac{m_p}{m_1} = \frac{m_p}{m_{p-1}} \frac{m_{p-1}}{m_{p-2}} \cdots \frac{m_3}{m_2} \frac{m_2}{m_1} \ge \frac{p}{p-1} \frac{p-1}{p-2} \cdots \frac{2}{1} = p.$$

Thus if the sequence (m_p^*) is nondecreasing, we have $m_p \ge pm_1$ for each $p \in \mathbb{N}$, which yields $m_p \to \infty$ as $p \to \infty$. Using these facts we prove the following.

Lemma 5.2.3. *Let the sequence (M_p) satisfy condition $(M.1)$ and suppose that the sequence (m_p^*) is nondecreasing. For every $t > m_1$ we have*

$$\frac{m(\lambda)}{\lambda} > \frac{m(t)}{2t}, \qquad m_1 \le \lambda < t. \tag{5.46}$$

Proof. Since the sequence (m_p^*) is nondecreasing, it follows from (5.35) that

$$(p+1)/m_{p+1} \le p/m_p, \qquad p \in \mathbb{N}. \tag{5.47}$$

Let $t > m_1$ be arbitrary but fixed. There is an integer $p \ge 1$ such that $m_p \le t < m_{p+1}$, and therefore $m(t) = p$. First assume that $m_p < t < m_{p+1}$ and λ satisfies the inequalities $m_p \le \lambda < t$. In this case, we have $m(\lambda) = p$ and

$$\frac{m(\lambda)}{\lambda} = \frac{p}{\lambda} > \frac{p}{t} = \frac{m(t)}{t} > \frac{m(t)}{2t}. \tag{5.48}$$

This is the desired result in the case where $m_p \le \lambda < t < m_{p+1}$ for the value of p such that $m_p < t < m_{p+1}$.

The remaining cases are covered by considering λ satisfying $m_1 \le \lambda < m_p$ for the index p for which $m_p \le t < m_{p+1}$. In this case, we have $m_k \le \lambda < m_{k+1}$ for some $k = 1, \dots, p-1$. Taking into account that $k \ge (k+1)/2$ for $k \in \mathbb{N}$ and $m_1 \le \lambda < m_p$ for the p such that $m_p \le t < m_{p+1}$ and using (5.47) repetitively, we get

$$\frac{m(\lambda)}{\lambda} = \frac{k}{\lambda} > \frac{k}{m_{k+1}} \ge \frac{k+1}{2m_{k+1}} \ge \frac{k+2}{2m_{k+2}} \ge \dots \ge \frac{p}{2m_p} \tag{5.49}$$

and

$$\frac{m(t)}{t} = \frac{p}{t} \le \frac{p}{m_p}. \tag{5.50}$$

Combining (5.49) and (5.50), we obtain

$$\frac{m(\lambda)}{\lambda} > \frac{p}{2m_p} \ge \frac{m(t)}{2t} \tag{5.51}$$

for the λ satisfying $m_1 \le \lambda < m_p$ for the p for which $m_p \le t < m_{p+1}$. Combining (5.48) and (5.51), we conclude (5.46). $\square$

In view of Lemma 5.2.1, the sequence (m_p^*) is nondecreasing if and only if (M_p^*) satisfies condition $(M.1)$. Thus the hypothesis that the sequence (m_p^*) is nondecreasing can be replaced in Lemmas 5.2.2 and 5.2.3 by the assumption that the sequence (M_p^*) satisfies condition $(M.1)$. This is the case also in the statements of Lemmas 5.2.4 and 5.2.5.

Lemma 5.2.4. *Let (M_p) satisfy condition $(M.1)$ and let the sequence (m_p^*) be nondecreasing. For $s := 2(m_1 + 1)$ we have*

$$M(t) > \frac{m(t)}{s}, \qquad t \geq m_1 + 1. \tag{5.52}$$

Proof. Since $m(t) = 0$ for $0 \leq t < m_1$, we see from (5.42) that

$$M(t) = \int_{m_1}^{t} m(\lambda)/\lambda d\lambda, \qquad t > m_1. \tag{5.53}$$

From (5.53) and the result (5.46) of Lemma 5.2.3, we deduce

$$M(t) > \frac{m(t)(t - m_1)}{2t} = \frac{m(t)}{2} - \frac{m(t)m_1}{2t}, \qquad t > m_1.$$

Consequently,

$$M(t) > \frac{m(t)}{2} - \frac{m(t)m_1}{2(m_1 + 1)} = \frac{m(t)}{2}\left(1 - \frac{m_1}{m_1 + 1}\right) = \frac{m(t)}{2(m_1 + 1)}$$

for $t \geq m_1 + 1$, which is (5.52) for $s = 2(m_1 + 1)$. $\square$

Lemma 5.2.5. *Let the sequence (M_p) satisfy condition $(M.1)$ and let (m_p^*) be nondecreasing. For $s := 2(m_1 + 1)$ we have*

$$M^*\left(\frac{t}{2sM(t)}\right) \leq M(t) + A, \qquad t \geq m_1 + 1, \tag{5.54}$$

for some constant A, where M and M^* are the associated functions defined in (2.8) and (2.9), respectively.

Proof. Fix $t \geq m_1 + 1$. It follows from inequality (5.52) in Lemma 5.2.4 that $sM(t) > m(t)$, where $s := 2(m_1 + 1)$. Hence $t' < t''$, where

$$t' := \frac{t}{2sM(t)}, \qquad t'' := \frac{t}{2m(t)}.$$

As m^* is a nondecreasing function of λ and $t' < t''$, we conclude from (5.44) in Lemma 5.2.2 that

$$m^*(t') \leq m^*(t'') \leq m(t), \qquad t \geq m_1 + 1. \tag{5.55}$$

Due to (5.41), we have $m^*(\lambda) = 0$ for $0 < \lambda < m_1$. Hence, in view of (5.43), we obtain

$$M^*(t) = \int_{m_1}^{t} \frac{m^*(\lambda)}{\lambda} d\lambda, \qquad t \geq m_1 + 1. \tag{5.56}$$

By a straightforward chain rule calculation and by (5.53), we have

$$\frac{dM^*(t')}{dt} = \frac{m^*(t')}{t'} \cdot \frac{1}{2s} \cdot \left(\frac{1}{M(t)} - \frac{m(t)}{(M(t))^2}\right) \tag{5.57}$$

for $t \geq m_1 + 1$. Using (5.57), (5.55), (5.42) and denoting

$$s' := \frac{m_1 + 1}{2sM(m_1 + 1)} = \frac{1}{4M(s/2)}, \qquad \mu(\lambda) := \frac{\lambda}{2sM(\lambda)},$$

we have for $t \geq m_1 + 1$ that

$$M^*(t') - M^*(s') = \int_{m_1+1}^{t} \frac{dM^*(\mu(\lambda))}{d\lambda} d\lambda$$

$$= \int_{s/2}^{t} \frac{m^*(\mu(\lambda))}{\mu(\lambda)} \cdot \frac{1}{2s} \cdot \left(\frac{1}{M(\lambda)} - \frac{m(\lambda)}{(M(\lambda))^2}\right) d\lambda$$

$$\leq \int_{s/2}^{t} \frac{m^*(\mu(\lambda))}{2s\mu(\lambda)M(\lambda)} d\lambda = \int_{s/2}^{t} \frac{m^*(\mu(\lambda))}{\lambda} d\lambda \leq \int_{s/2}^{t} \frac{m(\lambda)}{\lambda} d\lambda$$

$$= \int_{m_1}^{t} \frac{m(\lambda)}{\lambda} d\lambda - \int_{m_1}^{s/2} \frac{m(\lambda)}{\lambda} d\lambda = M(t) - M(s/2). \tag{5.58}$$

Now (5.54) follows from (5.58) with $A := M^*(s') - M(s/2)$ and the proof is finished. $\square$

We now proceed to prove several other lemmas which are needed to prove our boundary value results. The proof of the following lemma is similar to that of [113], Lemma 10.

Lemma 5.2.6. *Suppose that the sequence (M_p) satisfies conditions $(M.1)$ and $(M.2)$ and the sequence (m_p^*) is nondecreasing. Let C be an open connected cone in $\mathbb{R}^n$, and let $\nu > 0$ and $k > 0$ be arbitrary constants. There exist positive constants K and u such that*

$$\int_{C} \exp\left[-\nu\langle y, t\rangle - M^*(k/|y|)\right] dy \geq K \exp\left[-M(u|t|)\right], \quad t \in C^* \setminus \partial C^*,$$

$$\tag{5.59}$$

where M and M^* are the associated functions defined in (2.8) and (2.9), respectively, and ∂C^* is the boundary of the dual cone C^* of C.

Proof. Denote

$$I(t) := \int_C \exp\left[-\nu\langle y, t\rangle - M^*(k/|y|)\right] dy.$$

Let $s := 2(m_1 + 1)$. Let $t \in \mathbb{R}^n$ be arbitrary but fixed such that $t \in C^* \setminus \partial C^*$ and $|t| > m_1 + 1 = s/2$. For such t put

$$A(t) := \left\{ y \in \mathbb{R}^n : |y| \in [a(|t|), b(|t|)] \right\},$$

where

$$a(\theta) := 2skM(\theta)/\theta; \quad b(\theta) := (2skM(\theta) + 1)/\theta \qquad (5.60)$$

for $\theta > 0$, and

$$C(t) := A(t) \cap C.$$

Since the function M^* defined in (5.43) is nondecreasing, we conclude from Lemma 5.2.5 that

$$M^*(k/|y|) \leq M^*(t')$$

where

$$t' := |t|/(2sM(|t|));$$

and, consequently,

$$\exp\left[-M^*(k/|y|)\right] \geq \exp\left[-M^*(t')\right] \geq \exp\left[-M(|t|) - A\right] \qquad (5.61)$$

for $y \in C(t)$ and $|t| > s/2$, where the constant A from (5.54) depends on the sequence (M_p) through the functions M and M^*. For $y \in C(t)$, we have

$$\exp\left[-\nu|y||t|\right] \geq \exp\left[-2sk\nu M(|t|) - \nu\right]. \qquad (5.62)$$

Recall that $\langle y, t\rangle \geq 0$, $y \in C$, $t \in C^*$. Combining (5.61) and (5.62) we have, for $t \in C^* \setminus \partial C^*$ and $|t| > s/2$, that

$$I(t) \geq \int_{C(t)} \exp\left[-\nu|y||t| - M^*(k/|y|)\right] dy$$

$$\geq \exp\left[-2sk\nu M(|t|) - \nu\right] \exp\left[-M(|t|) - A\right] \int_{C(t)} 1 \, dy$$

$$= \exp\left[-(\nu + A)\right] \exp\left[-(1 + 2sk\nu)M(|t|)\right] \int_{C(t)} 1 \, dy. \qquad (5.63)$$

As before, let $\mathrm{pr}(C)$ denote the projection of C which is the intersection of C with the unit sphere in $\mathbb{R}^n$. We have

$$\int_{C(t)} 1 \, dy = \int_{\mathrm{pr}(C)} \int_{a(|t|)}^{b(|t|)} r^{n-1} \, dr \, d\sigma$$

$$= n^{-1} \int_{\mathrm{pr}(C)} 1 \, d\sigma \cdot [(b(|t|))^n - (a(|t|))^n] \geq S_C \, n^{-1} |t|^{-n},$$

$$(5.64)$$

where S_C is the surface area of $\mathrm{pr}(C)$. Recall from (5.42) that M is an increasing function. Thus

$$\frac{\exp[M(|t|)]}{|t|^n} \to \infty \qquad \text{as } |t| \to \infty.$$

For $t \in C^* \backslash \partial C^*$ and $|t| > s/2$ we can choose a constant $Q > 0$, independent of t, such that

$$|t|^{-n} \geq Q \exp[-M(|t|)], \qquad t \in C^* \backslash \partial C^*, \ |t| > s/2,$$

which we use in (5.64) to obtain

$$\int_{C(t)} 1 \, dy \geq Q S_C \, n^{-1} \exp[-M(|t|)], \qquad t \in C^* \backslash \partial C^*, \ |t| > s/2. \quad (5.65)$$

(Equivalently, it follows directly from the definition of the function M that

$$\exp[M(|t|)] \geq (M_0/M_n) |t|^n,$$

and this can also be used in (5.64) to obtain (5.65) with $Q := M_0/M_n$.)

Combining (5.63) and (5.65), we see that

$$I(t) \geq Q S_C \, n^{-1} \exp[-(\nu + A)] \exp[-(2 + 2sk\nu)M(|t|)]$$

$$= B \exp[-(2 + 2sk\nu)M(|t|)] \qquad (5.66)$$

for $t \in C^* \backslash \partial C^*$ and $|t| > s/2$, where the constant

$$B := Q S_C \, n^{-1} \exp[-(\nu + A)]$$

depends on ν, m_1, n, and the cone C, but not on t.

Now fix $t \in C^* \backslash \partial C^*$ such that $|t| \leq m_1 + 1 = s/2$. Put

$$C'(t) := A'(t) \cap C,$$

where

$$A'(t) := \{ y \in \mathbb{R}^n : \ |y| \in [a'(t), b'(t)] \}$$

with

$$a'(t) := a(|t| + m_1 + 1); \quad b'(t) := b(|t| + m_1 + 1),$$

according to the previous notation adopted in (5.60). Proceeding similarly as in (5.61), (5.62) and (5.64) and denoting

$$s' := \frac{2m_1 + 2}{2sM(m_1 + 1)}, \qquad t'' := \frac{|t| + m_1 + 1}{2sM(|t| + m_1 + 1)},$$

we now deduce that

$$\exp\left[-M^*(k/|y|)\right] \geq \exp\left[-M^*(t'')\right] \geq \exp\left[-M^*(s')\right]; \qquad (5.67)$$

$$\exp\left[-\nu|y||t|\right] \geq \exp\left[-\nu(2skM(s) + 1)\right]; \qquad (5.68)$$

$$\int_{C'(t)} 1 \, dy \geq S_C \, n^{-1} s^{-n}, \qquad (5.69)$$

for $y \in C'(t)$ and $|t| \leq s/2$, where again S_C is the surface area of $\mathrm{pr}(C)$. Combining (5.67), (5.68) and (5.69), we see that

$$I(t) \geq \int_{C'(t)} \exp\left[-\nu|y||t| - M^*(k/|y|)\right] dy$$

$$\geq \frac{S_C}{ns^n} \exp\left[-\nu(2skM(s) + 1) - M^*(s')\right] =: R \qquad (5.70)$$

for $t \in C^* \setminus \partial C^*$ and $|t| \leq s/2$, where the constant R depends on n, the sequence (M_p) through the functions M and M^*, and the cone C as well as on the given positive constants ν and k, but does not depend on t.

We have thus proved that (5.66) holds for each $t \in C^* \setminus \partial C^*$ such that $|t| > s/2$ and (5.70) holds for each $t \in C^* \setminus \partial C^*$ such that $|t| \leq s/2$. Combining (5.66) and (5.70), we can find a positive constant K_1 such that

$$I(t) \geq K_1 \exp\left[-2(1 + sk\nu)M(|t|)\right], \qquad t \in C^* \setminus \partial C^*. \qquad (5.71)$$

The sequence (M_p) is assumed to satisfy conditions $(M.1)$ and $(M.2)$ here. Thus (2.13) from Lemma 2.1.3 holds for the constant $L := (2 + 2sk\nu) > 1$. Hence the existence of constants $K > 0$ and $u > 0$ such that (5.59) holds follows from (5.71) and (2.13) in Lemma 2.1.3. The constants are independent of $t \in C^* \setminus \partial C^*$ but are dependent upon the dimension n, k, ν, the sequence (M_p), through the functions M and M^*, and the cone C. The proof is complete. $\square$

The sequence (M_p) given by $M_p := (p!)^s$ for $p \in \mathbb{N}_0$ and $s > 1$ is an example of a sequence which satisfies the hypothesis of the previous lemma.

We use Lemma 5.2.6 to prove the next lemma.

Lemma 5.2.7. *Assume that the sequence (M_p) satisfies conditions $(M.1)$ and $(M.2)$ and the sequence (m_p^*) is nondecreasing. Let C be a regular cone* *in* $\mathbb{R}^n$ *and* $1 \leq s < \infty$. *Let* g *be a measurable function on* $\mathbb{R}^n$ *such that* $\operatorname{supp} g \subseteq C^*$ *almost everywhere and*

$$\int_{\mathbb{R}^n} |(e_y g)(t)|^s \, dt \leq R \exp\left[M^*(k|y|)\right], \qquad y \in C, \tag{5.72}$$

for some constants $R > 0$ *and* $k > 0$, *where* e_y *has the meaning of (5.3). There exists a constant* $d > 0$ *such that*

$$\|g \exp\left[-M(d|\cdot|)\right]\|_{L^s} < \infty, \tag{5.73}$$

where the symbols M *and* M^* *here denote, as previously, the associated functions defined in (2.8) and (2.9), respectively.*

Proof. We need here the following additional notation:

$$e_y^s(t) := e^{-2\pi s \langle y, t \rangle}, \qquad y \in \mathbb{R}^n,$$

for $1 \leq s < \infty$. This means we have, in particular, $e_y^1 = e_y$ for $y \in \mathbb{R}^n$, according to (5.3).

Since $\operatorname{supp} g \subseteq C^*$ almost everywhere, we conclude from (5.72) that

$$\exp\left[-M^*(k/|y|)\right] \int_{C^*} |g(t)|^s e_y^s(t) \, dt \leq R, \qquad y \in C. \tag{5.74}$$

Fix $t_0 \in \operatorname{pr}(C^*) \setminus \partial C^*$ with ∂C^* denoting as before the boundary of C^*. Multiplying (5.74) by $e_y^s(t_0) = \exp\left[-2\pi s \langle y, t_0 \rangle\right]$ for arbitrary $y \in C$ and integrating over C we get

$$\int_C \left(\int_{C^*} (|g|^s e_y)(t) \, dt \right) \exp\left[-M^*(k/|y|\right] e_y^s(t_0) \, dy$$

$$\leq R \int_C \exp\left[-2\pi s \langle y, t_0 \rangle\right] \, dy. \tag{5.75}$$

As in the proof of Lemma 5.2.6, the integral on the right of (5.75) is finite for $t_0 \in \operatorname{pr}(C^*) \setminus \partial C^*$. By Fubini's theorem, interchanging the order of integration on the left of (5.75), we obtain

$$\int_{C^*} |g(t)|^s \int_C e_y^s(t + t_0) \exp\left[-M^*(k/|y|)\right] \, dy \, dt < \infty. \tag{5.76}$$

Hence, since $1 \leq s < \infty$ and the cone C is regular (and thus convex), so that $t \in C^* \setminus \partial C^*$ and $t_0 \in \operatorname{pr}(C^*) \setminus \partial C^*$ imply that $t + t_0 \in C^* \setminus \partial C^*$, it follows by Lemma 5.2.6 and (5.76) that there exist positive constants K and m such that

$$K \int_{C^*} |g(t)|^s \exp\left[-sM(m|t + t_0|)\right] \, dt$$

$$\leq K \int_{C^* \setminus \partial C^*} |g(t)|^s \exp\left[-M(m|t + t_0|)\right] \, dt$$

$$\leq \int_{C^* \setminus \partial C^*} |g(t)|^s \int_C e_y^s(t + t_0) \exp\left[-M^*(k/|y|)\right] \, dy \, dt < \infty$$

$$\tag{5.77}$$

and this proves that the integral on the left of (5.77) is finite. Now, since $t_0 \in \mathrm{pr}(C^*) \setminus \partial C^*$, we have $|t + t_0| \leq |t| + |t_0| = 1 + |t|$; and applying the fact that M is a nondecreasing function of $\rho > 0$ (see [127], p. 65) and the property (2.10) fom Lemma 2.1.3, we obtain

$$M(m|t + t_0|) \leq M(m|t| + m) \leq M(2m|t|) + M(2m). \tag{5.78}$$

Therefore, applying (5.78) in (5.77), we have

$$Ke^{-sM(2m)} \int_{C^*} |g(t)|^s e^{-sM(2m|t|)}\, dt < \infty, \tag{5.79}$$

since $\exp[sM(2m)]$ is finite. This proves (5.73) with $d := 2m$. $\square$

Lemma 5.2.7 is used to prove the next lemma which, in turn, will be used later in this section to obtain the ultradistribution boundary value results.

Lemma 5.2.8. *Assume that the sequence (M_p) satisfies conditions $(M.1)$ and $(M.2)$ and the sequence (m_p^*) is nondecreasing. Let C be a regular cone in $\mathbb{R}^n$ and $1 \leq s < \infty$. Let g be a measurable function on $\mathbb{R}^n$ such that $\mathrm{supp}\, g \subseteq C^*$ almost everywhere and*

$$\|e_y g\|_{L^s} \leq K \exp[M^*(k/|y|)], \qquad y \in C, \tag{5.80}$$

for some positive constants K and k with the notation (5.3) used. There is a constant $b > 0$ such that

$$\|g \exp[-M(b|\cdot|)]\|_{L^s} < \infty, \tag{5.81}$$

where M and M^* above denote the associated functions defined in (2.8) and (2.9), respectively.

Proof. From (5.80) we have

$$\int_{\mathbb{R}^n} |e_y g|^s\, dt \leq K^s \exp[sM^*(k/|y|)], \qquad y \in C. \tag{5.82}$$

Under the assumptions $(M.1)$ and $(M.2)$ on the sequence (M_p), it follows that the sequence $(M_p/p!)$ satisfies condition $(M.2)$; and $(M_p/p!)$ satisfies condition $(M.1)$ by Lemma 5.2.1. Note that

$$M^*(\rho) = \sup_p \log_+[\rho^p p! M_0/M_p] = \sup_p \log_+[\rho^p M_0/(M_p/p!)]. \tag{5.83}$$

Now, applying the proof of [111], Lemma 1.7 (b), pp. 140-141 (see also the proof of [82], Proposition 3.6, p. 51) to the sequence $(M_p/p!)$, which satisfies conditions $(M.1)$ and $(M.2)$, and taking into account (2.13) from

Lemma 2.1.3, we conclude that there exist a positive constant B and a constant $Q_s > 0$ depending on s such that

$$sM^*(k/|y|) \leq M^*(B^{s-1}k/|y|) + Q_s, \qquad 1 \leq s < \infty. \qquad (5.84)$$

Using (5.84) in (5.82), we have

$$\int_{\mathbb{R}^n} |(e_y g)(t)|^s \, dt \leq K^s \exp(Q_s) \exp[M^*(B^{s-1}k/|y|)], \qquad y \in C. \qquad (5.85)$$

The conclusion (5.81) now follows from (5.85) and the assumptions on g and (M_p) by applying Lemma 5.2.7. The proof is complete. $\square$

Recall the spaces $\mathcal{F}\mathcal{D}(*, L^r)$ of Section 2.4.

Lemma 5.2.9. *Assume that the sequence (M_p) satisfies conditions $(M.1)$ and $(M.2)$. Let $\varphi \in \mathcal{D}((M_p), L^1)$. We have*

$$\sup_{x \in R^n} |\widehat{\varphi}(x) \exp[M(h|x|)]| < \infty \qquad (5.86)$$

for every $h > 0$, where M is the associated function defined in (2.8).

Proof. Since $\varphi \in \mathcal{D}((M_p), L^1)$ implies $\widehat{\varphi} \in \mathcal{F}\mathcal{D}((M_p), L^1)$, for every $k > 0$, every n-tuple α of nonnegative integers, and each $x \in \mathbb{R}^n$ we have

$$\frac{|x^\alpha \widehat{\varphi}(x)|}{k^\alpha M_\alpha} \leq N \qquad (5.87)$$

where N is a positive constant which is independent of k, α and $x \in \mathbb{R}^n$. Applying (2.5) in (5.87), we get

$$\frac{1}{(M_0)^n B} \sup_{\alpha_1} \left[\frac{M_0}{M_{\alpha_1}} \left(\frac{|x_1|}{kE} \right)^{\alpha_1} \right] \cdots \sup_{\alpha_n} \left[\frac{M_0}{M_{\alpha_n}} \left(\frac{|x_n|}{kE} \right)^{\alpha_n} \right] |\widehat{\varphi}(x)| \leq N$$

and, by the definition of the function M in (2.8),

$$\frac{1}{(M_0)^n B} \exp[M(|x_1|/(kE)) + \ldots + M(|x_n|/(kE))] |\widehat{\varphi}(x)| \leq N \qquad (5.88)$$

for some positive constants B and E, and for all $x = (x_1, \ldots, x_n) \in \mathbb{R}^n$. Since $|x| \leq n(|x_1| + \ldots + |x_n|)$ and the function M is nondecreasing on $(0, \infty)$, we have

$$\begin{aligned}
M(|x|/(knE)) &\leq M(|x_1|/(kE) + \ldots + |x_n|/(kE)) \\
&\leq M(n|x_1|/(kE)) + \ldots + M(n|x_n|/(kE)) \\
&\leq (3n/2)(M(|x_1|/(kE)) + \ldots + M(|x_n|/(kE))) + K, \qquad (5.89)
\end{aligned}$$

by (2.10) and (2.12) from Lemma 2.1.3, where $K > 0$ is a constant. From (5.89) we have

$$(2/(3n))M(|x|/(knE)) \leq M(|x_1|/(kE)) + \ldots + M(|x_n|/(kE)) + K_1 \quad (5.90)$$

for $x \in \mathbb{R}^n$, where $K_1 > 0$ is a constant. Using (5.90) in (5.88) we have

$$\frac{\exp[-K_1]}{(M_0)^n B} \exp[(2/(3n))M(|x|/(knE))] \, |\widehat{\varphi}(x)| \leq N, \qquad x \in \mathbb{R}^n,$$

and hence

$$\frac{\exp[-K_1]}{(M_0)^n B} \sup_{x \in \mathbb{R}^n} \left(\exp[(2/(3n))\, M(|x|/(knE))] \, |\widehat{\varphi}(x)| \right) \leq N \quad (5.91)$$

for all $k > 0$. For arbitrary $h > 0$ and a positive integer q chosen such that $(2/(3n))2^q > 1$, select k such that

$$\frac{1}{knE} = H^q h, \quad (5.92)$$

where H is the constant from condition $(M.2)$. By (2.11) from Lemma 2.1.3, used repeatedly, there exists a constant $G > 0$ such that

$$\begin{aligned}
M(H^q h |x|) = M(H(H^{q-1}h|x|)) &\geq 2M(H^{q-1}h|x|) - G \\
&\geq 2^2 M(H^{q-2}h|x|) - 2G - G \\
&\geq 2^q M(h|x|) - (1 + 2 + \ldots + 2^{q-1})G. \quad (5.93)
\end{aligned}$$

From the choice of q, (5.92) and (5.93), we have

$$\begin{aligned}
\exp[(2/(3n))M(|x|/(knE))] &= \exp[(2/(3n))M(H^q h|x|)] \\
&\geq \exp[(2/3n)2^q M(h|x|)] \exp[-(2/(3n))(1 + 2 + \ldots + 2^{q-1})G] \\
&\geq \exp[-(2/(3n))(1 + 2 + \ldots + 2^{q-1})G] \exp[M(h|x|)] \quad (5.94)
\end{aligned}$$

for all $h > 0$. (5.91) and (5.94) now combine to prove our asssertion (5.86). $\square$

The proof of the following result can be obtained in a similar way as the proof of Lemma 5.2.9, and we omit the details.

Corollary 5.2.1. *If (φ_λ) is a net of elements in $\mathcal{D}((M_p), L^1)$ which converges to zero in $\mathcal{D}((M_p), L^1)$ as $\lambda \to \infty$, where the sequence (M_p) satisfies $(M.1)$ and $(M.2)$, we have*

$$\lim_{\lambda \to \infty} \sup_{x \in \mathbb{R}^n} |\exp[M(h|x|)]\, \widehat{\varphi}_\lambda(x)| = 0$$

for all $h > 0$, where M is the associated function defined in (2.8).

The following result will be used in the boundary value analysis.

Lemma 5.2.10. *Assume that the sequence (M_p) satisfies conditions $(M.1)$ and $(M.2)$. Let $\psi \in \mathcal{FD}((M_p), L^r)$ with $1 \leq r \leq 2$. We have, for arbitrary $k > 0$,*

$$\| \psi \exp [M(k| \cdot |)] \|_{L^s} < \infty, \tag{5.95}$$

with $1/r + 1/s = 1$, where M is the associated function defined in (2.8).

Proof. The proof is obtained by the calculation in [113], p. 205. For $k > 0$ and $t \in \mathbb{R}^n$, we have

$$\frac{\exp [M(k|t|)]}{M_0} \leq \sup_{p \in \mathbb{N}_0} \frac{(kn)^p(|t_1| + \ldots + |t_n|)^p}{M_p} \leq \sup_\alpha \frac{(kn^2)^\alpha |t^\alpha|}{M_\alpha},$$

where α is an n-tuple of nonnegative integers. Thus, if $\psi \in \mathcal{FD}((M_p), L^r)$ with $1 \leq r \leq 2$ and $1/r + 1/s = 1$, we have

$$\frac{\| \psi \exp [M(k| \cdot |)] \|_{L^s}}{M_0} \leq \left\| \psi \sup_\alpha \frac{(kn^2)^\alpha |t^\alpha|}{M_\alpha} \right\|_{L^s} \leq \sum_\alpha \frac{(kn^2)^\alpha}{M_\alpha} \| t^\alpha \psi \|_{L^s}. \tag{5.96}$$

From (2.50),

$$\sup_\alpha \frac{\| t^\alpha \psi \|_{L^s}}{h^\alpha M_\alpha} < \infty$$

for all $h > 0$. Putting $h := (2kn^2)^{-1}$, we obtain from (5.96)

$$\frac{\| \psi \exp [M(k| \cdot |)] \|_{L^s}}{M_0} \leq \sum_\alpha \frac{\| t^\alpha \psi \|_{L^s}}{2^\alpha h^\alpha M_\alpha} \leq \sup_\alpha \frac{\| t^\alpha \psi \|_{L^s}}{h^\alpha M_\alpha} \sum_\alpha \left(\frac{1}{2} \right)^\alpha < \infty,$$

which proves (5.95) as desired. $\square$

The following result is proved using the details of the proof of Lemma 5.2.10 just as the proof of Corollary 5.2.1 followed from the details of the proof of Lemma 5.2.9.

Corollary 5.2.2. *If (ψ_λ) is a net of elements in $\mathcal{FD}((M_p), L^r)$, with $1 \leq r \leq 2$, which converges to zero in $\mathcal{FD}((M_p), L^r)$ as $\lambda \to \infty$, where the sequence (M_p) satisfies $(M.1)$ and $(M.2)$, we have*

$$\lim_{\lambda \to \infty} \| \psi_\lambda \exp [M(k| \cdot |)] \|_{L^s} = 0$$

for all $k > 0$, where s satisfies the equality $1/r + 1/s = 1$ and M is the associated function defined in (2.8).

Using the lemmas and corollaries proved to this point of this section, we can now obtain boundary value results. In our proofs we also use properties obtained in Section 5.1.

Thus, throughout the remainder of this section we will assume that (M_p) is a sequence of positive numbers which satisfies conditions $(M.1)$, $(M.2)$, and $(M.3')$ of Section 2.1 and the sequence (m_p^*), defined in (5.35), is nondecreasing.

Let C be a regular cone in $\mathbb{R}^n$. We consider functions f of the variable z which are analytic in $T^C = \mathbb{R}^n + iC$ and which satisfy the following norm growth:

$$\|f(\cdot + iy)\|_{L^r} \leq K \exp\left[M^*(T/|y|)\right], \qquad y \in C, \tag{5.97}$$

where $K > 0$ and $T > 0$ are constants which are independent of $y \in C$ and M^* is the associated function of the sequence (M_p) defined in (2.9). Note that the norm growth (5.97) considered in this section is that in (5.1) with $m = 0$ or $q = 0$ there, and the functions f are certain elements in $H^r_{(M_p)}(T^C)$ as defined in Section 5.1. We first prove that elements of $H^r_{(M_p)}(T^C)$ with $1 < r \leq 2$, which satisfy (5.97), obtain ultradistributional boundary values in $\mathcal{D}'((M_p), L^1)$.

Theorem 5.2.1. *Let f be analytic in T^C and satisfy (5.97) with $1 < r \leq 2$. There exists $U \in \mathcal{D}'((M_p), L^1)$ such that*

$$\lim_{y \to 0, y \in C} f(\cdot + iy) = U \tag{5.98}$$

in $\mathcal{D}'((M_p), L^1)$.

Proof. Let $\varphi \in \mathcal{D}((M_p), L^1)$. From Section 2.3 we see that $\mathcal{D}((M_p), L^1) \subset \mathcal{D}((M_p), L^s)$ for all $s, 1 \leq s < \infty$.

By (5.97), we have $f(\cdot + iy) \in L^r$ for $y \in C$ with $1 < r \leq 2$. Thus $\langle f(\cdot + iy), \varphi \rangle$ is well defined for $y \in C$. By Corollary 5.1.1 (see the proof of Theorem 5.1.1) and the assumed inequality (5.97), there exists a measurable function g on $\mathbb{R}^n$ such that $\operatorname{supp} g \subseteq C^*$ almost everywhere,

$$\|e_y g\|_{L^s} \leq K \exp\left[M^*(T/|y|)\right], \qquad y \in C, \tag{5.99}$$

with s satisfying $1/r + 1/s = 1$, and

$$f(x + iy) = \mathcal{F}[e_y g](x), \qquad z = x + iy \in T^C, \tag{5.100}$$

where e_y is defined in (5.3) and M^* is the associated function of the sequence (M_p) defined in (2.9). The above Fourier transform is meant both in the sense of L^1 and L^r (see (5.26)). By (5.100) and the Parseval equality, we have

$$\langle f(\cdot + iy), \varphi \rangle = \langle \mathcal{F}[e_y \, g], \varphi \rangle = \langle e_y g, \hat{\varphi} \rangle \qquad (5.101)$$

for $\varphi \in \mathcal{D}((M_p), L^1)$ and $y \in C$.

We now want to show that $g\,\hat{\varphi} \in L^1$. To do so fix arbitrary constants $k_1, k_2 \geq 1$ and find, according to (2.12) and (2.13) from Lemma 2.1.3, constants $K_1, K_2, K_3 > 0$ such that

$$\exp\left[M(k_1|t|) + M(k_2|t|)\right] \leq K_1 \exp\left[(3/2)(k_1 + k_2)M(|t|)\right]$$
$$\leq K_2 \exp\left[M(K_3|t|)\right], \qquad (5.102)$$

where M is the associated function defined in (2.8).

Denoting $e^M_{-j}(t) := \exp\left[-M(k_j|t|)\right]$ for $j = 1, 2$ and $e^M_3(t) := \exp\left[M(K_3|t|)\right]$, we conclude from (5.102) and the Hölder inequality that

$$\int\limits_{\mathbb{R}^n} |g(t)\hat{\varphi}(t)|\, dt \leq K_2 \int\limits_{\mathbb{R}^n} |g(t)\,\hat{\varphi}(t)| e^M_{-1}(t)\, e^M_{-2}(t)\, e^M_3(t)\, dt$$

$$\leq K_2 \, \|e^M_3 \, \hat{\varphi}\|_{L^\infty} \, \|g e^M_{-1} e^M_{-2}\|_{L^1}$$

$$\leq K_2 \, \|e^M_3 \, \hat{\varphi}\|_{L^\infty} \, \|g e^M_{-1}\|_{L^s} \, \|e^M_{-2}\|_{L^r}. \qquad (5.103)$$

Clearly, $e^M_{-2} \in L^r$, so $\|e^M_{-2}\|_{L^r} < \infty$. Moreover, since $k_1 \geq 1$ was fixed arbitrarily, we may choose k_1 to be equal to b in (5.81). Hence, by Lemma 5.2.8, we have $\|g\, e^M_{-1}\|_{L^s} < \infty$. Finally, since the assertion of Lemma 5.2.9 holds for every $h > 0$, we can put $h := K_3$ in (5.86) to obtain $\|e^M_3 \hat{\varphi}\|_{L^\infty} < \infty$. Therefore the right side of (5.103) is finite. Consequently, $g\,\hat{\varphi} \in L^1$ by virtue of (5.103).

Since $\langle y, t \rangle \geq 0$ for $y \in C$, $t \in C^*$, and $\operatorname{supp} g \subseteq C^*$ almost everywhere, we have

$$|e_y(t)g(t)\hat{\varphi}(t)| \leq |g(t)\hat{\varphi}(t)|$$

for almost all $t \in \mathbb{R}^n$; and, in view of the preceding paragraph, $g\,\hat{\varphi} \in L^1$. Thus by the Lebesgue dominated convergence theorem,

$$\lim_{y \to 0, y \in C} \int_{\mathbb{R}^n} e_y(t)g(t)\hat{\varphi}(t)\, dt = \int_{\mathbb{R}^n} g(t)\hat{\varphi}(t)\, dt. \qquad (5.104)$$

We are now in a position to define U by

$$\langle U, \varphi \rangle := \langle g, \hat{\varphi} \rangle, \qquad \varphi \in \mathcal{D}((M_p), L^1). \qquad (5.105)$$

If (φ_λ) is a net in $\mathcal{D}((M_p), L^1)$ converging to zero in $\mathcal{D}((M_p), L^1)$ as $\lambda \to \infty$, similar arguments to those used in the proof of the inequalities (5.103) and of Corollary 5.2.1 show that

$$\lim_{\lambda \to \infty} \langle U, \varphi_\lambda \rangle = \lim_{\lambda \to \infty} \langle g, \hat{\varphi}_\lambda \rangle = 0.$$

Hence U is continuous on $\mathcal{D}((M_p), L^1)$ and the linearity of U is obvious. Thus $U \in \mathcal{D}'((M_p), L^1)$. Returning now to (5.101) and using (5.104) and the definition (5.105), we obtain

$$\lim_{y \to 0, y \in C} \langle f(\cdot + iy), \varphi \rangle = \lim_{y \to 0, y \in C} \langle e_y g, \widehat{\varphi} \rangle = \langle g, \widehat{\varphi} \rangle = \langle U, \varphi \rangle \qquad (5.106)$$

for $\varphi \in \mathcal{D}((M_p), L^1)$, which proves (5.98) as desired. $\square$

The following result is a dual theorem to Theorem 5.2.1, and boundary value results are obtained in $\mathcal{D}'((M_p), L^r)$, $1 < r \leq 2$.

Recall the spaces $\mathcal{F}\mathcal{D}((M_p), L^r)$ and $\mathcal{F}'\mathcal{D}((M_p), L^r)$ which were defined in Section 2.4.

Theorem 5.2.2. *Let $1 < r \leq 2$. Let g be a measurable function on $\mathbb{R}^n$ such that*

$$\|e_y g\|_{L^r} \leq K \exp\left[M^*(T/|y|)\right], \qquad y \in C, \qquad (5.107)$$

where M^* is the associated function of the sequence (M_p) defined in (2.9), e_y is defined in (5.3), and K and T are positive constants which are independent of $y \in C$. The function f defined by

$$f(z) := \int_{\mathbb{R}^n} g(t) e^{2\pi i \langle z, t \rangle} \, dt = \mathcal{F}[e_y g](x), \qquad z = x + iy \in T^C, \qquad (5.108)$$

is analytic in T^C, satisfies the inequality

$$\|f(\cdot + iy)\|_{L^s} \leq K \exp\left[M^*(T/|y|)\right], \qquad y \in C, \qquad (5.109)$$

where $1/r + 1/s = 1$, and there is an ultradistribution $U \in \mathcal{D}'((M_p), L^r)$, namely $U := \mathcal{F}^{-1}[\tilde{g}]$, where $\tilde{g}(t) := g(-t)$ for $t \in \mathbb{R}^n$, such that

$$\lim_{y \to 0, y \in C} f(\cdot + iy) = U \qquad (5.110)$$

in $\mathcal{D}'((M_p), L^r)$.

Proof. By Theorem 5.1.2 with $q := 0$, the function f is analytic in T^C and satisfies (5.109), i.e. (5.97) with L^r replaced by L^s, $1/r + 1/s = 1$. By the proof of Lemma 5.1.1, the Fourier transform $\mathcal{F}[e_y g]$ on the right of (5.108) can be interpreted in both the L^1 and L^r sense. Further, by Lemma 5.1.2, $\operatorname{supp} g \subseteq C^*$ almost everywhere.

Now fix an arbitrary $\varphi \in \mathcal{D}((M_p), L^r)$ with $1 < r \leq 2$ and let $\psi := \mathcal{F}[\varphi] = \widehat{\varphi}$ and $\tilde{\psi}(t) = \psi(-t)$. Hence $\psi \in \mathcal{F}\mathcal{D}((M_p), L^r)$. By the proof of Lemma 5.2.10, we have

$$\|\psi \exp\left[M(h|\cdot|)\right]\|_{L^s} < \infty, \qquad (5.111)$$

for arbitrary $h > 0$, with $1/r + 1/s = 1$. By Hölder's inequality, (5.111), and the proof of Lemma 5.2.8, there exists a constant $b > 0$ such that

$$\int_{\mathbb{R}^n} |g(t)\psi(t)| \, dt \leq \|g \exp[-M(b|\cdot|)]\|_{L^r} \|\psi \exp[M(b|\cdot|)]\|_{L^s} < \infty, \quad (5.112)$$

in view of inequalities (5.81) and (5.111). By (5.112) and Corollary 5.2.2, we have $g \in \mathcal{F}'\mathcal{D}((M_p), L^r)$ with $1 < r \leq 2$.

Now define $U := \mathcal{F}^{-1}[\tilde{g}]$ by means of the formula (2.52); that is,

$$\langle U, \varphi \rangle = \langle \mathcal{F}^{-1}[\tilde{g}], \varphi \rangle := \langle \tilde{g}, \tilde{\hat{\varphi}} \rangle = \langle g, \hat{\varphi} \rangle, \quad \varphi \in \mathcal{D}((M_p), L^r). \quad (5.113)$$

Since $\tilde{g} \in \mathcal{F}'\mathcal{D}((M_p), L^r)$, we have $U \in \mathcal{D}'((M_p), L^r)$ with $1 < r \leq 2$ by Lemma 2.4.2. Thus, by (5.108) and (5.113), we have

$$\langle f(\cdot + iy) - U, \varphi \rangle = \langle e_y g, \hat{\varphi} \rangle - \langle g, \hat{\varphi} \rangle = \langle g(e_y - 1), \psi \rangle \quad (5.114)$$

for our fixed $\varphi \in \mathcal{D}((M_p), L^r)$ and $\psi = \hat{\varphi} \in \mathcal{F}\mathcal{D}((M_p), L^r)$. Since $\text{supp} \, g \subseteq C^*$ almost everywhere and $\langle y, t \rangle \geq 0$ for $y \in C$, $t \in C^*$, we have

$$|g(t)(e_y(t) - 1)\,\psi(t)| \leq 2|g(t)\psi(t)|$$

for almost all $t \in \mathbb{R}^n$; and $g\psi \in L^1$ due to (5.112). By the Lebesgue dominated convergence theorem,

$$\lim_{y \to 0, y \in C} \langle g(e_y - 1), \psi \rangle = \lim_{y \to 0, y \in C} \int_{\mathbb{R}^n} g(t)(e_y(t) - 1)\,\psi(t)\, dt = 0. \quad (5.115)$$

Now (5.110) is obtained by combining (5.114) and (5.115), since the fixed function $\varphi \in \mathcal{D}((M_p), L^r)$ was chosen arbitrarily. The proof is complete. $\square$

Corollary 5.2.3. *Let f be analytic in T^C and satisfy (5.97) with $r = 2$. There is a $U \in \mathcal{D}'((M_p), L^2)$ such that*

$$\lim_{y \to 0, y \in C} f(\cdot + iy) = U \quad (5.116)$$

in $\mathcal{D}'((M_p), L^2)$ and

$$f(z) = \langle U, K(z - \cdot) \rangle, \quad z \in T^C. \quad (5.117)$$

Proof. By Theorem 5.1.1 and its proof, Corollary 5.1.1, and (5.97), there exists a measurable function g of the variable $t \in \mathbb{R}^n$ such that $\text{supp} \, g \subseteq C^*$ almost everywhere, the estimate (5.107) holds with $r = 2$, and f has the representation (5.108). Since $\text{supp} \, g \subseteq C^*$ almost everywhere, the representation (5.108) can be written in the form:

$$f(z) = \int_{C^*} g(t) E_z(t)\, dt = \langle g, I_{C^*} E_z \rangle, \quad (5.118)$$

where I_{C^*} is the characteristic function of the set C^*.

By Theorem 5.2.2, the ultradistribution $U \in \mathcal{D}'((M_p), L^2)$ defined by
$$U := \mathcal{F}^{-1}[\tilde{g}] \qquad (5.119)$$
in the sense of formula (2.52) satisfies (5.116); so it remains to prove (5.117).

Note that the Cauchy kernel $K(z - t)$, $z \in T^C$, $t \in \mathbb{R}^n$, defined in (1.15) satisfies the relation $K(z - \cdot) \in \mathcal{D}((M_p), L^2)$ for $z \in T^C$ by Theorem 4.1.1. Thus $\langle U, K(z - \cdot) \rangle$ is well defined for $z \in T^C$. Moreover, by (1.15) and (1.16),
$$K(z - t) = \int_{C^*} E_{z-t}(\eta)\, d\eta = \mathcal{F}^{-1}[h_z], \qquad z \in T^C, \qquad (5.120)$$
where $h_z := I_{C^*} E_z$. By (5.119), (5.120), the definition (5.113) of the inverse Fourier transform of $\tilde{g}$, and (5.118), we have
$$\langle U, K(z - \cdot) \rangle = \langle \mathcal{F}^{-1}[\tilde{g}], \mathcal{F}^{-1}[h_z] \rangle = \langle \tilde{g}, \tilde{h}_z \rangle = \langle g, I_{C^*} E_z \rangle = f(z) \quad (5.121)$$
for $z \in T^C$. Hence (5.117) is obtained and the proof is complete. $\square$

If C_1 is an arbitrary regular cone such that $C^* \cap C_1^*$ is a set of Lebesgue measure zero, where C is the cone from Corollary 5.2.3, then the calculation in the proof of Corollary 5.2.3, in (5.121), shows that
$$\langle U, K(z - t) \rangle = 0, \qquad z \in T^{C_1},$$
where U is the boundary value of f in (5.116) and $K(z - t)$ is the Cauchy kernel in (5.120).

Theorem 5.2.3 below is a companion theorem to the boundary value results presented previously in this section. The proof techniques for the following result are the same as those for these previous results, and hence the proof is omitted.

Theorem 5.2.3. *Let f be an analytic function of the variable $z = x + iy \in T^C$ and be the Fourier transform of a function in L^r with $1 < r \leq 2$. Suppose that there exist constants $K > 0$ and $T > 0$ which are independent of $y \in C$ and such that*
$$\|\mathcal{F}^{-1}[f(\cdot + iy)]\|_{L^r} \leq K \exp\left[M^*(T/|y|)\right], \qquad y \in C,$$
where M^* is the associated function of the sequence (M_p) defined in (2.9). There is an ultradistribution $U \in \mathcal{D}'((M_p), L^r)$ such that
$$\lim_{y \to 0, y \in C} f(\cdot + iy) = U$$
in $\mathcal{D}'((M_p), L^r)$.

Boundary value results for $\mathcal{D}'(\{M_p\}, L^r)$ similar to those contained in this section need to be proved; we leave this for future investigations. We also desire in the future to extend the boundary value results of this section to analytic functions in tubes T^C for which (5.97) holds for $1 < r < \infty$.

5.3 Case $2 < r < \infty$

We will extend results of Sections 5.1 and 5.2, where possible, for values of r in (5.1) for which $2 < r < \infty$. The results of this section will concern tubes T^C defined by special types of cones C which will be considered here.

Let $u \in \Theta$ with Θ defined in (1.10) and (4.2). Thus let $u = (u_1, \ldots, u_n)$ be any of the 2^n n-tuples whose components are -1 or 1; for such a u, we defined in (1.11) and (4.1) the n-rant C_u in $\mathbb{R}^n$ as the subset of $\mathbb{R}^n$ of the form

$$C_u := \{y \in \mathbb{R}^n : \ u_j y_j > 0, \ \ j = 1, \ldots, n\}.$$

The n-rants C_u for $u \in \Theta$ are open convex cones with the property that $C_u^* = \overline{C}_u$. We will denote the first n-rant as C_0; it is the n-rant $C_0 = \{y \in \mathbb{R}^n : y_j > 0, \ j = 1, \ldots, n\}$.

Now let C be the interior of the convex hull of n linearly independent rays meeting at $0 \in \mathbb{R}^n$ (see [137], p. 118). Select vectors $a_1, \ldots, a_n$ in the direction of these rays; then C can be written as

$$C = \{y \in \mathbb{R}^n : \ y = y_1 a_1 + \ldots + y_n a_n, \ y_1 > 0, \ldots, y_n > 0\}.$$

The set C is an open convex cone in $\mathbb{R}^n$. We call such a C an *n-rant cone*. There is a nonsingular linear transformation L mapping the standard basis vectors e_j for $j = 1, \ldots, n$ in $\mathbb{R}^n$ (i.e. e_j is the vector with 1 in the j-th component and 0 in the other $n - 1$ components) one-one and onto the vectors a_j, $j = 1, \ldots, n$. The transformation L is then a one-one mapping from C_0 onto C and the boundary of C_0 is mapped to the boundary of C. Further, an arbitrary cone contained in C_0 is mapped in a one-one and onto fashion to a cone contained in C with boundary being mapped to boundary. We extend L to $\mathbb{C}^n$ by putting

$$L(u + iv) := L(u) + iL(v), \qquad u + iv \in \mathbb{C}^n;$$

then L maps the tube $T^{C_0} := \mathbb{R}^n + iC_0$ one-one and onto the tube $T^C := \mathbb{R}^n + iC$. If f is an analytic function in the tube T^C, the function $g := f \circ L$ is analytic in the tube T^{C_0} (see [137], p. 118); and the same is true for the tubes corresponding to open convex cones which are proper subsets of C_0 and C and which are mapped to one another by L and L^{-1}. Therefore analyticity in these tubes is preserved under the transformation L. Similar statements to the above can be made for L^{-1}, the inverse of L, since L is nonsingular. We call a cone C described in this paragraph a *n-rant cone* because of its identification with C_0 by the linear transformations L and

L^{-1}. Note that Rudin in [128] used the analytic invariance of tubes under nonsingular linear transformations to prove edge of the wedge theorems.

A cone C is called a *polygonal cone* (see [137], p. 118) if it is the interior of the convex hull of a finite number of rays meeting the origin $0 \in \mathbb{R}^n$ among which there are n (at least n) that are linearly independent. Thus a polygonal cone C is a finite union of n-rant cones C_j, $j = 1, \ldots, m$ (see [137], p. 118).

Recall that every n-rant cone is an open convex cone as is every polygonal cone. Thus the n-rant cones C_j, $j = 1, \ldots, m$, whose union is C possess a very important intersection property which we describe now. No boundary point of any of the C_j, $j = 1, \ldots, m$, is an element of that C_j. Thus if $y \in C$ such that y is on the boundary of some C_j, then y must belong to one or more of the other C_j. Because of this property (i.e. because of the convexity of the polygonal cone C), the n-rant cones C_j, $j = 1, \ldots, m$, whose union is the polygonal cone C must overlap as they cover C in the following sense:

$1°$ given the n-rant cone C_1 there is another one (call it C_2) of the C_j, $j \neq 1$, such that $C_1 \cap C_2 \neq \emptyset$;

$2°$ given $C_1 \cup C_2$ there is another one (call it C_3) of the C_j, $j \neq 1, 2$, such that at least one of the intersections $C_1 \cap C_3$ and $C_2 \cap C_3$ is not empty;

$3°$ given $C_1 \cup C_2 \cup C_3$ there is another one (call it C_4) of the C_j, $j \neq 1, 2, 3$, such that at least one of the intersections $C_1 \cap C_4$, $C_2 \cap C_4$ and $C_3 \cap C_4$ is not empty;

. . . .

$m°$ given $\bigcup_{j=1}^{m-1} C_j$ the remaining n-rant cone C_m intersects at least one of the C_j, $j = 1, \ldots, m - 1$.

The above intersection properties of the n-rant cones whose union is a given polygonal cone will be important in our proof of a result below; we collectively refer to the above described intersections of the n-rant cones C_j, $j = 1, \ldots, m$, whose union is a given polygonal cone C as the *intersection property* of the n-rant cones.

Let us also note that the intersection of two open convex cones is an open convex cone. Thus each of the nonempty intersections in the intersection property of the n-rant cones above is itself an open convex cone which is contained in an n-rant cone.

The notion of polygonal cone of Stein and Weiss is closely associated with the notion of a cone containing an admissible set of vectors in the sense of Vladimirov (see [148], p. 930). A cone with an admissible set of vectors can be a polygonal cone.

A regular cone is an open convex cone C in $\mathbb{R}^n$ such that $\overline{C}$ does not contain any entire straight line. Every regular cone C is properly contained in an open convex cone $\Gamma \subset \mathbb{R}^n$ such that $\overline{C} \subset \Gamma \cup \{0\}$. The proof of Theorem 5.5 from [137], p. 118, shows that there is a finite number of polygonal cones Λ_j, $j = 1, \ldots, k$, and the polygonal cone Λ which is the convex hull of $\bigcup\limits_{j=1}^{k} \Lambda_j$ such that

$$C \subset \bigcup_{j=1}^{k} \Lambda_j \subset \Lambda \subset \Gamma; \qquad \overline{C} \subset \bigcup_{j=1}^{k} \Lambda_j \cup \{0\} \subset \Lambda \cup \{0\} \subset \Gamma \cup \{0\}.$$

According to the preceding paragraph, each polygonal cone Λ and Λ_j, $j = 1, \ldots, k$, is the union of a finite number of n-rant cones. Hence $C \subset \bigcup\limits_{j=1}^{m} C_j$ for $m \geq k$, where the C_j, $j = 1, \ldots, m$, are n-rant cones. If $C \cap C_{j_0} = \emptyset$ for some $j_0 \in \{1, \ldots, m\}$, we delete such an n-rant cone C_{j_0} from the family $C_1, \ldots, C_m$ of n-rant cones. Consequently, we obtain $C \subset \bigcup\limits_{j=1}^{r} C_j$ for $r \leq m$ where $C \cap C_j \neq \emptyset$ for $j = 1, \ldots, r$. (The inclusion $C \subset \bigcup\limits_{j=1}^{r} C_j$ can be obtained equally well from the inclusion $C \subset \Lambda$ since Λ is a polygonal cone.) Now we have

$$C = \bigcup_{j=1}^{r} (C \cap C_j). \tag{5.122}$$

Using the same arguments as in the proof of the intersection property for the polygonal cone case in the preceding paragraph, we conclude that (5.122) is the representation of a regular cone C in terms of the open convex cones $C \cap C_j$, $j = 1, \ldots, r$, which satisfy the stated intersection property and each of these cones is contained in or is an n-rant cone (namely, in C_j). We will need this conclusion in one of the results below.

Analysis similar to that in [23] leads to the Fourier-Laplace integral representation of analytic functions in tubes T^C which satisfy (5.1) for $2 < r < \infty$ where C is an n-rant, an n-rant cone, or a polygonal cone. We begin with the following result.

Theorem 5.3.1. *Let C be an open convex cone which is contained in or is any of the 2^n n-rants C_u in $\mathbb{R}^n$. Let f be an analytic function of the* *variable $z \in T^C$ which satisfies (5.1) for $2 < r < \infty$. There exists a measurable function g of the variable $t \in \mathbb{R}^n$ such that $\operatorname{supp} g \subseteq C^*$ almost everywhere and*

$$\|e_y g\|_{L^2} \leq M[1 + (d(y))^{-m}]^q \exp[M^*(T/|y|)], \qquad y \in C, \qquad (5.123)$$

where e_y is defined in (5.3), M^* is the associated function of the sequence (M_p) defined in (2.9), and $M > 0$, $T > 0$, $m \geq 0$, and $q \geq 0$ are constants which are independent of $y \in C$; and

$$f(z) = W(z) \int_{\mathbb{R}^n} g(t) E_z(t) \, dt, \qquad z \in T^C, \qquad (5.124)$$

where E_z is defined in (5.2) and W is a polynomial of the variable $z \in T^C$.

Proof. For C being contained in or being the n-rant C_u put

$$W(z) := \prod_{j=1}^{n} [1 - iu_j z_j]^{n+2}, \qquad z = x + iy \in T^C. \qquad (5.125)$$

We have

$$|1/W(x+iy)| \leq \prod_{j=1}^{n} (1 + x_j^2)^{-1-n/2}, \qquad z \in T^C. \qquad (5.126)$$

The function F defined by

$$F(z) := f(z)/W(z), \qquad z \in T^C, \qquad (5.127)$$

is analytic in T^C. By (5.127), Hölder's inequality, (5.1), and (5.126), we have

$$\int_{\mathbb{R}^n} |F(x+iy)|^2 \, dx \leq \||f(\cdot+iy)|^2\|_{L^{r/2}} \, \||1/W(\cdot+iy)|^2\|_{L^{r/(r-2)}}$$

$$= (\|f(\cdot+iy)\|_{L^r})^2 \left(\int_{\mathbb{R}^n} |1/W(x+iy)|^{2r/(r-2)} \, dx \right)^{(r-2)/r}$$

$$\leq \left(M \, (1 + (d(y))^{-m})^q \exp[M^*(T/|y|)] \right)^2, \qquad (5.128)$$

where

$$M := K \left(\int_{\mathbb{R}^n} \prod_{j=1}^{n} (1 + x_j^2)^{-2r/(r-2)-nr/(r-2)} \, dx \right)^{(r-2)/2r}$$

with the constant K from inequality (5.1). Due to (5.128) and the fact that the function F is analytic in T^C, we can apply Corollary 5.1.1 to find a function g of the variable $t \in \mathbb{R}^n$, with the inclusion supp $g \subseteq C^*$ satisfied almost everywhere, such that the estimate (5.123) holds and

$$F(z) = \int_{\mathbb{R}^n} g(t)e^{2\pi i\langle z,t\rangle}\,dt = \int_{\mathbb{R}^n} g(t)E_z(t)\,dt, \qquad z \in T^C. \qquad (5.129)$$

Equality (5.124) follows from (5.127) and (5.129). The proof is complete. □

We ask if the representation (5.124) can be rewritten in the form $f = \langle V, E_z \rangle$, $z \in T^C$, for some ultradistribution V? If so, g will have to possess sufficient properties to allow for $\langle V, E_z \rangle$ to be well defined.

If $m = 0$ or $q = 0$ in (5.1), the Fourier-Laplace integral in (5.124) obtains an ultradistributional boundary value as $y = \operatorname{Im} z \to 0$, $y \in C$, by Theorem 5.2.2, since g satisfies (5.123) with $m = 0$ or $q = 0$. Can this fact be used along with (5.124) to prove that f also obtains an ultradistributional boundary value?

Recall the concept of n-rant cone given above. We extend Theorem 5.3.1 to the case where C is contained in or is an n-rant cone.

Theorem 5.3.2. *Let C be an open convex cone that is contained in or is a n-rant cone in $\mathbb{R}^n$. Let f be an analytic function in T^C and satisfy (5.1) for $2 < r < \infty$. There exists a measurable function g on $\mathbb{R}^n$ and a nonsingular linear transformation L of $\mathbb{R}^n$ onto $\mathbb{R}^n$ such that* supp $g \subseteq (L^{-1}(C))^*$ *almost everywhere and, denoting $v := L^{-1}(y)$,*

$$\|e_v\,g\|_{L^r} \le M[k + (d\,(v))^{-m}]^q \exp\left[M^*(R/|v|)\right], \qquad y \in C, \qquad (5.130)$$

where M^* is the associated function of the sequence (M_p) defined in (2.9) and $M > 0$, $R > 0$, $k > 0$, $m \ge 0$, $q \ge 0$ are constants which are independent of $y \in C$; and

$$f(z) = W(u + iv) \int_{\mathbb{R}^n} g(t)E_{u+iv}(t)\,dt, \qquad (5.131)$$

for $z = x + iy \in \mathbb{R}^n + iC$, where W is a polynomial of variable $u + iv := L^{-1}(x) + iL^{-1}(y)$.

Proof. Let Γ denote the n-rant cone that C is contained in or is. There exists a nonsingular linear transformation L (with domain and range being $\mathbb{R}^n$) which maps the first n-rant C_0 onto Γ in a one-one manner such that the boundary of C_0 is mapped to the boundary of Γ. Further, if C is properly contained in Γ, then $L^{-1}(C)$ is an open convex cone which is contained in

C_0 and L maps $L^{-1}(C)$ one-one and onto C with the boundary of $L^{-1}(C)$ being mapped to the boundary of C. (If $C = \Gamma$, then $L^{-1}(C) = C_0$.) For $u + iv := L^{-1}(x) + iL^{-1}(y) \in \mathbb{R}^n + iL^{-1}(C)$, put

$$G(u + iv) := f(L(u) + iL(v)) = f(x + iy), \qquad (5.132)$$

where $u + iv \in \mathbb{R}^n + iL^{-1}(C)$ and $x + iy \in \mathbb{R}^n + iC$. The function G of the variable $u + iv$ is analytic in $\mathbb{R}^n + iL^{-1}(C)$. By (5.1), we have

$$\int_{\mathbb{R}^n} |G(u + iv)|^r \, du = \frac{1}{|\det(L)|} \int_{\mathbb{R}^n} |f(x + iL(v))|^r \, dx$$

$$\leq \frac{1}{|\det(L)|} \left(K[1 + (d(L(v)))^{-m}]^q \, \exp\left[M^*(T/|L(v)|) \right] \right)^r$$

$$(5.133)$$

for $y = L(v) \in C$. Recalling that the boundary of $L^{-1}(C)$ is mapped by L to the boundary of C, we have

$$d(L(v)) = \inf_{y' \in \partial C} |L(v) - y'| = \inf_{v' \in \partial L^{-1}(C)} |L(v) - L(v')|$$

$$= \inf_{v' \in \partial L^{-1}(C)} |L(v - v')| \qquad (5.134)$$

for $v \in L^{-1}(C)$, where ∂C denotes the boundary of C. Corresponding to the nonsingular linear transformation L there exist constants $a > 0$ and $b > 0$ such that

$$a|w| \leq |L(w)| \leq b|w|, \qquad w \in \mathbb{R}^n, \qquad (5.135)$$

with a and b being independent of $w \in \mathbb{R}^n$ (see [23], p. 93). Using (5.135) in (5.134), we have

$$d(L(v)) = \inf_{v' \in \partial L^{-1}(C)} |L(v - v')|$$

$$\geq \inf_{v' \in \partial L^{-1}(C)} a|v - v'| = a\, d(v), \qquad v \in L^{-1}(C).$$

$$(5.136)$$

Applying (5.136) and (5.135) in (5.133), we obtain

$$|\det(L)| \int_{\mathbb{R}^n} |G(u + iv)|^r \, du$$

$$\leq \left(K(1 + (a\, d(v))^{-m})^q \exp\left[M^*(T/a|v|) \right] \right)^r$$

$$= \left((K/a^{mq})(a^m + (d(v))^{-m})^q \exp\left[M^*(T/a|v|) \right] \right)^r$$

$$(5.137)$$

for $v \in L^{-1}(C)$, because M^* is an increasing function. Since G is an analytic function of the variable $u + iv \in \mathbb{R}^n + iL^{-1}(C)$, it follows from Theorem 5.3.1 that there exists a measurable function g on $\mathbb{R}^n$ such that $\operatorname{supp} g \subseteq (L^{-1}(C))^*$ almost everywhere and

$$\|e_v g\|_{L^2} \leq M[k + (d(v))^{-m}]^q \exp[M^*(R/|v|)], \qquad v \in L^{-1}(C), \quad (5.138)$$

for some $M > 0$, $k > 0$, $m \geq 0$, and $R > 0$, i.e. (5.130) holds with $L^{-1}(y) = v$; and

$$G(u+iv) = W(u+iv) \int_{\mathbb{R}^n} g(t) E_{u+iv}(t)\, dt, \quad u+iv \in \mathbb{R}^n + iL^{-1}(C), \quad (5.139)$$

where W is a polynomial. Thus, by (5.132) and (5.139), we get (5.131). The proof is complete. $\square$

If $m = 0$ or $q = 0$ in (5.1), can we obtain an ultradistributional boundary value for f in Theorem 5.3.2 using (5.132)? We could if we knew an ultradistributional boundary value existed in Theorem 5.3.1 for $m = 0$ or $q = 0$ there.

Now recall the concept of a polygonal cone given above and the fact of the intersection property for the n-rant cones C_j, $j = 1, \ldots, m$, whose union is the polygonal cone. Let us extend Theorems 5.3.1 and 5.3.2 to the case that C can be a polygonal cone.

For C being a polygonal cone, $C = \bigcup_{j=1}^{m} C_j$, where the C_j are n-rant cones which have the intersection property described above. Let f be an analytic function in T^C satisfying (5.1) for $2 < r < \infty$. Note that $y \in C$ implies $y \in C_j$ for some $j = 1, \ldots, m$ and that the distance from y to the boundary of C is greater or equal to the distance from y to the boundary of C_j. Thus f is analytic in $\mathbb{R}^n + iC_j$ and satisfies (5.1) for $y \in C_j$ and $j = 1, \ldots, m$. Thus, by Theorem 5.3.2, for each n-rant cone C_j, $j = 1, \ldots, m$, there is a nonsingular linear transformation L_j which maps C_0 one-one and onto C_j and a function g_j with $\operatorname{supp} g_j \subseteq (L_j^{-1}(C_j))^* = C_0^*$ almost everywhere and there is a polynomial W_j such that, denoting $v_j := L_j^{-1}(y)$,

$$\|e_{v_j}(y) g_j\|_{L^2} \leq M[k + (d(v_j))^{-m}]^q \exp[M^*(R/|v_j|)] \quad (5.140)$$

for $y \in C_j$ and

$$f(x + iy) = W_j(u_j + iv_j) \int_{\mathbb{R}^n} g_j(t) E_{u_j+iv_j}(t)\, dt \quad (5.141)$$

for $x + iy \in \mathbb{R}^n + iC_j$, where W_j is a polynomial of the variable $u_j + iv_j :=$ $L_j^{-1}(x) + iL_j^{-1}(y)$.

In this way, we have proved the following result.

Theorem 5.3.3. *Let C be a polygonal cone in $\mathbb{R}^n$ and let f be an analytic function in T^C satisfying (5.1) for $2 < r < \infty$. There exist n-rant cones $C_j, j = 1, \ldots, m$; nonsingular linear transformations L_j which map C_0 one-one and onto C_j; functions g_j satisfying the inclusions $\mathrm{supp}\, g_j \subseteq \overline{C}_0$ almost everywhere and (5.140); and polynomials W_j such that (5.141) holds with*

$$C := \bigcup_{j=1}^{m} C_j.$$

A result similar to Theorem 5.3.3 can be proved for C being a regular cone.

Can an ultradistributional boundary value be obtained for f in Theorem 5.3.3 and for the corresponding result for C being a regular cone?

5.4 Boundary values via almost analytic extensions

In this section we give another approach which is based on almost analytic extensions. This concept gives for $\mathbb{R}^n$ the most general results, although the cases $p = \infty$ and $p = 1$ are still open. Here we will consider the case where the space dimension is $n = 1$; the case $n > 1$ is considered in [32]. We refer to papers [31], [32], [113] - [115] and [122].

We continue to assume conditions $(M.1)$, $(M.2)$, $(M.3')$ and that (m_p^*) is a nondecreasing sequence.

Using the same method as in [113], 2.2. Proposition, and the Minkowski inequality one can prove the following lemma.

Lemma 5.4.1. *Let $r > 1$ and $h > 0$. There exists an $H > 0$ such that for every $\varphi \in \mathcal{D}((M_p), h, L^r)$ there are a $\phi \in C^1(\mathbb{C})$ and a $C > 0$ such that $\phi|\mathbb{R} = \varphi$ and*

$$\sup_{y \in \mathbb{R}} e^{M^*(hH/|y|)} \left\| \left(\frac{\partial}{\partial \bar{z}} \phi \right)(\cdot + iy) \right\|_{L^r} < C \|\varphi\|_{L^r, h},$$

where M^* is the associated function of the sequence (M_p) defined in (2.9), and

$$\sup_{y \in \mathbb{R}} \|(\phi^{(j)})(\cdot + iy)\|_{L^r} < C \|\varphi\|_{L^r, h}, \qquad j = 0, 1,$$

with the convention $(\partial/\partial \bar{z}) \phi (x + i0) = 0$.

The estimate for $\varphi'(\cdot + iy)$ is added in Lemma 5.4.1 in order to have a symmetric assertion to the assertion of Lemma 5.4.4 below.

For the main results of this section we need the following three lemmas.

Lemma 5.4.2. *Let F be an analytic function on $\mathbb{C}\backslash\mathbb{R}$ such that*
1° *(in the Beurling case) there are a $k > 0$ and a $C > 0$,*
2° *(in the Roumieu case) for every $k > 0$ there is a $C > 0$,*
for which

$$\|F(\cdot + iy)\|_{L^s} \le Ce^{M^*(k/|y|)}, \qquad y \ne 0, \tag{5.142}$$

where M^* is the associated function of the sequence (M_p) defined in (2.9). We have
1° *(in the Beurling case) for every compact set $K \subset \mathbb{R}$ there are a $p > 0$ and a $B > 0$,*
2° *(in the Roumieu case) for every $p > 0$ there is a $B > 0$,*
such that

$$\sup_{x \in K}\{|F(x + iy)|\} \le Be^{M^*(p/|y|)}, \qquad y \ne 0.$$

Proof. We shall prove the assertion only for the Beurling case, since the proof for the Roumieu case is similar.

Let $\alpha \in \mathcal{D}((M_p), \mathbb{R})$, $\operatorname{supp}\alpha \subset [-a, a]$ and $\alpha \equiv 1$ in a neighborhood of K. Fix $y \ne 0$. Let

$$K_{x,t} := \{z = x + iy\colon |z - t - iy| = \frac{|y|}{4}\}$$

for a given $x \in K$ and $t \in [-a, a]$ and let $s := r/(r-1)$. Since

$$F(x + iy) = \alpha(x)F(x + iy) = \int_{-\infty}^{x} [\alpha(t)F(t + iy)]'\, dt,$$

we have, by Cauchy's formula,

$$|F(x + iy)| \le \frac{1}{2\pi}[G_1(y) + G_2(y)],$$

where

$$G_1(y) := \int_{-a}^{a} |\alpha'(t)| \left| \int_{z \in K_{x,t}} \frac{F(z)}{z - t - iy}\, dz \right| dt$$

and

$$G_2(y) := \int_{-a}^{a} |\alpha(t)| \left| \int_{z \in K_{x,t}} \frac{F(z)\, dz}{(z - t - iy)^2} \right| dt.$$

By Hölder's inequality, we have

$$G_1(y) \leq \left(\int_{-a}^{a} |\alpha'(t)|^r \, dt \right)^{1/r} \left(\int_{-a}^{a} \left| \int_{z \in K_{x,t}} \frac{F(z)}{z - t - iy} \, dz \right|^s dt \right)^{1/s}$$

$$\leq C_1 \left(\int_{-a}^{a} \left| \int_{0}^{2\pi} F(t + iy + e^{i\theta}|y|/4) \, d\theta \right|^s dt \right)^{1/s}$$

$$\leq C_1 \left[\int_{-a}^{a} \left(\int_{0}^{2\pi} 1 \, d\theta \right)^{s/r} \left(\int_{0}^{2\pi} |F(t + iy + e^{i\theta}|y|/4|^s \, d\theta \right) dt \right]^{1/s}$$

$$\leq C_2 \left[\int_{0}^{2\pi} \left(\int_{-a}^{a} |F(t + iy + e^{i\theta}|y|/4)|^s \, dt \right) d\theta \right]^{1/s}$$

and, similarly,

$$G_2(y) \leq \left(\int_{-a}^{a} |\alpha(t)|^r \, dt \right)^{1/r} \left(\int_{-a}^{a} \left| \int_{z \in K_{x,t}} \frac{F(z)}{(z - t - iy)^2} \, dz \right|^s dt \right)^{1/s}$$

$$\leq \frac{D_1}{|y|} \left(\int_{-a}^{a} \left| \int_{0}^{2\pi} [F(t + iy + e^{i\theta}|y|/4]e^{-i\theta} \, d\theta \right|^s dt \right)^{1/s}$$

$$\leq \frac{D_2}{|y|} \left[\int_{0}^{2\pi} \left(\int_{-a}^{a} |F(t + iy + e^{i\theta}|y|/4)|^s \, dt \right) d\theta \right]^{1/s}$$

for suitable positive constants C_1, C_2, D_1, D_2 and for arbitrary $x \in K$. Since the function M^* is increasing and $t \leq C_0 M^*(t)$ for some constant $C_0 > 0$ and all $t > 0$ (see (2.8)), we conclude from the assumption (5.142) that

$$\sup_{x \in K} \{ |F(x + iy)| \}$$

$$\leq B_0 \left(1 + 1/|y| \right) \left(\int_{0}^{2\pi} (\|F(\cdot + iy + e^{i\theta}|y|/4)\|_{L_s})^s \, d\theta \right)^{1/s}$$

$$\leq B_1 \left(1 + 1/|y| \right) \exp \left[M^* \left(k/|y + \sin \theta|y|/4| \right) \right]$$

$$\leq B_1 \left(1 + 1/|y| \right) \exp \left[M^* \left(\frac{4k}{3|y|} \right) \right] \leq B_2 \exp \left[M^* \left(\frac{2k}{|y|} \right) \right]$$

with suitable positive constants B_0, B_1, B_2. The lemma is proved. $\square$

Using Sobolev's lemma, one can easily prove the following assertion.

Lemma 5.4.3. *Let* $r > 1$ *and* $\varphi \in \mathcal{D}((M_p), L^r)$ *(respectively,* $\varphi \in \mathcal{D}(\{M_p\}, L^r)$ *). For every compact set* $K \subset \mathbb{R}$ *and every* $h > 0$ *(respectively, for some* $h > 0$ *there are* $C > 0$ *and* $k > 0$ *) we have*

$$\sup_{\substack{x \in K \\ p \in \mathbb{N}_0}} \left\{ \frac{h^p}{(M_p)} |\varphi^{(p)}(x)| \right\} \leq C \|\varphi\|_{k, L^r}.$$

Lemma 5.4.4. *Fix* $\delta > 0$ *and let* $\Delta := \{z : |\operatorname{Im} z| < \delta\}$. *Suppose that* $\phi \in C^1(\Delta)$ *and* $\phi^{(j)}(\cdot + iy) \in L^r$ *with* $r > 1$ *for* $j = 0, 1$ *and* $|y| < \delta$. *Moreover, assume that for every* $h > 0$ *(respectively, for some* $h > 0$*), we have*

$$D_{j,h} := \sup_{0 < |y| < \delta} \|\phi^{(j)}(\cdot + iy)\|_{L^r} < \infty, \qquad j = 0, 1, \tag{5.143}$$

and

$$D_{2,h} := \sup_{0 < |y| < \delta} e^{M^*(h/|y|)} \|\frac{\partial}{\partial \bar{z}} \phi(\cdot + iy)\|_{L^r} < \infty, \tag{5.144}$$

where M^* *is the associated function of the sequence* (M_p) *defined in (2.9). We have that the function* $\varphi := \phi_{|\mathbb{R}}$ *is in* $\mathcal{D}((M_p), L^r)$ *(respectively, in* $\mathcal{D}(\{M_p\}, L^r)$*) and, for every* $h > 0$ *(respectively, for some* $h > 0$*) there is a* $C > 0$ *such that*

$$\|\varphi\|_{L^r, h} \leq C D_h,$$

where $D_h := \max\{D_{j,h} : j = 0, 1, 2\}$.

Proof. We denote

$$\Gamma_{1,a}^{\pm} := \{\zeta : \zeta = t \pm i\delta, |t| < a\}, \quad \Gamma_1^{\pm} := \{\zeta : \zeta = t \pm i\delta, t \in \mathbb{R}\},$$

$$\Gamma_{2,a}^{\pm} := \{\zeta : \zeta = \pm a + it, |t| < \delta\}, \quad \Delta_a := \{\zeta : |\operatorname{Im} \zeta| < \delta, |\operatorname{Re} \zeta| < a\}.$$

This notation will be used later, as well.

Fix $p \in \mathbb{N}$ and let $x \in \mathbb{R}$. By Cauchy's formula, for sufficiently large a, we have

$$\varphi^{(p)}(x) = \frac{p!}{2\pi i}(I_{1,a}^- - I_{1,a}^+ + I_{2,a}^- - I_{2,a}^+ + I_a), \tag{5.145}$$

where

$$I_{j,a}^{\pm} := \int_{\Gamma_{j,a}^{\pm}} \frac{\phi(\zeta)\, d\zeta}{(\zeta - x)^{p+1}}, \quad j = 1, 2; \qquad I_a := \int_{\Delta_a} \frac{(\partial/\partial\bar{\zeta})\phi(\zeta)\, d\zeta \wedge d\bar{\zeta}}{(\zeta - x)^{p+1}}.$$

Since

$$|\phi(x + iy)| = \left| \int_0^x \phi'(t + iy)\, dt \right| \leq |x|^{1/s} \left(\int_{-\infty}^{\infty} |\phi'(t + iy)|^r\, dt \right)^{1/r},$$

inequality (5.143) implies that $I_{2,a}^{\pm} \to 0$ as $a \to 0$.

Hence, by (5.145),

$$\varphi^{(p)}(x) = \frac{1}{2\pi i}(I_1^-(x) - I_1^+(x) + I(x)),$$

where

$$I_1^{\pm}(x) := p! \int_{\Gamma_1^{\pm}} \frac{\phi(\zeta)\, d\zeta}{(\zeta - x)^{p+1}}; \qquad I(x) := p! \int_{\Delta} \frac{(\partial/\partial\bar{\zeta})\phi(\zeta)\, d\zeta \wedge d\bar{\zeta}}{(\zeta - x)^{p+1}}.$$

Let us estimate $\|I_1^-\|_{L^r}$. We have

$$|I_1^-(x)|^r \leq (p!)^r \left(\int_{-\infty}^{\infty} \frac{|\phi(t + x - i\delta)|\, dt}{|t - i\delta|^{p+1}} \right)^r$$

$$\leq \frac{A(p!)^r}{\delta} \int_{-\infty}^{\infty} \frac{|\phi(t + x - i\delta)|^r\, dt}{|t - i\delta|^{pr}},$$

where

$$A := \left(\int_{-\infty}^{\infty} \frac{dt}{(1 + t^2)^{s/2}} \right)^{r/s}.$$

By Hölder's inequality, Fubini's theorem and (5.143), we have

$$\int_{-\infty}^{\infty} |I_1^-(x)|^r\, dx \leq \frac{A(p!)^r}{\delta} \int_{-\infty}^{\infty} \frac{dt}{|t - i\delta|^2} \int_{-\infty}^{\infty} \frac{|\phi(t + x - i\delta)|^r}{|t - i\delta|^{pr-2}}\, dx$$

$$\leq \frac{A(p!)^r}{\delta} \int_{-\infty}^{\infty} \frac{dt}{t^2 + \delta^2} \int_{-\infty}^{\infty} \frac{|\phi(t + x - i\delta)|^r\, dx}{\delta^{pr-2}}$$

$$\leq \frac{A\pi(p!)^r}{2\delta^2} \frac{(D_{0,h})^r}{\delta^{pr-2}}.$$

Since $p! \prec M_p$, we obtain, for suitable $\tilde{A} > 0$, the inequality

$$\|I_1^-\|_{L^r} = \left(\int_{-\infty}^{\infty} |I_1^-(x)|^r\, dx \right)^{1/r} \leq \frac{\tilde{A}D_h p!}{\delta^p e^{M^*(h/\delta)}} \leq \tilde{A}D_h h^{-p} M_p.$$

An analogous inequality holds for $\|I_1^+\|_{L^r}$.

Let us estimate $\|I\|_{L^r}$. We have

$$\frac{|I(x)|^r}{(p!)^r} = \left| \int_{\Delta} \frac{(\partial/\partial\bar{\zeta})\phi(\zeta)\, d\zeta \wedge d\bar{\zeta}}{(\zeta - x)^{p+1}} \right|^r \leq \left(\int_{-\infty}^{\infty} \int_{-\delta}^{\delta} \frac{|(\partial/\partial\bar{\zeta})\phi(\xi + i\eta)|\, d\eta\, d\xi}{|\xi + i\eta - x|^{p+1}} \right)^r$$

$$= \left(\int_{-\infty}^{\infty} \int_{-\delta}^{\delta} \frac{|(\partial/\partial\bar{\zeta})\phi(\xi + i\eta)|}{|\eta|^{1/s}|\xi + i\eta - x|^{p+1-(2/s)}} \frac{|\eta|^{1/s} d\eta d\xi}{|\xi + i\eta - x|^{2/s}} \right)^r$$

$$\leq B^{r/s} \int_{-\infty}^{\infty} \int_{-\delta}^{\delta} \frac{|(\partial/\partial\bar{\zeta})\phi(\xi + i\eta)|^r\, d\eta\, d\xi}{|\eta|^{r/s}|\xi + i\eta - x|^{(p+1-(2/s))r}},$$

where

$$B := \int_{-\infty}^{\infty} \int_{-\delta}^{\delta} \frac{|\eta| d\eta d\xi}{|\xi + i\eta - x|^2} = 2 \int_0^{\delta} \left(\int_{-\infty}^{\infty} \frac{d(\xi/\eta)}{((\xi - x)/\eta)^2 + 1} \right) d\eta$$

$$= 2\pi\delta.$$

This implies that

$$(p!)^{-r} \ (2\pi\delta)^{-r/s} \int_{-\infty}^{\infty} |I(x)|^r dx$$

$$\leq \int_{-\infty}^{\infty} \left(\int_{-\infty}^{\infty} \int_{-\delta}^{\delta} \frac{|(\partial/\partial\bar{\zeta})\phi(\xi + i\eta)|^r}{|\eta|^{(p+1-(1/s))r-2}} \frac{d\xi d\eta}{|\xi + i\eta - x|^2} \right) dx$$

$$\leq \int_{-\infty}^{\infty} \int_{-\delta}^{\delta} \left(\int_{-\infty}^{\infty} |\frac{\partial}{\partial\bar{\zeta}}\phi(\xi + x + i\eta)|^r dx \right) \frac{|\eta|^{((1/r)-p)r}}{\xi^2 + \eta^2} d\xi d\eta.$$

Hence, by (5.144),

$$\|I\|_{L^r} \leq (2\pi\delta)^{1/s} p! D_{2,h} \left(\int_{-\delta}^{\delta} \int_{-\infty}^{\infty} \left(\frac{|\eta|^{1/r}}{|\eta|^p e^{M^*(h/|\eta|)}} \right)^r \frac{d\xi d\eta}{\xi^2 + \eta^2} \right)^{1/r}$$

$$\leq D_h (2\pi\delta)^{1/s} h^{-p} M_p \left(\int_{-\infty}^{\infty} \int_{-\delta}^{\delta} \frac{|\eta| d\xi d\eta}{\xi^2 + \eta^2} \right)^{1/r} \leq A D_h \delta h^{-p} M_p.$$

Minkowski's inequality implies that for every $h > 0$ (respectively, for some $h > 0$) there is a constant $C > 0$ such that

$$\|\varphi^{(p)}\|_{L^r} \leq C D_h h^{-p} M_p, \qquad p \in \mathbb{N}_0.$$

This implies the assertion. $\square$

Let $s \in [1, \infty]$. Denote by $\mathcal{H}((M_p), L^s)$ (respectively, by $\mathcal{H}(\{M_p\}, L^s)$) the space of all functions f for which there exists a $\delta = \delta_f > 0$ such that f is analytic in $\Delta \setminus \mathbb{R}$, where $\Delta = \Delta_f := \{x + iy : x \in \mathbb{R}, |y| < \delta\}$, and which satisfies the following estimate: for some $k > 0$ and some $C > 0$ (respectively, for every $k > 0$ there exists a $C > 0$) such that

$$\|f(\cdot + iy)\|_{L^s} \leq C e^{M^*(k/|y|)}, \qquad |y| < \delta, \ y \neq 0,$$

where M^* is the associated function of the sequence (M_p) defined in (2.9). The common notation for both spaces is $\mathcal{H}(*, L^s) = \mathcal{H}_{L^s}^*$.

We denote by $\mathcal{H}(L^s, \mathbb{R}) = \mathcal{H}_{L^s}$ the space of functions f analytic in the corresponding Δ_f and satisfying the estimate

$$\|f(\cdot + iy)\|_{L^s} < C_f, \qquad |y| < \delta_f.$$

Let f be an analytic function in $\Delta_f \setminus \mathbb{R}$ with $\Delta_f := \{z: \ |\operatorname{Im} z| < \delta_f\}$ for the respective $\delta_f > 0$. If for every $\varphi \in \mathcal{D}(*, L^r)$ the limit

$$\langle Tf, \varphi \rangle := \lim_{\varepsilon \to 0} \int_{\mathbb{R}} \varphi(x)(f(x + i\varepsilon) - f(x - i\varepsilon))\, dx \tag{5.146}$$

exists, we call Tf the boundary value of f in $\mathcal{D}'(*, L^r)$.

Theorem 5.4.1. *Let $r > 1$ and let $f \in \mathcal{H}^*_{L^s}$. The boundary value Tf of f defined in (5.146) can be represented for every $\varphi \in \mathcal{D}(*, L^r)$ in the following form:*

$$\langle Tf, \varphi \rangle = \int_{\Delta} f(z) \frac{\partial}{\partial z}\phi(z)\, dz \wedge d\bar{z} - \int_{\Gamma_\delta^-} f(z)\phi(z)\, dz + \int_{\Gamma_\delta^-} f(z)\phi(z)\, dz,$$

where φ is as defined in Lemma 5.4.4, i.e. $\varphi := \phi_{|\mathbb{R}}$. Moreover, Tf belongs to $\mathcal{D}'(*, L^r)$.

Proof. Fix δ_0 and put

$$\Delta_a^+ := \{z: \ \operatorname{Im} z \in (0, \delta), \ |\operatorname{Re} z| < a\}; \quad \Delta^+ := \{z: \ \operatorname{Im} z \in (0, \delta)\};$$

$$\Delta_a^- := \{z: \ \operatorname{Im} z \in (-\delta, 0), \ |\operatorname{Re} z| < a\}; \quad \Delta^- := \{z: \ \operatorname{Im} z \in (-\delta, 0)\}$$

for $\delta \in (0, \delta_0)$. Let $\varepsilon < (\delta_0 - \delta)/2$. Lemmas 5.4.2 and 5.4.3 enable us to apply Stokes' theorem which implies

$$\int_{\Delta_a^+} f(x + i(y + \varepsilon)) \frac{\partial}{\partial \bar{z}}\phi(z)\, d\bar{z} \wedge dz = \int_{\partial \Delta_a^+} f(x + i(y + \varepsilon))\phi(z)\, dz.$$

Since $y + \varepsilon \in (\varepsilon, \varepsilon + \delta) \subset (0, \delta_0)$ for $y \in (0, \delta)$, we obtain

$$\|f(\cdot + i(y + \varepsilon))\|_{L^s} \leq C e^{M^*(k/\varepsilon)}.$$

This fact and [143] (p. 125, Lemma) imply $f(x + i(y + \varepsilon)) \to 0$ as $|x| \to \infty$, uniformly for $y \in (0, \delta)$. Thus, by Lemma 5.4.3 and by letting $a \to \infty$, we obtain

$$\int_{\Delta^+} f(x + i(y + \varepsilon)) \frac{\partial}{\partial \bar{z}}\phi(z)\, d\bar{z} \wedge dz$$

$$= \int_{\mathbb{R}} f(x + i\varepsilon)\varphi(x)\, dx - \int_{\mathbb{R}} f(x + i(\varepsilon + \delta))\phi(x + i\delta)\, dx. \tag{5.147}$$

Similarly,

$$\int_{\Delta^-} f(x + i(y - \varepsilon)) \frac{\partial}{\partial z}\phi(z)\, d\bar{z} \wedge dz$$

$$= \int_{\mathbb{R}} f(x - i(\varepsilon + \delta))\phi(x - i\delta)\, dx - \int_{\mathbb{R}} f(x - i\varepsilon)\varphi(x)\, dx. \tag{5.148}$$

We have (with a suitable $C > 0$)

$$\left| \iint_{\Delta^+} f(x + i(y + \varepsilon)) \frac{\partial}{\partial \bar{z}} \phi(z) \, d\bar{z} \wedge dz \right|$$

$$= 2 \left| \int_0^\delta dy \left(\int_{-\infty}^\infty f(x + i(y + \varepsilon)) \frac{\partial}{\partial \bar{z}} \phi(x + iy) \, dx \right) \right|$$

$$\leq 2 \int_0^\delta dy \left(\int_{-\infty}^\infty |f(x + i(y + \varepsilon))|^s \, dx \right)^{1/s}$$

$$\bullet \left(\int_{-\infty}^\infty \left| \frac{\partial}{\partial \bar{z}} \phi(x + iy) \right|^r dx \right)^{1/r}$$

$$\leq C \int_0^\delta e^{M^*(k/(y+\varepsilon)) - M^*(k/y)} \, dy < \infty.$$

The same holds for the integral over Δ^-. Since the integrands in (5.147) and (5.148) converge pointwise to the corresponding integrable functions, as $\varepsilon \to 0$, we obtain

$$\langle Tf, \varphi \rangle = \int_\Delta f(z) \frac{\partial}{\partial \bar{z}} \phi(z) \, d\bar{z} \wedge dz - \int_{\partial \Delta} f(z) \phi(z) \, dz,$$

which proves the first part of the assertion.

Using Hölder's inequality, the estimate for f and Lemma 5.4.1, we conclude that for some $h > 0$ and some $C > 0$ (respectively, for every $h > 0$ there is $C > 0$) such that

$$|\langle Tf, \varphi \rangle| \leq C\|\varphi\|_{h, L^r}$$

which completes the proof of Theorem 5.4.1. $\square$

Now let

$$\mathbb{C}_0 := \{z = x + iy \in \mathbb{C} \colon x \in \mathbb{R}, \ y \neq 0\} \tag{5.149}$$

and, for every $z = x + iy \in \mathbb{C}_0$, denote by χ_z the complex valued function defined on $\mathbb{R}$ given by

$$\chi_z(t) := \frac{1}{t - z}, \qquad t \in \mathbb{R}. \tag{5.150}$$

The following estimate will be needed in the proof of the next theorem.

Lemma 5.4.5. *There is a constant $B > 0$ such that for every $y \neq 0$ and every $g \in L^s$ with $s > 1$ the following inequality holds:*

$$\|\langle g, \chi_{\cdot + iy} \rangle\|_{L^s} = \left(\int_{\mathbb{R}} \left| \int_{\mathbb{R}} \frac{g(t) \, dt}{t - x - iy} \right|^s dx \right)^{1/s} \leq B\|g\|_s. \tag{5.151}$$

Proof. The assertion can be obtained by combining Theorem 1.4, Lemma 1.5 (Ch. IV)and Theorems 3.10 and 3.7 (Ch. II) in [143]. □

Theorem 5.4.2. *Let $r > 1$ and let $s := r/(r-1)$. The mapping $T : \mathcal{H}(*, L^s) \to \mathcal{D}'(*, L^s)$ defined in (5.146) is surjective. Its kernel is $\mathcal{H}_{L^s}$.*

Proof. We shall prove the assertion only for the Beurling case, because the Roumieu case can be proved similarly.

Let $f \in \mathcal{D}'((M_p), L^s)$ be of the form

$$f = \sum_{p=0}^{\infty} (-1)^p f_p^{(p)} \tag{5.152}$$

with $f_p \in L^s (p \in \mathbb{N}_0)$ such that

$$\sum_{p=0}^{\infty} \frac{M_p}{K^p} \|f_p\|_{L^s} < \infty. \tag{5.153}$$

By Theorem 4.1.1, the function χ_z defined in (5.150) belongs to $\mathcal{D}((M_p), L^r)$ for every $z \in \mathbb{C}_0$, the set defined in (5.149). We shall prove that the function $\tilde{f} \colon \mathbb{C}_0 \to \mathbb{C}$ defined by

$$\tilde{f}(z) := -\langle f, \chi_z \rangle, \qquad z \in \mathbb{C}_0,$$

belongs to $\mathcal{H}((M_p), L^s)$.

Fix $y \neq 0$. By inequality (5.152) from Lemma 5.4.5, Minkowski's inequality and the continuity of the L^s norm, we have

$$\|\tilde{f}(\cdot + iy)\|_{L^s} \leq \sum_{p=0}^{\infty} \left\| \langle f_p^{(p)}, \chi_{\cdot + iy} \rangle \right\|_{L^s} = \sum_{p=0}^{\infty} p! \, \|F_p\|_{L^s}, \tag{5.154}$$

where

$$F_p(x) := \int_{\mathbb{R}} \frac{f_p(t)}{[t - (x+iy)]^{p+1}} \, dt, \qquad p \in \mathbb{N}_0, \ x \in \mathbb{R}. \tag{5.155}$$

Consider $p \geq 1$ and recall that $r = s/(s-1)$. Since $|t - (x+iy)| \geq |y|$, we have

$$|F_p(x)| \leq \frac{\bar{F}_p(x)}{|y|^{p-1}}, \qquad x \in \mathbb{R}, \tag{5.156}$$

where

$$\bar{F}_p(x) := \int_{\mathbb{R}} \frac{|f_p(t)| \, dt}{|t - (x+iy)|^2}, \qquad x \in \mathbb{R}.$$

Using the substitution $u := (t-x)/y$ in the integral below, we have

$$\int_{\mathbb{R}} \frac{dt}{|t-z|^{1+r/2}} = \frac{1}{|y|^{r/2}} \int_{\mathbb{R}} \frac{du}{|u-i|^{1+r/2}} = \frac{A_r}{y^{r/2}} \tag{5.157}$$

(note that the integral on the left side does not depend on x), where

$$A_r := \int_{\mathbb{R}} \frac{du}{|u-i|^{1+r/2}} = \int_{\mathbb{R}} \frac{dv}{|v+i|^{1+r/2}} < \infty.$$

By Hölder's inequality,

$$\begin{aligned}
|\bar{F}_p(x)| &\leq \int_{\mathbb{R}} \frac{|f_p(t)|}{|t-z|^{3/2-1/r}} \cdot \frac{1}{|t-z|^{1/2+1/r}} \, dt \\
&\leq \left(\int_{\mathbb{R}} \frac{|f_p(t)|^s}{|t-z|^{1+s/2}} \, dt \right)^{1/s} \cdot \left(\int_{\mathbb{R}} \frac{dt}{|t-z|^{1+r/2}} \right)^{1/r} \\
&= \frac{A_r^{1/r}}{|y|^{1/2}} \left(\int_{\mathbb{R}} \frac{|f_p(t)|^s}{|t-z|^{1+s/2}} \, dt \right)^{1/s}
\end{aligned}$$

and hence

$$|\bar{F}_p(x)|^s \leq \frac{A_r^{s/r}}{|y|^{s/2}} \int_{\mathbb{R}} \frac{|f_p(t)|^s}{|t-z|^{1+s/2}} \, dt. \tag{5.158}$$

Consequently, by (5.156), (5.158), the Fubini theorem and (5.157) applied with t replaced by x and r replaced by s, we obtain

$$\begin{aligned}
\|F_p\|_{L^s} &\leq \frac{1}{|y|^{p-1}} \left(\int_{\mathbb{R}} |\bar{F}_p(x)|^s \, dx \right)^{1/s} \\
&\leq \frac{A_r^{1/r}}{|y|^{p+1/2}} \left[\int_{\mathbb{R}} \left(\int_{\mathbb{R}} \frac{|f_p(t)|^s}{|t-z|^{1+s/2}} \, dt \right) dx \right]^{1/s} \tag{5.159} \\
&= \frac{A_r^{1/r}}{|y|^{p+1/2}} \left[\int_{\mathbb{R}} |f_p(t)|^s \left(\int_{\mathbb{R}} \frac{dx}{|t-z|^{1+s/2}} \right) dt \right]^{1/s} \\
&= \frac{A_r^{1/r}}{|y|^{p+1/2}} \left(\frac{A_s}{|y|^{s/2}} \right)^{1/s} \|f_p\|_{L^s} = \frac{A_r^{1/r} A_s^{1/s}}{|y|^{p+1}} \|f_p\|_{L^s}.
\end{aligned}$$

By (5.154), (5.151) and (5.159),

$$\|\tilde{f}(\cdot + iy)\|_{L^s} \leq B\|f_0\|_{L^s} + A_r^{1/r} A_s^{1/s} \sum_{p=1}^{\infty} \frac{p!}{|y|^{p+1}} \|f_p\|_{L^s}$$

$$\leq A_1 \sum_{p=0}^{\infty} \frac{p!}{|y|^p} \|f_p\|_{L^s},$$

where $A_1 = A_1(y) := \max\{B, A_r^{1/r} A_s^{1/s}/|y|\}$. By (5.153), this implies that, for arbitrary $y \neq 0$,

$$\|\tilde{f}(\cdot + iy)\|_{L^s} \leq A_1 \sup_p \left[\frac{k^p p!}{M_p |y|^p}\right] \sum_{p=0}^{\infty} \frac{M_p}{k^p} \|f_p\|_{L^s} \leq A_2 e^{M^*(k/|y|)}$$

for a suitable constant A_2 depending on y. Consequently, $\tilde{f} \in \mathcal{H}((M_p), L^s)$.

We shall show that $f = T\tilde{f}$. Let $\varphi \in \mathcal{D}((M_p), L^r)$ and ϕ be its almost analytic extension. For $z \in \mathbb{C}$ put

$$\phi_1(z) := \frac{1}{2\pi i} \int_\Delta \frac{(\partial/\partial\bar\zeta)\phi(\zeta)}{\zeta - z}\, d\zeta \wedge d\bar\zeta, \qquad (5.160)$$

$$\phi_2(z) := \frac{1}{2\pi i} \int_{\Gamma_{\delta-}} \frac{\phi(\zeta)}{\zeta - z}\, d\zeta, \qquad (5.161)$$

$$\phi_3(z) := -\frac{1}{2\pi i} \int_{\Gamma_{\delta+}} \frac{\phi(\zeta)}{\zeta - z}\, d\zeta \qquad (5.162)$$

and let $\varphi_j := \phi_{j|\mathbb{R}}$ for $j = 1, 2, 3$, i.e. φ_1, φ_2 and φ_3 are given by formulas (5.160), (5.161) and (5.162), respectively, with the variable $z \in \mathbb{C}$ replaced by $x \in \mathbb{R}$. We have $\varphi(x) = \varphi_1(x) + \varphi_2(x) + \varphi_3(x)$ for $x \in \mathbb{R}$.

By the same arguments as in Lemma 5.4.4, it follows that the functions φ_2 and φ_3 are in $\mathcal{D}((M_p), L^r)$. Thus the function φ_1 is in $\mathcal{D}((M_p), L^r)$.

Applying formulas (5.160), (5.161) and (5.162) with $z \in \mathbb{C}$ replaced by $x \in \mathbb{R}$, we obtain

$$2\pi i \langle f, \varphi_1 \rangle = -\int_\Delta \langle f, \chi_\zeta \rangle \frac{\partial}{\partial\bar\zeta}\phi(\zeta)\, d\zeta \wedge d\bar\zeta = \int_\Delta \tilde{f}(\zeta) \frac{\partial}{\partial\bar\zeta}\phi(\zeta)\, d\zeta \wedge d\bar\zeta,$$
$$(5.163)$$

$$2\pi i \langle f, \varphi_2 \rangle = -\int_{\Gamma_{\delta-}} \langle f, \chi_\zeta \rangle \phi(\zeta)\, d\zeta = \int_{\Gamma_{\delta-}} \tilde{f}(\zeta) \phi(\zeta)\, d\zeta \qquad (5.164)$$

$$2\pi i \langle f, \varphi_3 \rangle = \int_{\Gamma_{\delta+}} \langle f, \chi_\zeta \rangle \phi(\zeta)\, d\zeta = -\int_{\Gamma_{\delta+}} \tilde{f}(\zeta) \phi(\zeta)\, d\zeta, \qquad (5.165)$$

where χ_ζ denotes the function defined in (5.150), under the assumption that the change of the order of the functional f and the respective integrals is allowed. In this case, we may conclude from (5.163)-(5.165) that

$$\langle f, \varphi \rangle = \langle f, \varphi_1 \rangle + \langle f, \varphi_2 \rangle + \langle f, \varphi_3 \rangle = \langle T\tilde{f}, \varphi \rangle.$$

The interchange of f and integrals applied above is allowed, because it is allowed in case the sets $\int_\Delta$ and $\int_{\Gamma_{\delta\pm}}$ under the integrals above are replaced by $\int_{\Delta_a}$ and $\int_{\Gamma_{a,\delta\pm}}$, respectively, for arbitrary $a > 0$, and because

$$\int_{\Delta_a} \frac{(\partial/\partial\bar\zeta)\varphi(\zeta)}{\zeta - \cdot} \, d\zeta \wedge d\bar\zeta \to \int_{\Delta} \frac{(\partial/\partial\bar\zeta)\varphi(\zeta)}{\zeta - \cdot} \, d\zeta \wedge d\bar\zeta,$$

$$\int_{\Gamma_{a,\delta\pm}} \frac{\varphi(\zeta)}{\zeta - \cdot} \, d\zeta \to \int_{\Gamma_{\delta\pm}} \frac{\varphi(\zeta)}{\zeta - \cdot} \, d\zeta$$

as $a \to \infty$ in the sense of convergence in $\mathcal{D}((M_p), L^r)$. By similar arguments as in the proof of Theorem 3.3 in [110] one can prove that $\operatorname{Ker} T = \mathcal{H}(L^s, \mathbb{R})$, and the assertion of Theorem 5.4.2 is proved. $\square$

5.5 Cases $s = \infty$ and $s = 1$

The method used in the previous section cannot be applied for $s = \infty$ and $s = 1$ because the function

$$\mathbb{R} \ni t \mapsto \frac{1}{t - x - iy} \in \mathbb{C}, \qquad x + iy \in \mathbb{C}, \quad y \neq 0,$$

is not in L^1. Note that this function belongs to $\dot{\mathcal{B}}((M_p), \mathbb{R})$, but we have not succeeded in proving that for an $f \in \mathcal{D}'(*, L^\infty)$ or $f \in \mathcal{D}'(*, L^1)$ there exists the corresponding $F(z)$ in $\mathcal{H}(*, L^\infty)$ or $\mathcal{H}(*, L^1)$ which converges to f in $\mathcal{D}'(*, L^\infty)$ or $\mathcal{D}'(*, L^1)$. We shall prove the converse assertion, i.e., that elements in $\mathcal{H}(*, L^\infty)$ and $\mathcal{H}(*, L^1)$ determine elements in $\mathcal{D}'(*, L^\infty)$ and $\mathcal{D}'(*, L^1)$ as boundary values, but assuming the stronger condition (5.151) instead of (5.148). This condition enables us to follow the method of Komatsu used in the proof of Theorem 11.5 in [82]. The following lemma from [82] is needed.

Lemma 5.5.1. *Assume that the sequence* (N_p) *satisfies* (M.1), (M.2), (M.3') *and, denoting* $n_p := N_p/N_{p-1}$, *consider*

$$P(\zeta) := (1 + \zeta)^2 \prod_{p=1}^{\infty} \left(1 + \frac{\zeta}{n_p}\right), G(\zeta) := \frac{1}{2\pi} \int_0^{\infty} P(t)^{-1} e^{i\zeta t} \, dt, \zeta \in \mathbb{C}.$$

Then the function G *is analytic, can be continued analytically to the Riemann domain* $\{z: \ -\pi < \arg z < 2\pi\}$ *on which* $P(D)G(z) = -(2\pi i z)^{-1}$ *and is bounded on the domain* $\{z: \ -\frac{\pi}{2} \leq \arg z \leq \frac{3\pi}{2}\}$. *Moreover, defining*

$$g(y) := G_+(-iy) - G_-(-iy), \qquad y > 0,$$

where G_+ *and* G_- *are the branches of* G *on* $\{z: \ -\pi < \arg z \leq 0\}$ *and* $\{z: \ \pi \leq \arg z < 2\pi\}$, *respectively, we have*

$$|g(y)| \leq A\sqrt{y}e^{M^*(L/y)}, \qquad y > 0,$$

for some $A > 0$, where M^* is the associated function of the sequence (M_p) defined in (2.9).

Theorem 5.5.1. *Assume that the sequence (M_p) satisfies conditions $(M.1)$, $(M.2)$, $(M.3')$ and the sequence (m_p^*) is nondecreasing. If $F \in \mathcal{H}(*, L^\infty)$ (respectively, if $F \in \mathcal{H}(*, L^1)$), then $F(\cdot + iy) \to F(\cdot + i0) \in \mathcal{D}'(*, L^\infty)$ (respectively, $F(\cdot + iy) \to F(\cdot + i0)) \in \mathcal{D}'(*, L^1))$ as $y \to 0^+$ in the sense of convergence in $\mathcal{D}'(*, L^\infty)$ (respectively, in $\mathcal{D}'(*, L^1))$.*

Proof. We shall prove the theorem only for the Roumieu case $* = \{M_p\}$ which is more complicated. We shall use the construction from [82], Theorem 5.4.3 (see also [114]). Our aim is to prove that, for every $\varphi \in \mathcal{D}(\{M_p\}, L^1)$ (respectively, $\varphi \in \dot{\mathcal{B}}(\{M_p\}, \mathbb{R}))$, the set

$$\{\langle F(\cdot + iy), \varphi \rangle \colon \, 0 < y < \delta_0\}$$

is bounded and, moreover, $\langle F(\cdot + iy), \varphi \rangle$ converges as $y \to 0^+$ for every $\varphi \in \mathcal{D}(\{M_p\}, \mathbb{R})$. Since $\mathcal{D}(\{M_p\}, \mathbb{R})$ is dense in $\mathcal{D}(\{M_p\}), L^1)$ and in $\dot{\mathcal{B}}(\{M_p\}, \mathbb{R}))$, this will imply the assertion in Theorem 5.5.1.

Assume first that $F \in \mathcal{H}(\{M_p\}, L^\infty)$ and let $\varphi \in \mathcal{D}(\{M_p\}, L^1)$ be such that $\|\varphi\|_{L^1, h_0} < \infty$ for $h_0 > 0$.

Let $I_k := (k - 2, k + 2)$ for $k \in \mathbb{Z}$, the set of all integers, and let ψ_k be a partition of unity in $\mathcal{D}(\{M_p\}, \mathbb{R})$ such that, for some $R > 0$ which does not depend on k,

$$\mathrm{supp}\,\psi_k \subseteq I_k, \quad \|\psi_k\|_{L^1, h_0} \le R, \qquad k \in \mathbb{Z}.$$

We have

$$\int_{-\infty}^{\infty} F(x + iy)\varphi(x)\,dx = \sum_{k \in \mathbb{Z}} \int_{I_k} F(x + iy)\varphi(x)\psi_k(x)\,dx, \qquad 0 < y < \delta_0.$$

We will construct an ultradifferential operator P of class $\{M_p\}$ of the form

$$P(D) = (1 + D)^2 \prod_{p=1}^{\infty} \left(1 + \frac{D}{n_p}\right) \tag{5.166}$$

such that the equations

$$P(D)H_k(x + iy) = F(x + iy), \qquad k \in \mathbb{Z},$$

have the solutions $H_k(x + iy)$ which are analytic in

$$\Pi_k := \{x + iy \colon \, x \in I_k, \,\, 0 < y < \delta_0/2\}$$

and bounded in some neighborhood of I_k for $k \in \mathbb{Z}$.

As in [82], pp. 98–99, one can show that there is a sequence (N_p) such that the operator (5.166) is of class $\{M_p\}$, $M_p \prec N_p$, and

$$\|F(\cdot + iy)\|_{L^\infty} < Ce^{N^*(1/y)}, \qquad |y| < \delta_0.$$

Conditions $(M.1)$, $(M.2)$, $(M.3')$ imply that if the operator $P(D)$ is of the form (5.166), then it is of the class $\{M_p\}$; and in this case, the condition that the sequence (m_p^*) is nondecreasing could not be replaced by $(M.1)$.

Fix k and denote $z_k^0 := k + i\delta$ with $\delta_0/2 < \delta < \delta_0$. Let

$$H_k(z) := \int_\Gamma G(z - \omega)F(\omega)\,d\omega, \qquad z = x + iy, \ \ x \in I_k, \ \ 0 < y < \delta_0/2,$$

where $G(z)$ is the Green kernel of $P(D)$ given in Lemma 5.5.1 and Γ_k is a simple closed curve lying in $\{x + iy\colon x \in I_k, y \in (0,\delta)\}$ starting at z_k^0 and encircling counterclockwise a slit connecting z_k^0 and z. We deform the path Γ_k to the union of segments joining z_k^0 and $z_k^1 = x + i\delta_0/2$, a segment joining z_k^1 and z, a segment joining z and z_k^1, and a segment joining z_k^1 and z_k^0. This is possible because G is bounded in the set $\{z\colon -\pi/2 \le \arg z \le 3\pi/2\}$. By the same arguments as in [82], we have $P(D)H_k(z) = F(z)$ for $z \in \Pi_k$ and, consequently,

$$\left|\int_{-\infty}^\infty F(x+iy)\varphi(x)\,dx\right| \le \sum_{k\in\mathbb{Z}}\int_{I_k}|F(x+iy)\psi_k(x)\varphi(x)|\,dx$$

$$\le \sum_{k\in\mathbb{Z}}\int_{I_k}\left|\left[\int_{\Gamma_k}G(z-\omega)F(\omega)\,d\omega\right]P(D)[\psi_k(x)\varphi(x)]\right|\,dx$$

for $0 < y < \delta_0/2$. Denote the part of γ_k from z_k^1 to z and from z to z_k^1 by Γ_k^1 and the rest by Γ_k^0 for $k \in \mathbb{Z}$. We have

$$\int_{I_k}\left|\left[\int_{\Gamma_k^1}G(z-\omega)F(\omega)\,d\omega\right]P(D)[\varphi(x)\psi_k(x)]\right|\,dx$$

$$\le \sup_{x\in I_k}\left|\int_{\Gamma_k^1}G(z-\omega)F(\omega)\,d\omega\right|\int_{I_k}|P(D)[\varphi(x)\psi_k(x)]|\,dx. \qquad (5.167)$$

Denote by A_k the first and by B_k the second factor on the right side of (5.167). Since $P(D) = \sum_\alpha D^\alpha$ is of class $\{M_p\}$, it follows from (5.166) with $2r < h_0^2$ and from the inequalities $M_{\alpha-j}M_j \le M_\alpha$ for $j, \alpha \in \mathbb{N}_0^n$, $j \le \alpha$, that

$$\sum_{k\in\mathbb{Z}}B_k \le \sum_{k\in\mathbb{Z}}\sum_{\alpha=0}^\infty a_\alpha \sum_{j=0}^\alpha \binom{\alpha}{j}\int_{I_k}|\varphi^{(\alpha-j)}(x)\psi_k^{(j)}(x)|\,dx$$

$$\leq C \sum_{\alpha=0}^{\infty} \frac{r^{\alpha} a_{\alpha}}{(h_0^2)^{\alpha}} \sum_{j=0}^{\alpha} \binom{\alpha}{j} \|\varphi\|_{L^1, h_0} \|\psi_k\|_{L^{\infty}, h_0}$$

$$\leq CR \|\varphi\|_{L^1, h_0} \sum_{\alpha=0}^{\infty} \frac{r^{\alpha}}{(h_0^2)^{\alpha}} < \infty.$$

On the other hand,

$$A_k = \sup_{x \in I_k} \left| \int_0^{\delta-y} g(t) F(x + iy + it) \, dt \right|$$

$$\leq A\sqrt{y} \sup_{x \in I_k} \int_0^{\delta-y} e^{-N^*(\frac{1}{t+y})} e^{N^*(\frac{1}{t})} \, dt < \infty.$$

This implies that $\sum_{k \in \mathbb{Z}} A_k B_k < \infty$. Consider the path Γ_k^0. We have

$$\left| \int_{I_k} \left| \left[\int_{\Gamma_k^0} G(z-\omega) F(\omega) \, d\omega \right] P(D)[\varphi(x)\psi_k(x)] \right| dx \right|$$

$$\leq \sup_{x \in I_k} \left| \int_{\Gamma_k^0} G(z-\omega) F(\omega) \, d\omega \right| \int_{I_k} |P(D)[\varphi(x)\psi_k(x)]| \, dx = D_k B_k.$$

Since, for $z \in \Pi_k$ and $\omega \in \Gamma_k^0$, the function $G(z-\omega)$ is uniformly bounded by a constant which does not depend on k, we obtain $\sum_{k \in \mathbb{Z}} D_k B_k < \infty$. This implies

$$\left| \int_{-\infty}^{\infty} F(x+iy)\varphi(x) \, dx \right| \leq \sum_{k \in \mathbb{Z}} (D_k + A_k) B_k < \infty.$$

The proof that there is $F(\cdot + i0) \in \mathcal{D}'(\{M_p\}, \mathbb{R})$ such that

$$\langle F(x+iy), \varphi \rangle \to \langle F(x+i0), \varphi \rangle, \qquad y \to 0^+,$$

for every $\varphi \in \mathcal{D}(\{M_p\}, \mathbb{R})$, is given in [82] and [110]. We conclude that

$$F(x+iy) \to F(x+i0) \in \mathcal{D}'(\{M_p\}, L^{\infty}), \qquad y \to 0^+,$$

which finishes the proof in case $F \in \mathcal{H}(\{M_p\}, L^{\infty})$.

The proof of Theorem 5.5.1 for $F \in \mathcal{H}(\{M_p\}, L^1)$ is analogous to the previous case. The partition of unity (ψ_k) and the constructed sequence of functions $H_k(z)$, $z \in \Pi_k$, lead us to the proof that for every $\varphi \in \mathcal{D}(\{M_p\}, L^{\infty})$ the set

$$\{\langle F(\cdot + iy), \varphi \rangle : \ 0 < y < \delta_0/2\}$$

is bounded. So we have to prove that $\langle F(\cdot + iy), \varphi \rangle$ converges as $y \to 0^+$ for each $\varphi \in \mathcal{D}(\{M_p\}, \mathbb{R})$.

Let I be a bounded open interval and

$$\Pi_I := \{x + iy\colon x \in I, y \in (0, \delta_0/2)\}.$$

As in the first part of the proof, we construct $P(D)$ of the form (5.166) and of class $\{M_p\}$ and H_I such that

$$P(D)H_I(x + iy) = F(x + iy), \qquad x + iy \in \Pi_I.$$

We put

$$H_I(x + iy) := \int_{\Gamma^1} G(z - \omega)F(\omega)\, d\omega + \int_{\Gamma^0} G(z - \omega)F(\omega)\, d\omega,$$

where $\Gamma := \Gamma^1 \cup \Gamma^2$ is a path constructed in the same way as Γ_k with I instead of I_k and $z^0 := x^0 + i\delta$ (x^0 is the middle point of I) instead of z_k^0. By using Hölder's inequality, we obtain $H(\cdot + iy) \in L^1(I)$ for every y, $0 < y < \delta_0/2$, and

$$\|H(\cdot + iy)\|_{L^1} < C, \qquad 0 < y < \delta_0/2.$$

This implies that

$$H_I(\cdot + iy) \to H_I(\cdot + i0) \in L^1, \qquad y \to 0^+,$$

and, consequently,

$$H_I(x + iy) \to H_I(x + i0), \qquad y \to 0^+,$$

in $\mathcal{D}'(\{M_p\}, \mathbb{R})$.

Consequently,

$$\langle F(x + iy), \varphi \rangle \to \langle F(x + i0), \varphi \rangle, \qquad y \to 0^+,$$

for every $\varphi \in \mathcal{D}(\{M_p\}, \mathbb{R})$, and the proof is completed. $\square$

Chapter 6

Convolution of Ultradistributions

6.1 Introduction

In the theory of generalized functions (distributions, ultradistributions, hyperfunctions) the convolution of two generalized functions is usually defined only in the case where one of them has compact support. On the basis of such a definition of the convolution of ultradistributions, Braun, Meise, Taylor, Voigt and their collaborators (see [103] and references there) extensively studied convolution equations in the space of ultradistributions.

However, it is natural and very important in many situations (e.g. convolution equations, convolutional algebras) to have the convolution of generalized functions defined without any restrictions on their supports.

In the space $\mathcal{D}'$ of Schwartz's distributions (and in certain of its subspaces, e.g. in the spaces $\mathcal{S}'$ of tempered distributions and $K\{M_p\}'$ of Gelfand-Shilov) a variety of such general definitions of the convolution have appeared in many books and papers (see e.g. [36, 132, 57, 156, 135, 149, 151, 2, 45, 69]), but almost thirty years passed until it became clear that most of them are equivalent (see [135, 45, 69]).

First two general definitions of the convolution for ultradistributions were introduced in [119]. The equivalence of analogues of the main classical definitions of the convolution of ultradistributions was proved in [70]. In [70], several definitions of the convolution in the space $\mathcal{S}'^{(M_p)}$ of tempered ultradistributions (analogous to the notion of $\mathcal{S}'$-convolution of tempered distributions) were formulated, but without proof of their equivalence. The proof will be given in this chapter. This proof is not a simple modification of the proofs known for distributions. New methods are necessary, because ultradistributions are not finite but infinite sums of derivatives of functions on bounded open sets (so e.g. the Leibniz formula cannot be used). It is interesting that several sequential definitions of convolution are equivalent to those introduced in the functional analysis approach to the theory in which the space of integrable ultradistributions plays a crucial role. This gives the possibility of further generalizations of the definition by selecting suitably narrow classes of unit-sequences. We restrict our study to convolution of ultradistributions of Beurling type and follow the approach of Denjoy, Carleman and Komatsu to the theory (see [82]-[84]), but the results can be obtained by other approaches, such as those developed by Cioranescu-Zsidó, Beurling, Björck, Grudzinski and Braun-Meise-Taylor (see [44, 7, 8, 56, 11]).

Recall that if ω is a function on $\mathbb{R}^n$, then the symbol $\omega^\triangle$ traditionally denotes the function on $\mathbb{R}^{2n}$ given by

$$\omega^\triangle(x, y) := \omega(x + y), \qquad x, y \in \mathbb{R}^n. \tag{6.1}$$

Evidently, if $\omega \in \mathcal{E} = \mathcal{E}(\mathbb{R}^n)$, then $\omega^\triangle \in \mathcal{E}(\mathbb{R}^{2n})$ and

$$\frac{\partial^{\alpha+\beta}}{\partial x^\alpha \partial y^\beta}[\omega^\triangle(x, y)] = \left[\frac{\partial^{\alpha+\beta}}{\partial x^\alpha \partial y^\beta}\omega(x, y)\right]^\triangle, \qquad \alpha, \beta \in \mathbb{N}^n.$$

We denote this common value in abbreviation by $(\omega^\triangle)^{(\alpha+\beta)}(x, y)$.

As before, the norm in $L^r = L^r(\mathbb{R}^n)$, $r \in [1, \infty]$, is denoted by $\|\cdot\|_{L^r}$.

We introduce now certain classes of sequences of functions (see [45], [68], [119]). A sequence (η_j) of elements of $\mathcal{D}^{(M_p)}(\mathbb{R}^n)$ is said to be an (M_p)-*unit-sequence* or, briefly, a *unit-sequence* [in symbols: $(\eta_j) \in \mathbf{1}^{(M_p)}(\mathbb{R}^n)$] if it converges to 1 in $\mathcal{E}^{(M_p)}(\mathbb{R}^n)$ and if there exists $h > 0$ such that

$$\sup_{j \in \mathbb{N}} \sup_{\alpha \in \mathbb{N}^n} \left(\frac{h^\alpha}{M_\alpha} \|\eta_j^{(\alpha)}\|_{L^\infty}\right) < \infty. \tag{6.2}$$

If moreover, for every compact set $K \subset \mathbb{R}^n$, there exists $j_0 \in \mathbb{N}$ such that

$$\eta_j(x) = 1 \qquad \text{for } x \in K, \ j \geq j_0,$$

then (η_j) is called a *strong* (M_p)-*unit-sequence* or, briefly, a *strong unit-sequence* [in symbols: $(\eta_j) \in \bar{\mathbf{1}}^{(M_p)}(\mathbb{R}^n)$].

A sequence (δ_j) of elements of $\mathcal{D}^{(M_p)}(\mathbb{R}^n)$ is said to be an (M_p)-*delta-sequence* or, briefly, a *delta-sequence* [in symbols: $(\delta_j) \in \mathbf{\Delta}^{(M_p)}(\mathbb{R}^n)$] if

$$\delta_j \geq 0, \quad \int \delta_j = 1, \quad \text{supp } \delta_j \subseteq [-\alpha_j, \alpha_j]^n \qquad \text{for } j \in \mathbb{N},$$

where $\alpha_j > 0$ for $j \in \mathbb{N}$ and $\alpha_j \to 0$ as $j \to \infty$.

A locally convex space F is said to be a *space of ultradistributions* if

(I) $\mathcal{D}^{(M_p)} \subset F \subset \mathcal{D}'$ and the injections $\mathcal{D}^{(M_p)} \to F$ and $F \to \mathcal{D}'$ are continuous;

(II) $\mathcal{D}^{(M_p)}$ is dense in F .

Following [57] we say that the space of ultradistributions of the class (M_p) is *permitted* if $(\eta_j f) * \delta_j \to f$ and $\eta_j(f * \delta_j) \to f$ in $\mathcal{D}'(\mathbb{R}^n)$ as $j \to \infty$ for arbitrary $f \in F$, $(\eta_j) \in \mathbf{1}^{(M_p)}(\mathbb{R}^n)$ and $(\delta_j) \in \mathbf{\Delta}^{(M_p)}(\mathbb{R}^n)$.

A space F of ultradistributions is said to have *property* $(C_{(M_p)})$ (see [119]) if for each barrelled space E every linear mapping $\tau : E \to F$ is continuous as a mapping $\tau : E \to \mathcal{D}'^{(M_p)}$. For example, L^1 and $\mathcal{D}'^{(M_p)}_{L^1}$ have this property.

6.2 Definitions of $\mathcal{D}'^{(M_p)}$−convolution

In this section we give a list of general definitions of the convolution of ultra-distributions, analogous to the definitions given in the case of distributions by Schwartz [132] (Definition 6.2.1), Chevalley [36], pp. 67, 112 (Definitions 6.2.2, 6.2.3, and 6.2.4), Vladimirov [149], p. 138 (Definition 6.2.5), Dierolf and Voigt [45] (Definition 6.2.6), and Kamiński [69] (Definitions 6.2.7 and 6.2.8). In the next section we will prove their equivalence.

Recall that for a given ultradistribution $T \in \mathcal{D}'^{(M_p)}(\mathbb{R}^n)$ the symbol $\tilde{T}$ is meant in the sense of formulas (1.5)-(1.6) and the symbol $\omega^\triangle$ for a function $\omega \in \mathcal{D}^{(M_p)}$ is defined in (6.1).

We give now several definitions (Definitions 6.2.1-6.2.8) of the $\mathcal{D}'^{(M_p)}$−convolution of ultradistributions denoting the convolution introduced in Definition 6.2.k by the corresponding symbol $\overset{k}{*}$ for $k = 1, 2, 3, 4, 5, 6, 7_0, 7_1, 7_2, 8_0, 8_1, 8_2$. In the next section, we will prove that all of these definitions are equivalent and for the $\mathcal{D}'^{(M_p)}$−convolution of ultradistributions the common symbol $*$ will be used further on.

In all definitions of the $\mathcal{D}'^{(M_p)}$−convolution below S and T mean two fixed ultradistributions, i.e. elements of the space $\mathcal{D}'^{(M_p)} := \mathcal{D}'^{(M_p)}(\mathbb{R}^n)$.

Definition 6.2.1. The convolution $S \overset{1}{*} T$ is defined by the formula

$$\langle S \overset{1}{*} T, \varphi \rangle := \langle (S \otimes T)\varphi^\triangle, 1 \rangle, \qquad \varphi \in \mathcal{D}^{(M_p)},$$

whenever

$$\overset{1}{(*)} \qquad (S \otimes T)\varphi^\triangle \in \mathcal{D}'^{(M_p)}_{L^1}(\mathbb{R}^{2n}) \qquad \text{for} \quad \varphi \in \mathcal{D}^{(M_p)}.$$

Definition 6.2.2. The convolution $S \overset{2}{*} T$ is defined by the formula

$$\langle S \overset{2}{*} T, \varphi \rangle := \langle (S * \varphi)\tilde{T}, 1 \rangle, \qquad \varphi \in \mathcal{D}^{(M_p)},$$

whenever

$(\overset{2}{*})$ $\qquad (S * \varphi)\tilde{T} \in \mathcal{D}'^{(M_p)}_{L^1}$ $\qquad$ for $\quad \varphi \in \mathcal{D}^{(M_p)}$.

Definition 6.2.3. The convolution $S \overset{3}{*} T$ is defined by the formula

$$\langle S \overset{3}{*} T, \varphi \rangle := \langle S(\tilde{T} * \varphi), 1 \rangle, \qquad \varphi \in \mathcal{D}^{(M_p)},$$

whenever

$(\overset{3}{*})$ $\qquad S(\tilde{T} * \varphi) \in \mathcal{D}'^{(M_p)}_{L^1}$ $\qquad$ for $\quad \varphi \in \mathcal{D}^{(M_p)}$.

Definition 6.2.4.

The convolution $S \overset{4}{*} T$ is defined by the formula

$$\langle (S \overset{4}{*} T) * \varphi, \psi \rangle := \langle (S * \varphi)(\tilde{T} * \psi), 1 \rangle, \qquad \varphi \in \mathcal{D}^{(M_p)}, \ \psi \in \mathcal{D}^{(M_p)},$$

whenever

$(\overset{4}{*})$ $\qquad (S * \varphi)(\tilde{T} * \psi) \in L^1$ $\qquad$ for $\quad \varphi \in \mathcal{D}^{(M_p)}, \ \psi \in \mathcal{D}^{(M_p)}$.

As in the theory of distributions, one can prove that the mapping

$$\mathcal{D}^{(M_p)}(\mathbb{R}^n) \ni \varphi \mapsto A_\varphi \in \mathcal{D}'^{(M_p)}(\mathbb{R}^n),$$

where

$$\langle A_\varphi, \psi \rangle := \int_{\mathbb{R}^n} (S * \varphi)(x)(\tilde{T} * \psi)(x)dx, \qquad \psi \in \mathcal{D}^{(M_p)},$$

is a linear, continuous and translation invariant mapping from $\mathcal{D}^{(M_p)}(\mathbb{R}^n)$ into $\mathcal{E}^{(M_p)}(\mathbb{R}^n)$. This implies that there exists a unique ultradistribution $V \in \mathcal{D}'^{(M_p)}(\mathbb{R}^n)$ such that $V * \varphi = A_\varphi$ for $\varphi \in \mathcal{D}^{(M_p)}(\mathbb{R}^n)$. Consequently, Definition 6.2.4 is consistent and $S \overset{4}{*} T = V$.

Definition 6.2.5. The convolution $S \overset{5}{*} T$ is defined by the formula

$$\langle S \overset{5}{*} T, \varphi \rangle := \lim_{j \to \infty} \langle S \otimes T, \eta_j \varphi^\triangle \rangle, \qquad \varphi \in \mathcal{D}^{(M_p)},$$

whenever

$(\overset{5}{*})$ $\qquad$ the limit exists for all $\varphi \in \mathcal{D}^{(M_p)}$ and for all $(\eta_j) \in \bar{\mathbf{1}}^{(M_p)}(\mathbb{R}^{2n})$.

Definition 6.2.6. The convolution $S \overset{6}{*} T$ is defined by the formula

$$\langle S \overset{6}{*} T, \varphi \rangle := \lim_{j \to \infty} \langle S \otimes T, \eta_j \, \varphi^{\triangle} \rangle, \qquad \varphi \in \mathcal{D}^{(M_p)},$$

whenever

$\left(\overset{6}{*} \right)$ the limit exists for all $\varphi \in \mathcal{D}^{(M_p)}$ and for all $(\eta_j) \in \mathbf{1}^{(M_p)}(\mathbb{R}^{2n})$.

Definition 6.2.7.

a) The convolution $S \overset{7_0}{*} T$ is defined by the formula

(d_0) $$S \overset{7_0}{*} T := \lim_{j \to \infty} (\eta_j S) * (\tilde{\eta}_j T) \quad \text{in } \mathcal{D}'^{(M_p)},$$

whenever

$\left(\overset{7_0}{*} \right)$ the limit in (d_0) exists for all $(\eta_j) \in \bar{\mathbf{1}}^{(M_p)}$ and for all $(\tilde{\eta}_j) \in \bar{\mathbf{1}}^{(M_p)}$.

b) The convolution $S \overset{7_1}{*} T$ is defined by the formula

(d_1) $$S \overset{7_1}{*} T := \lim_{j \to \infty} (\eta_j S) * T \quad \text{in } \mathcal{D}'^{(M_p)},$$

whenever

$\left(\overset{7_1}{*} \right)$ the limit in (d_1) exists for all $(\eta_j) \in \bar{\mathbf{1}}^{(M_p)}$.

c) The convolution $S \overset{7_2}{*} T$ is defined by the formula

(d_2) $$S \overset{7_2}{*} T := \lim_{j \to \infty} S * (\eta_j T) \quad \text{in } \mathcal{D}'^{(M_p)},$$

whenever

$\left(\overset{7_2}{*} \right)$ the limit in (d_2) exists for all $(\eta_j) \in \bar{\mathbf{1}}^{(M_p)}$.

Definition 6.2.8. For every $k = 0, 1, 2$, the convolution $S \overset{8_k}{*} T$ is defined by formula (d_k), whenever

$\left(\overset{8_k}{*} \right)$ the limit in (d_k) exists for all $(\eta_j) \in \mathbf{1}^{(M_p)}$ *(and for all* $(\tilde{\eta}_j) \in \mathbf{1}^{(M_p)})$.

In the case of distributions, the equivalence of the above listed definitions of convolution follows from several papers of different authors (see
[135], [45], [69], [154]).

In the case of ultradistributions, it has been proved in [119] that Definitions 6.2.1 and 6.2.5 are equivalent and that they imply Definitions 6.2.2
and 6.2.3.

The full proof of the equivalence of all the above mentioned definitions
was given in [73], and we reproduce it in the next section.

6.3 Equivalence of definitions of $\mathcal{D}'^{(M_p)}$−convolution

In this section we are going to prove the equivalence of all the definitions
of the $\mathcal{D}'^{(M_p)}$−convolution of ultradistributions listed in the preceding section. The form of the theorem concerning this equivalence that we are going
to prove is an extension of the known theorem by Shiraishi (see [135]) on
the equivalence of four definitions of the convolution of distributions, but
our theorem includes also the equivalence of analogues of several definitions
given later by other authors (see [45] and [69]). Our proof is similar to those
given in [135] and [69], but it requires new techniques. Traditional methods fail in the case of ultradistributions. For example the Leibniz formula
cannot be used since ultradistributions are infinite sums of derivatives of
corresponding continuous functions on a bounded open set.

In the proof we will need three assertions. The first one is the equivalence of the three conditions $(i) - (iii)$ which was proved for distributions
in [45]. For ultradistributions, only the equivalence of conditions (i) and
(ii) was shown formally in [119], but condition (iii) is equivalent to (i) and
(ii), as well.

Lemma 6.3.1. (see [45], [119]) *Let $f \in \mathcal{D}'^{(M_p)}$. The following conditions
are equivalent:*

(i) *f is continuous on the space $\dot{\mathcal{B}}^{(M_p)}$, i.e., $f \in \mathcal{D}'^{(M_p)}_{L^1}$;*

(ii) *the sequence $(< f, \eta_k >)$ is convergent for arbitrary $(\eta_k) \in \overline{\mathbf{1}}^{(M_p)}$;*

(iii) *the sequence $(< f, \eta_k >)$ is convergent for arbitrary $(\eta_k) \in \mathbf{1}^{(M_p)}$.*

The second assertion needed in the proof is the following result of Komatsu (see [82]; for the proof see also [43]):

Lemma 6.3.2. (see [82]) *Let K be a compact neighborhood of zero and $r > 0$. There are $u \in \mathcal{D}_{K,r/2}^{(M_p)}(\mathbb{R}^n)$ and $v \in \mathcal{D}_K^{(M_p)}(\mathbb{R}^n)$ such that*

$$P_r(D)u = \delta + v. \tag{6.3}$$

Finally, we shall also need the following lemma, proved in [119].

Lemma 6.3.3. (see [119]) *If the convolution $S \overset{1}{*} T$ of $S, T \in \mathcal{D}'(\mathbb{R}^n)$ exists and $P(D)$ is an ultradifferential operator of the class (M_p), then*

$$P(D)(S * T) = S * (P(D)T).$$

The main result of this section is the following theorem.

Theorem 6.3.1. *Let $S, T \in \mathcal{D}'^{(M_p)}$. The conditions $\left(\overset{j}{*}\right)$ for $j = 1, 2, 3, 4, 5, 6$ and $\left(\overset{j_k}{*}\right)$ for $j = 7, 8; k = 0, 1, 2$ listed in Definitions 6.2.1-6.2.8 of the existence of $\mathcal{D}'^{(M_p)}$–convolution of ultradistributions are equivalent and the respective convolutions are equal.*

Proof. The equivalence of the conditions $\left(\overset{1}{*}\right)$, $\left(\overset{5}{*}\right)$ and $\left(\overset{6}{*}\right)$ follow from Lemma 6.3.1 (see [119]) and the implications $\left(\overset{1}{*}\right) \Rightarrow \left(\overset{2}{*}\right)$, $\left(\overset{1}{*}\right) \Rightarrow \left(\overset{3}{*}\right)$ have been proved in [119]. We shall begin here by proving the two implications: $\left(\overset{4}{*}\right) \Rightarrow \left(\overset{1}{*}\right)$ and $\left(\overset{3}{*}\right) \Rightarrow \left(\overset{4}{*}\right)$. Since the proof of the implication $\left(\overset{2}{*}\right) \Rightarrow \left(\overset{4}{*}\right)$ is similar to that of the latter, we shall have already proved in this way the equivalence of conditions $\left(\overset{1}{*}\right) - \left(\overset{6}{*}\right)$.

Next we shall observe that the following implications hold: $\left(\overset{5}{*}\right) \Rightarrow \left(\overset{7_0}{*}\right)$, $\left(\overset{6}{*}\right) \Rightarrow \left(\overset{8_0}{*}\right)$, $\left(\overset{7_0}{*}\right) \Rightarrow \left(\overset{7_1}{*}\right)$, $\left(\overset{8_0}{*}\right) \Rightarrow \left(\overset{8_1}{*}\right)$, $\left(\overset{7_0}{*}\right) \Rightarrow \left(\overset{7_2}{*}\right)$, $\left(\overset{8_0}{*}\right) \Rightarrow \left(\overset{7_2}{*}\right)$, $\left(\overset{7_1}{*}\right) \Rightarrow \left(\overset{3}{*}\right)$, $\left(\overset{8_1}{*}\right) \Rightarrow \left(\overset{3}{*}\right)$, $\left(\overset{7_2}{*}\right) \Rightarrow \left(\overset{2}{*}\right)$, $\left(\overset{8_2}{*}\right) \Rightarrow \left(\overset{2}{*}\right)$. This will complete the proof of the equivalence of all the conditions listed in the assertion. $\left(\overset{1}{*}\right) - \left(\overset{6}{*}\right)$, $\left(\overset{7_0}{*}\right) - \left(\overset{7_2}{*}\right)$ and $\left(\overset{8_0}{*}\right) - \left(\overset{8_2}{*}\right)$.

Suppose that $S, T \in \mathcal{D}'^{(M_p)} = \mathcal{D}'^{(M_p)}(\mathbb{R}^n)$.

Proof of $\left(\overset{4}{*}\right) \Rightarrow \left(\overset{1}{*}\right)$. Fix $\phi \in \mathcal{D}^{(M_p)} = \mathcal{D}^{(M_p)}(\mathbb{R}^n)$. The mapping

$$\mathcal{D}^{(M_p)} \ni \psi \mapsto (S * \phi)(\tilde{T} * \psi) \in \mathcal{D}'^{(M_p)} \tag{6.4}$$

is continuous. Since the space L^1 has property $(C_{(M_p)})$, it follows that the mapping $\mathcal{D}^{(M_p)} \to L^1(\mathbb{R}^n)$ given by (6.4) is continuous, which implies the continuity of the mapping

$$\mathbb{R}^n \ni x \mapsto [(S * \phi)(x)(\tilde{T} * \psi)(x - \cdot)] \in L^1(\mathbb{R}^n),$$

where $\phi, \psi \in \mathcal{D}^{(M_p)}$ are fixed. Hence, for a fixed $\omega \in \mathcal{D}^{(M_p)}$, the function given by

$$f(x) := \int_{\mathbb{R}^n} |(S * \phi)(x)(\tilde{T} * \psi)(x - y)\omega(y)| \, dy, \qquad x \in \mathbb{R}^n,$$

belongs to $L^1(\mathbb{R}^n)$. The Fubini theorem implies that, for arbitrary $\phi, \psi, \omega \in \mathcal{D}^{(M_p)}$, the function

$$g(x, y) := (S * \phi)(x)(\tilde{T} * \psi)(x - y)\omega(y), \quad (x, y) \in \mathbb{R}^{2n},$$

belongs to $L^1(\mathbb{R}^{2n})$. Changing variables, we conclude that

$$[(S * \phi) \otimes (T * \tilde{\psi})] \omega^\triangle \in L^1(\mathbb{R}^{2n})$$

for arbitrary $\phi, \psi, \omega \in \mathcal{D}^{(M_p)}$.

Now let Q and K be compact neighborhoods of zero in $\mathbb{R}^n$ such that Q is a subset of the interior of K. Without loss of generality we may assume that $K = -K$ and $Q = -Q$. The mapping

$$\mathcal{D}_K^{(M_p)} \times \mathcal{D}_K^{(M_p)} \times \mathcal{D}_K^{(M_p)} \ni (\phi, \psi, \omega) \mapsto [(S * \phi) \otimes (T * \tilde{\psi})] \omega^\triangle \in L^1(\mathbb{R}^{2n}),$$

is separately continuous and thus, since $\mathcal{D}_K^{(M_p)}$ is a Fréchet space, continuous. Hence there exist $r > 0$ and $C > 0$ such that

$$\| [(S * \phi) \otimes (T * \tilde{\psi})] \, \omega^\triangle \|_{L^1(\mathbb{R}^{2n})}$$
$$\leq C \left[\| \phi \|_{\mathcal{D}_{K,r}^{(M_p)}} + \| \psi \|_{\mathcal{D}_{K,r}^{(M_p)}} + \| \omega \|_{\mathcal{D}_{K,r}^{(M_p)}} \right] \tag{6.5}$$

for $\phi, \psi, \omega \in \mathcal{D}_K^{(M_p)}$. Let $\phi, \psi, \omega \in \mathcal{D}_{Q,r}^{(M_p)}$. Since $\mathcal{D}^{(M_p)}$ is permitted, we can find sequences (ϕ_j), (ψ_j) and (ω_j) of elements of $\mathcal{D}_K^{(M_p)}$ such that $\phi_j \to \phi$, $\psi_j \to \psi$ and $\omega_j \to \omega$ in $\mathcal{D}_{K,r}^{(M_p)}$. Notice that

$$[(S * \phi_j) \otimes (T * \tilde{\psi}_j)] (\omega_j)^\triangle \rightarrow [(S * \phi) \otimes (T * \tilde{\psi})] \omega^\triangle$$

as $j \to \infty$ in $\mathcal{D}'(\mathbb{R}^{2n})$ (see [82]) and in $L^1(\mathbb{R}^{2n})$. Hence, replacing in (6.5) ϕ, ψ, ω by ϕ_j, ψ_j, ω_j, respectively, and passing to the limit, we get

$$\| [(S * \phi) \otimes (T * \tilde{\psi})] \omega^\triangle \|_{L^1(\mathbb{R}^{2n})}$$
$$\leq C \left[\| \phi \|_{\mathcal{D}_{K,r}^{(M_p)}} + \| \psi \|_{\mathcal{D}_{K,r}^{(M_p)}} + \| \omega \|_{\mathcal{D}_{K,r}^{(M_p)}} \right]$$
$$= C \left[\| \phi \|_{\mathcal{D}_{Q,r}^{(M_p)}} + \| \psi \|_{\mathcal{D}_{Q,r}^{(M_p)}} + \| \omega \|_{\mathcal{D}_{Q,r}^{(M_p)}} \right] < \infty \tag{6.6}$$

for arbitrary $\phi, \psi, \omega \in \mathcal{D}_{K,r}^{(M_p)}$.

Lemma 6.3.2 implies that there exist $u \in \mathcal{D}_{Q,r}^{(M_p)}$ and $v \in \mathcal{D}_Q^{(M_p)}$ such that

$$\delta = P_{2r}(D)u + v. \tag{6.7}$$

Since $P_{2r}(z)$ is an ultrapolynomial of class (M_p), we have

$$P_{2r}(z) = \sum_{\alpha \in \mathbb{N}^n} c_\alpha z^\alpha, \tag{6.8}$$

where c_α satisfy inequality (2.14). Thus, for each $\varphi \in \mathcal{D}_K^{(M_p)}(\mathbb{R}^n)$, we have

$$
\begin{aligned}
(S \otimes T)\,\varphi^\triangle \\
&= [(S * P_{2r}(D)u + S * v) \otimes (T * P_{2r}(D)u + T * v)]\,\varphi^\triangle \\
&= [(S * P_{2r}(D)u) \otimes (T * P_{2r}(D)u) + (S * v) \otimes (T * P_{2r}(D)u) \\
&\quad + (S * P_{2r}(D)u) \otimes (T * v) + (S * v) \otimes (T * v)]\,\varphi^\triangle.
\end{aligned} \tag{6.9}
$$

By (6.6), it follows that $[(S * u) \otimes (T * u)]\,\varphi^\triangle$, $[(S * v) \otimes (T * u)]\,\varphi^\triangle$, $[(S * u) \otimes (T * v)]\,\varphi^\triangle$ and $[(S * v) \otimes (T * v)]\,\varphi^\triangle$ belong to $L^1(\mathbb{R}^{2n})$. We shall prove that $[(S * P_{2r}(D)u) \otimes (T * P_{2r}(D)u)]\,\varphi^\triangle$ belongs to $\mathcal{D}'^{(M_p)}_{L^1}(\mathbb{R}^{2n})$. In a similar manner one can prove that $[(S * v) \otimes (T * P_{2r}(D)u)]\,\varphi^\triangle$ and $[(S * P_{2r}(D)u) \otimes (T * v)]\,\varphi^\triangle$ belong to $\mathcal{D}'^{(M_p)}_{L^1}(\mathbb{R}^{2n})$. This will mean that all the terms on the right side of (6.9) are elements of $\mathcal{D}'^{(M_p)}_{L^1}$, i.e., the proof of the implication $\overset{4}{(*)} \Rightarrow \overset{1}{(*)}$ will be completed.

For simplicity, let us denote

$$V_w := (S * w) \otimes (T * w); \quad U_w := V_{P_{2r}(D)w}$$

for each $w \in \mathcal{D}_K^{(M_p)}(\mathbb{R}^n)$. By (6.8) and Lemma 6.3.3, we obtain

$$
\begin{aligned}
(U_u\,\varphi^\triangle)\,(x,y) &= [(S * P_{2r}(D)u)_x \otimes (T * P_{2r}(D)u)_y]\,\varphi(x+y) \\
&= [(P_{2r}(D_x)P_{2r}(D_y))\,((S * u)_x \otimes (T * u)_y)]\,\varphi(x,y) \\
&= \sum_{\alpha,\beta \in \mathbb{N}^n} \frac{c_\alpha c_\beta}{\imath^{\alpha+\beta}} \sum_{\substack{0 \le k \le \alpha \\ 0 \le l \le \beta}} (-1)^{k+l}\binom{\alpha}{k}\binom{\beta}{l}[V_u\,(\varphi^\triangle)^{(k+l)}]^{(\alpha-k,\beta-l)}(x,y).
\end{aligned}
$$

Therefore, for an arbitrary $\omega \in \mathcal{D}^{(M_p)}(\mathbb{R}^{2n})$, we have

$$
\begin{aligned}
\langle U_u\,\varphi^\triangle, \omega \rangle \\
&= \sum_{\alpha,\beta \in \mathbb{N}_0^n} (-\imath)^{\alpha+\beta} c_\alpha c_\beta \sum_{\substack{0 \le k \le \alpha \\ 0 \le l \le \beta}} \binom{\alpha}{k}\binom{\beta}{l}\langle V_u\,(\varphi^{(k+l)})^\triangle, \omega^{(\alpha-k,\beta-l)} \rangle.
\end{aligned}
$$

$$\tag{6.10}$$

Since, for a fixed $u \in \mathcal{D}^{(M_p)}(\mathbb{R}^n)$, the mapping

$$\mathcal{D}^{(M_p)}(\mathbb{R}^n) \ni \vartheta \mapsto V_u \vartheta^\triangle \in L^1(\mathbb{R}^{2n})$$

is continuous and, for an arbitrary constant $B > 0$, the set $\{B^\alpha \varphi^{(\alpha)}/M_\alpha : \alpha \in \mathbb{N}_0^n\}$ is bounded in $\mathcal{D}^{(M_p)}(\mathbb{R}^n)$, we infer that the set

$$\left\{ \frac{B^{k+l}}{M_{k+l}} V_u (\varphi^{(k+l)})^\triangle : k, l \in \mathbb{N}_0^n \right\}$$

is bounded in $L^1(\mathbb{R}^{2n})$ for an arbitrary constant $B > 0$. By $(M.2)$, there exist constants $h > 0$ and $B_0 > 0$ such that

$$\left| \left\langle \frac{B^{k+l}}{M_k M_l} V_u (\varphi^\triangle)^{(k+l)}, \omega \right\rangle \right| \leq B_0 H^{k+l} \sum_{r \in \mathbb{N}^{2n}} \frac{h^r}{M_r} \| \omega^{(r)} \|_{L^\infty(\mathbb{R}^{2n})} \quad (6.11)$$

for arbitrary $\omega \in \mathcal{D}^{(M_p)}(\mathbb{R}^{2n})$ and $k, l \in \mathbb{N}_0^n$.

Note that $(M.2)$ and (2.2) imply that there are constants $B_1, H > 1$ such that

$$M_k M_l M_{\alpha+\beta+p+q-k-l} \leq B_1 H^{2(\alpha+\beta+p+q-k-l)} M_\alpha M_\beta M_{p+q} \quad (6.12)$$

for arbitrary $p, q, \alpha, \beta, k, l \in \mathbb{N}_0^n$ such that $0 \leq k \leq \alpha, 0 \leq l \leq \beta$. By (6.10),

$$|\langle U_u \varphi^\triangle, \omega \rangle|$$
$$\leq \sum_{\alpha, \beta \in \mathbb{N}_0^n} \sum_{\substack{0 \leq k \leq \alpha \\ 0 \leq l \leq \beta}} \binom{\alpha}{k} \binom{\beta}{l} |c_\alpha c_\beta| \, |\langle V_u (\varphi^{(k+l)})^\triangle, \omega^{(\alpha-k, \beta-l)} \rangle|.$$

$$(6.13)$$

Applying (2.14), (6.11), and (6.12), setting $\lambda_r := \| \omega^{(r)} \|_{L^\infty(\mathbb{R}^{2n})}$ for $r \in \mathbb{N}^{2n}$ and denoting by $\mathcal{C}$ a positive constant, not necessarily the same at each occurrence, we have

$$|c_\alpha c_\beta| |\langle V_u (\varphi^{(k+l)})^\triangle, \omega^{(\alpha-k, \beta-l)} \rangle|$$
$$\leq \mathcal{C} \frac{A^{\alpha+\beta}}{M_\alpha M_\beta} |\langle V_u(\varphi^{(k+l)})^\triangle, \omega^{(\alpha-k, \beta-l)} \rangle|$$
$$\leq \mathcal{C} \frac{A^{\alpha+\beta-k-l} M_k M_l}{(4H)^{k+l} M_\alpha M_\beta} |\langle \frac{(4AH)^{k+l}}{M_k M_l} V_u (\varphi^{(k+l)})^\triangle, \omega^{(\alpha-k, \beta-l)} \rangle|$$
$$\leq \mathcal{C} \frac{(AH^2 h)^{\alpha+\beta+p+q-k-l}}{4^{k+l} M_{\alpha+\beta+p+q-k-l}} \lambda_{\alpha-k+p, \beta-l+q}$$

for arbitrary $p, q, \alpha, \beta, k, l \in \mathbb{N}_0^n$ such that $0 \leq k \leq \alpha, 0 \leq l \leq \beta$.

Hence, by (6.13),

$$|\langle U_u \varphi^\triangle, \omega \rangle|$$
$$\leq \mathcal{C} \sum_{\alpha, \beta, p, q \in \mathbb{N}_0^n} \sum_{\substack{0 \leq k \leq \alpha \\ 0 \leq l \leq \beta}} \binom{\alpha}{k} \binom{\beta}{l} 4^{-(\alpha+\beta+p+q)} \sum_{r \in \mathbb{N}^{2n}} \frac{(4AH^2 h)^r}{M_r} \lambda_r$$
$$< \infty.$$

But this implies that

$$U_u\varphi^\triangle \in \mathcal{D}'^{(M_p)}_{L^1}(\mathbb{R}^{2n}),$$

as desired.

Proof of $\overset{3}{(*)} \Rightarrow \overset{4}{(*)}$. The mappings

$$\mathcal{D}^{(M_p)} \ni \psi \mapsto S(\tilde{T} * \psi) \in \mathcal{D}'^{(M_p)}_{L^1}$$

and

$$\mathbb{R}^n \times \mathcal{D}'^{(M_p)}_{L^1} \ni (y, U) \mapsto U(\cdot - y) \in \mathcal{D}'^{(M_p)}_{L^1}$$

are continuous. This implies that, for every $\psi \in \mathcal{D}^{(M_p)}$, the mapping

$$\mathbb{R}^n \ni y \mapsto S(\cdot - y)(\tilde{T} * \psi)(\cdot) \in \mathcal{D}'^{(M_p)}_{L^1}$$

is also continuous. Therefore (see [10], p. 81)

$$V_{\varphi,\psi} \in \mathcal{D}'^{(M_p)}_{L^1}(\mathbb{R}^n), \qquad \varphi, \psi \in \mathcal{D}^{(M_p)}(\mathbb{R}^n),$$

where

$$V_{\varphi,\psi}(\cdot) := \int \varphi(y)[S(\cdot - y)(\tilde{T} * \psi)](\cdot)\, dy, \qquad \varphi, \psi \in \mathcal{D}^{(M_p)}(\mathbb{R}^n),$$

and the integral $\int_{\mathbb{R}^n} R(x, y) dy$ of an ultradistribution $R \in \mathcal{D}'^{(M_p)}_{L^1}(\mathbb{R}^{2n})$ is meant as the ultradistribution in $\mathcal{D}'^{(M_p)}(\mathbb{R}^n)$ defined by

$$\left\langle \int_{\mathbb{R}^n} R(x, y)\, dy,\, \omega(x) \right\rangle := \int_{\mathbb{R}^n} \langle R(x, y),\, \omega(x) \rangle\, dy, \qquad \omega \in \mathcal{D}^{(M_p)}(\mathbb{R}^n).$$

For an arbitrary $\omega \in \mathcal{D}^{(M_p)}$, we have

$$\left\langle \int \varphi(y) S(x - y)(\tilde{T} * \psi)(x)\, dy, \omega(x) \right\rangle$$

$$= \int \langle \varphi(y) S(x - y)(\tilde{T} * \psi)(x), \omega(x) \rangle\, dy$$

$$= \int \langle \varphi(y) S(x - y), (\tilde{T} * \psi)(x)\omega(x) \rangle\, dy$$

$$= \langle S * \varphi, (\tilde{T} * \psi)\omega \rangle = \langle (S * \varphi)(\tilde{T} * \psi), \omega \rangle.$$

Consequently, $(S * \varphi)(\tilde{T} * \psi) \in \mathcal{D}'^{(M_p)}_{L^1}$ for $\varphi, \psi \in \mathcal{D}^{(M_p)}$. Theorem 3 in [121] now implies that $[(S * \varphi)(\tilde{T} * \psi)] * \omega \in L^1$ for $\varphi, \psi, \omega \in \mathcal{D}^{(M_p)}$. On the other hand, it follows by Lemma 6.3.2 that

$$(S * \varphi)(\tilde{T} * \psi) = [(S * \varphi)(\tilde{T} * \psi)] * P_r(D)u + [(S * \varphi)(\tilde{T} * \psi)] * v,$$

and similar arguments to that in the proof of implication $\left(\overset{6}{*}\right) \Rightarrow \left(\overset{1}{*}\right)$ yield $(S * \varphi)\,(\tilde{T} * \psi) \in L^1$ for arbitrary $\varphi, \psi \in \mathcal{D}^{(M_p)}$.

Proof of $\left(\overset{5}{*}\right) \Rightarrow \left(\overset{7_0}{*}\right)$ *and* $\left(\overset{6}{*}\right) \Rightarrow \left(\overset{8_0}{*}\right)$. Since, for arbitrary $\omega \in \mathcal{D}^{(M_p)}$ and unit-sequences $(\eta_j), (\tilde{\eta}_j) \in \mathbf{1}^{(M_p)}(\mathbb{R}^n)$, we have

$$\langle (\eta_j S) * (\tilde{\eta}_j T), \omega \rangle = \langle (\eta_j S)_x \otimes (\tilde{\eta}_j T)_y, \omega(x+y) \rangle$$
$$= \langle (S_x \otimes T_y)\,\omega(x+y), \eta_j(x)\tilde{\eta}_j(y) \rangle,$$

it follows from $\left(\overset{6}{*}\right)$ that the limit (a) exists for all strong unit-sequences $(\eta_j), (\tilde{\eta}_j) \in \bar{\mathbf{1}}^{(M_p)}(\mathbb{R}^n)$, i.e., (a) is valid and $S \overset{7_0}{*} T = S \overset{5}{*} T$. Similarly, $\left(\overset{6}{*}\right)$ implies $\left(\overset{8_0}{*}\right)$ and $S \overset{8_0}{*} T = S \overset{6}{*} T$.

Proof of $\left(\overset{7_0}{*}\right) \Rightarrow \left(\overset{7_1}{*}\right)$, $\left(\overset{8_0}{*}\right) \Rightarrow \left(\overset{8_1}{*}\right)$, $\left(\overset{7_0}{*}\right) \Rightarrow \left(\overset{7_2}{*}\right)$ *and* $\left(\overset{8_0}{*}\right) \Rightarrow \left(\overset{7_2}{*}\right)$. Note that $\left(\overset{7_0}{*}\right)$ implies that for arbitrary strong unit-sequences $(\eta_j), (\tilde{\eta}_j) \in \bar{\mathbf{1}}^{(M_p)}(\mathbb{R}^n)$ and $\varphi \in \mathcal{D}^{(M_p)}$ we have

$$\lim_{p,q \to \infty} \langle (\eta_p S) * (\tilde{\eta}_q T), \varphi \rangle = \langle S \overset{7_2}{*} T, \varphi \rangle. \tag{6.14}$$

In fact, if $\left(\overset{7_1}{*}\right)$ were not true, there would exist $\varphi \in \mathcal{D}^{(M_p)}$, $\varepsilon > 0$ and increasing sequences (p_j) and (q_j) of positive integers such that

$$|\langle (\eta_{p_j}) S * (\tilde{\eta}_{q_j} T), \varphi \rangle - \langle S \overset{7_2}{*} T, \varphi \rangle| > \varepsilon.$$

But since (η_{p_j}) and $(\tilde{\eta}_{q_j})$ are again strong unit-sequences from $\bar{\mathbf{1}}^{(M_p)}(\mathbb{R}^n)$, the above inequality would contradict $\left(\overset{7_0}{*}\right)$. Notice that (6.14) yields

$$\langle S \overset{7_2}{*} T, \varphi \rangle = \lim_{p \to \infty} \lim_{q \to \infty} \langle (\eta_p S) * (\tilde{\eta}_q T), \varphi \rangle = \lim_{p \to \infty} \langle (\eta_p) * T, \varphi \rangle,$$

which implies condition $\left(\overset{7_1}{*}\right)$ and the identity $S \overset{7_2}{*} T = S \overset{7_0}{*} T$. In the same way one proves that $\left(\overset{8_0}{*}\right)$ implies $\left(\overset{8_1}{*}\right)$ and $S \overset{8_2}{*} T = S \overset{8_0}{*} T$. The remaining two implications and the respective identities of the convolutions follow in view of symmetry.

Proof of $\left(\overset{7_1}{*}\right) \Rightarrow \left(\overset{3}{*}\right)$, $\left(\overset{8_1}{*}\right) \Rightarrow \left(\overset{3}{*}\right)$, $\left(\overset{7_2}{*}\right) \Rightarrow \left(\overset{2}{*}\right)$ *and* $\left(\overset{8_2}{*}\right) \Rightarrow \left(\overset{2}{*}\right)$. Since

$$\langle (\eta_j S) * T, \varphi \rangle = \langle \eta_j S, \varphi * \tilde{T} \rangle = \langle S(\tilde{T} * \varphi), \eta_j \rangle,$$

we infer from $\left(\overset{7_1}{*}\right)$ or $\left(\overset{8_1}{*}\right)$ that $S(\tilde{T} * \varphi) \in \mathcal{D}'^{(M_p)}_{L^1}$ for $\varphi \in \mathcal{D}^{(M_p)}$. Consequently, $\left(\overset{3}{*}\right)$ holds and $S \overset{7_0}{*} T = S \overset{3}{*} T$ and $S \overset{3}{*} T = S \overset{7_1}{*} T$ in the respective cases. The remaining implications and the respective identities hold true, due to the symmetry argument.

Thus the equivalence of all conditions $\left(\overset{1}{*}\right) - \left(\overset{6}{*}\right)$, $\left(\overset{7_0}{*}\right) - \left(\overset{7_2}{*}\right)$ and $\left(\overset{8_0}{*}\right) - \left(\overset{8_2}{*}\right)$ is proved.

By standard arguments one can show that the remaining identities are satisfied:

$$S \overset{1}{*} T = S \overset{2}{*} T = S \overset{3}{*} T = S \overset{4}{*} T = S \overset{5}{*} T = S \overset{6}{*} T,$$

which completes the proof of Theorem 6.3.1. $\square$

6.4 Definitions of $\mathcal{S}'^{(M_p)}$−convolution

In this section, we introduce the notion of the $\mathcal{S}'^{(M_p)}$−convolution of two tempered ultradistributions, which is an analogue of the notion of $\mathcal{S}'$−convolution of tempered distributions. We formulate, similarly to the case of the $\mathcal{D}'^{(M_p)}$−convolution defined in Section 6.2, several definitions of the $\mathcal{S}'^{(M_p)}$−convolution of tempered ultradistributions and give the respective conditions for its existence.

The definitions of the $\mathcal{S}'^{(M_p)}$−convolution we present here are quite analogous to the definitions of the $\mathcal{D}'^{(M_p)}$−convolution of ultradistributions given in Section 6.2 and, of course, analogous to the definitions of the $\mathcal{D}'$−convolution of distributions and to the definitions of the $\mathcal{S}'$−convolution of tempered distributions; see [132] (Definition 6.4.1), [36] (Definitions 6.4.2, 6.4.3 and 6.4.4), [151] (Definition 6.4.5), [45] (Definition 6.2.6), [69] (Definitions 6.4.7 and 6.4.8). The only difference is that Definition 6.4.4 consists of three equivalent versions.

The $\mathcal{S}'^{(M_p)}$−convolution introduced in Definitions 6.4.1-6.4.4 will be denoted, in contrast to the symbol $*$ of the $\mathcal{D}'^{(M_p)}$−convolution, by the symbol $\overset{k}{\star}$ with the corresponding index; more precisely: the symbol $\overset{k}{\star}$ will mean the $\mathcal{S}'^{(M_p)}$−convolution introduced in Definition 6.4.k for $k = 1, 2, 3, 4_0, 4_1, 4_2,$ $5, 6, 7_0, 7_1, 7_2, 8_0, 8_1, 8_2$. In the next section, we will prove that all these definitions are equivalent; and for the $\mathcal{S}'^{(M_p)}$−convolution of tempered ultradistributions, the common symbol $\star$ will be used further on.

In all definitions of the $\mathcal{S}'^{(M_p)}$−convolution given below S and T will mean two fixed tempered ultradistributions, i.e. elements of the space $\mathcal{S}'^{(M_p)} := \mathcal{S}'^{(M_p)}(\mathbb{R}^n)$.

Definition 6.4.1. The convolution $S \overset{1}{\star} T$ is defined by the formula

$$\langle S \overset{1}{\star} T, \varphi \rangle := \langle (S \otimes T) \varphi^{\triangle}, 1 \rangle, \qquad \varphi \in \mathcal{S}^{(M_p)},$$

whenever

$(\overset{1}{\star})$ $\qquad (S \otimes T) \varphi^{\triangle} \in \mathcal{D}'^{(M_p)}_{L^1}(\mathbb{R}^{2n}) \qquad$ for $\varphi \in \mathcal{S}^{(M_p)}$.

Definition 6.4.2. The convolution $S \overset{2}{\star} T$ is defined by the formula

$$\langle S \overset{2}{\star} T, \varphi \rangle := \langle (S * \varphi)\tilde{T}, 1 \rangle, \qquad \varphi \in \mathcal{S}^{(M_p)},$$

whenever

$(\overset{2}{\star})$ $\qquad (S * \varphi)\tilde{T} \in \mathcal{D}'^{(M_p)}_{L^1} \qquad$ for $\varphi \in \mathcal{S}^{(M_p)}$.

Definition 6.4.3. The convolution $S \overset{3}{\star} T$ is defined by the formula

$$\langle S \overset{3}{\star} T, \varphi \rangle := \langle S(\tilde{T} * \varphi), 1 \rangle, \qquad \varphi \in \mathcal{S}^{(M_p)},$$

whenever

$(\overset{3}{\star})$ $\qquad S(\tilde{T} * \varphi) \in \mathcal{D}'^{(M_p)}_{L^1} \qquad$ for $\varphi \in \mathcal{S}^{(M_p)}$.

Definition 6.4.4.

a) The convolution $S \overset{4_0}{\star} T$ is defined by the formula

$$\langle (S \overset{4_0}{\star} T) * \varphi, \psi \rangle := \langle (S * \varphi)(\tilde{T} * \psi), 1 \rangle, \qquad \varphi \in \mathcal{S}^{(M_p)}, \ \psi \in \mathcal{S}^{(M_p)},$$

whenever

$(\overset{4_0}{\star})$ $\qquad (S * \varphi)(\tilde{T} * \psi) \in L^1 \qquad$ for $\varphi \in \mathcal{S}^{(M_p)}, \ \psi \in \mathcal{S}^{(M_p)}$.

b) The convolution $S \overset{4_1}{\star} T$ is defined by the formula

$$\langle (S \overset{4_1}{\star} T) * \varphi, \psi \rangle := \langle (S * \varphi)(\tilde{T} * \psi), 1 \rangle, \qquad \varphi \in \mathcal{S}^{(M_p)}, \ \psi \in \mathcal{D}^{(M_p)},$$

whenever

$(\overset{4_1}{\star})$ $\qquad (S * \varphi)(\tilde{T} * \psi) \in L^1 \qquad$ for $\varphi \in \mathcal{S}^{(M_p)}, \ \psi \in \mathcal{D}^{(M_p)}$.

c) The convolution $S \overset{4_2}{\star} T$ is defined by the formula

$$\langle (S \overset{4_2}{\star} T) * \varphi, \psi \rangle := \langle (S * \varphi)(\tilde{T} * \psi), 1 \rangle, \qquad \varphi \in \mathcal{D}^{(M_p)}, \ \psi \in \mathcal{S}^{(M_p)},$$

whenever

$(\overset{4_2}{\star})$ $\qquad (S * \varphi)(\tilde{T} * \psi) \in L^1 \qquad$ for $\varphi \in \mathcal{D}^{(M_p)}, \ \psi \in \mathcal{S}^{(M_p)}$.

As in Section 6.2 for the case of $\mathcal{D}'^{(M_p)}$−convolution, one can prove that the mapping

$$\mathcal{S}^{(M_p)}(\mathbb{R}^n) \ni \varphi \mapsto A_\varphi \in \mathcal{S}'^{(M_p)}(\mathbb{R}^n),$$

where

$$\langle A_\varphi, \psi \rangle := \int_{\mathbb{R}^n} (S * \varphi)(x)(\tilde{T} * \psi)(x)dx, \qquad \psi \in \mathcal{S}^{(M_p)},$$

is a linear, continuous and translation invariant mapping from $\mathcal{D}^{(M_p)}(\mathbb{R}^n)$ into $\mathcal{E}^{(M_p)}(\mathbb{R}^n)$. This implies that there exists a unique ultradistribution $V \in \mathcal{S}'^{(M_p)}(\mathbb{R}^n)$ such that $V * \varphi = A_\varphi$ for $\varphi \in \mathcal{S}^{(M_p)}(\mathbb{R}^n)$. Consequently, Definition 6.4.4, a) is consistent and $S \overset{4_0}{\star} T = V$. A similar reasoning can be applied to the variants b) and c) of Definition 6.4.4.

Definition 6.4.5. The convolution $S \overset{5}{\star} T$ is defined by the formula

$$\langle S \overset{5}{\star} T, \varphi \rangle := \lim_{j \to \infty} \langle S \otimes T, \eta_j \varphi^\triangle \rangle, \qquad \varphi \in \mathcal{S}^{(M_p)},$$

whenever

$\left(\overset{5}{\star} \right)$ the limit exists for all $\varphi \in \mathcal{S}^{(M_p)}$ and for all $(\eta_j) \in \bar{\mathbf{1}}^{(M_p)}(\mathbb{R}^{2n})$.

Definition 6.4.6. The convolution $S \overset{6}{\star} T$ is defined by the formula

$$\langle S \overset{6}{\star} T, \varphi \rangle := \lim_{j \to \infty} \langle S \otimes T, \eta_j \varphi^\triangle \rangle, \qquad \varphi \in \mathcal{S}^{(M_p)},$$

whenever

$\left(\overset{6}{\star} \right)$ the limit exists for all $\varphi \in \mathcal{S}^{(M_p)}$ and for all $(\eta_j) \in \mathbf{1}^{(M_p)}(\mathbb{R}^{2n})$.

Definition 6.4.7.
   a) The convolution $S \overset{7_0}{\star} T$ is defined by the formula

(b_0) $$S \overset{7_0}{\star} T := \lim_{j \to \infty} (\eta_j S) * (\tilde{\eta}_j T) \qquad \text{in } \mathcal{S}'^{(M_p)},$$

whenever

$\left(\overset{7_0}{\star} \right)$ the limit in (b_0) exists for all $(\eta_j) \in \bar{\mathbf{1}}^{(M_p)}$ and for all $(\tilde{\eta}_j) \in \bar{\mathbf{1}}^{(M_p)}$.

b) The convolution $S \overset{7_1}{\star} T$ is defined by the formula

(b_1) $$S \overset{7_1}{\star} T := \lim_{j \to \infty} (\eta_j S) * T \qquad \text{in } \mathcal{S}'^{(M_p)},$$

whenever

$\left(\overset{7_1}{\star} \right)$ the limit in (b_1) exists for all $(\eta_j) \in \bar{\mathbf{1}}^{(M_p)}$.

c) The convolution $S \overset{7_2}{\star} T$ is defined by the formula

(b_2) $$S \overset{7_2}{\star} T := \lim_{j \to \infty} S * (\eta_j T) \qquad \text{in } \mathcal{S}'^{(M_p)},$$

whenever

$\left(\overset{7_2}{\star} \right)$ the limit in (b_2) exists for all $(\eta_j) \in \bar{\mathbf{1}}^{(M_p)}$.

Definition 6.4.8. For every $k = 0, 1, 2$, the convolution $S \overset{8_k}{\star} T$ is defined by formula (b_k), whenever

$\left(\overset{8_k}{\star} \right)$ the limit in (b_k) exists for all $(\eta_j) \in \mathbf{1}^{(M_p)}$ *(and for all $(\tilde{\eta}_j) \in$* $\mathbf{1}^{(M_p)}$).

6.5   Equivalence of definitions of $\mathcal{S}'^{(M_p)}$−convolution

The purpose of this section is to prove Theorem 6.5.1 on equivalence of all conditions for the existence of $\mathcal{S}'^{(M_p)}$−convolution of ultradistributions listed in Definitions 6.4.1-6.4.8.

As in the case of Theorem 6.3.1, the proof of Theorem 6.5.1 applies methods shown in [135] and [69] (in particular, we need Lemma 6.3.1 as before). But also new techniques are used, because traditional ones fail for tempered ultradistributions.

A crucial role in the proof of the theorem will be played by the following assertion as proved both for the spaces of ultradistributions of Beurling and Roumieu type in [120], Proposition 4:

Lemma 6.5.1. (see [120], [146]) *The space $\mathcal{O}_M^{(M_p)}$ of multipliers of $\mathcal{S}'^{(M_p)}$ is nuclear.*

Now let us formulate and prove the equivalence of Definitions 6.4.1-6.4.8 of the $\mathcal{S}'^{(M_p)}$–convolution of ultradistributions.

Theorem 6.5.1. *Let* $S, T \in \mathcal{S}'^{(M_p)}$. *The conditions* $\left(\overset{j}{\star}\right)$ *for* $j = 1, 2, 3, 5, 6$ *and* $\left(\overset{j_k}{\star}\right)$ *for* $j = 4, 7, 8; k = 0, 1, 2$ *listed in Definitions 6.4.1-6.4.8 of the existence of* $\mathcal{S}'^{(M_p)}$ *– convolution of ultradistributions are equivalent and the respective convolutions are equal.*

Proof.

Proof of $\left(\overset{1}{\star}\right) \Leftrightarrow \left(\overset{5}{\star}\right) \Leftrightarrow \left(\overset{6}{\star}\right)$. The equivalence of conditions $\left(\overset{1}{\star}\right)$, $\left(\overset{5}{\star}\right)$ and $\left(\overset{6}{\star}\right)$ as well as the equalities $S \overset{1}{\star} T = S \overset{5}{\star} T = S \overset{6}{\star} T$ follow directly from Lemma 6.3.1.

Proof of $\left(\overset{1}{\star}\right) \Rightarrow \left(\overset{3}{\star}\right)$. Clearly,

$$\langle S \overset{1}{\star} T, \varphi \rangle = \langle S_x \otimes T_y, \varphi^\triangle \rangle = \lim_{k \to \infty} \langle S \otimes T, [\eta_k \otimes 1]\varphi^\triangle \rangle \tag{6.15}$$

for arbitrary $(\eta_k) \in \overline{\mathbf{1}}^{(M_p)}(\mathbb{R}^n)$, $\varphi \in \mathcal{D}^{(M_p)}(\mathbb{R}^n)$ and the constant function 1 on $\mathbb{R}^n$. On the other hand,

$$\langle S \otimes T, [\eta_k \otimes 1]\varphi^\triangle \rangle = \lim_{l \to \infty} \langle S \otimes T, [\eta_k \otimes \eta_l]\varphi^\triangle \rangle = \langle (\tilde{S} * \varphi)T, \eta_k \rangle.$$

By (6.15), the limit $\lim_{k \to \infty} \langle (\tilde{S} * \vartheta)T, \eta_k \rangle$ exists for all $(\eta_k) \in \overline{\mathbf{1}}$, i.e., $\left(\overset{3}{\star}\right)$ follows from the equivalence $\left(\overset{5}{\star}\right) \Leftrightarrow \left(\overset{1}{\star}\right)$ and $S \overset{3}{\star} T = S \overset{1}{\star} T$.

Proof of $\left(\overset{3}{\star}\right) \Rightarrow \left(\overset{4_2}{\star}\right)$. The proof of this implication and the identity $S \overset{3}{\star} T = S \overset{4}{\star} T$ proceeds in a similar manner as the proof of the respective implication in Theorem 6.5.1, which was proved in [70].

Proof of $\left(\overset{4_2}{\star}\right) \Rightarrow \left(\overset{1}{\star}\right)$. For arbitrary $\theta, \vartheta \in \mathcal{D}^{(M_p)}$, we have

$$\langle [(S * \varphi) \otimes \psi] T^\triangle, \theta \otimes \vartheta \rangle = \langle (S * \varphi)[\tilde{T} * (\vartheta \psi)]\theta, 1 \rangle. \tag{6.16}$$

Obviously, the mappings

$$\Theta \colon \mathcal{O}_M^{(M_p)} \ni \vartheta \mapsto \vartheta \psi \in \mathcal{S}^{(M_p)};$$

$$\Gamma \colon \mathcal{S}^{(M_p)} \ni \psi \mapsto (S * \varphi)(\tilde{T} * \psi) \in \mathcal{D}'^{(M_p)}$$

are continuous. But $(S * \varphi)(\tilde{T} * \psi) \in L^1$, due to $\left(\overset{4_2}{\star}\right)$, and L^1 has the property $(C_{(M_p)})$. Thus both Γ meant as the mapping from $\mathcal{S}^{(M_p)}$ into L^1 and the composition mapping $\Gamma \circ \Theta : \mathcal{O}_M^{(M_p)} \mapsto \mathcal{D}'^{(M_p)}$ are continuous; so $((S * \varphi) \otimes \psi)T^{\triangle}$ can be extended as a continuous mapping to the space $\dot{\mathcal{B}}^{(M_p)} \hat{\otimes}_{\pi} \mathcal{O}_M^{(M_p)}$. But, by Lemma 6.5.1,

$$\dot{\mathcal{B}}^{(M_p)} \hat{\otimes}_{\pi} \mathcal{O}_M^{(M_p)} = \dot{\mathcal{B}}^{(M_p)} \hat{\otimes}_{\varepsilon} \mathcal{O}_M^{(M_p)} \supset \dot{\mathcal{B}}^{(M_p)} \hat{\otimes}_{\varepsilon} \dot{\mathcal{B}}^{(M_p)} \supset \dot{\mathcal{B}}^{(M_p)}(\mathbb{R}^{2n}), \quad (6.17)$$

where $\hat{\otimes}_{\pi}$ and $\hat{\otimes}_{\varepsilon}$ denote the tensor products with the projective topology and the Grothendieck topology respectively, and the corresponding embeddings are continuous.

Assume that $\eta_k \to \eta$ in the space $\dot{\mathcal{B}}^{(M_p)}(\mathbb{R}^{2n})$, where

$$\eta_k = \sum_{j=1}^{m_k} \theta_{j,k} \otimes \vartheta_{j,k}; \qquad \theta_{j,k}, \vartheta_{j,k} \in \dot{\mathcal{B}}^{(M_p)}(\mathbb{R}^n), \quad j = 1, \ldots, m_k.$$

By (6.17), $\eta_k \to \eta$ also in $\dot{\mathcal{B}}^{(M_p)} \hat{\otimes}_{\varepsilon} \dot{\mathcal{B}}^{(M_p)}$ and thus

$$\lim_{k \to \infty} \langle [(S * \varphi) \otimes \psi] T^{\triangle}, \eta_k \rangle = \langle [(S * \varphi) \otimes \psi] T^{\triangle}, \eta \rangle;$$

so $((S * \varphi) \otimes \psi) T^{\triangle} \in \mathcal{D}'^{(M_p)}_{L^1}$ and $S \overset{4}{\star} T = S \overset{1}{\star} T$.

Proof of $\left(\overset{1}{\star}\right) \Rightarrow \left(\overset{2}{\star}\right) \Rightarrow \left(\overset{4_1}{\star}\right) \Rightarrow \left(\overset{1}{\star}\right)$. These implications as well as the identities $S \overset{1}{\star} T = S \overset{2}{\star} T = S \overset{4}{\star} T$ follow, by symmetry, from the chain of the implications $\left(\overset{1}{\star}\right) \Rightarrow \left(\overset{3}{\star}\right) \Rightarrow \left(\overset{4_2}{\star}\right) \Rightarrow \left(\overset{1}{\star}\right)$ proved above.

Proof of $\left(\overset{4_0}{\star}\right) \Rightarrow \left(\overset{4_1}{\star}\right)$; $\left(\overset{4_0}{\star}\right) \Rightarrow \left(\overset{4_2}{\star}\right)$. The implications are evident.

Proof of $\left(\overset{4_1}{\star}\right) \Rightarrow \left(\overset{4_0}{\star}\right)$; $\left(\overset{4_2}{\star}\right) \Rightarrow \left(\overset{4_0}{\star}\right)$. By symmetry, it suffices to prove the second implication. Suppose that condition $\left(\overset{4_2}{\star}\right)$ is satisfied. Analysis similar to that in [135] shows that $(S * \varphi)(\tilde{T} * \psi) \in \mathcal{D}'^{(M_p)}_{L^1}$ for arbitrary $\varphi, \psi \in \mathcal{S}^{(M_p)}$. As in the proof of the implication $(D3) \Rightarrow (D4)$ in Theorem 6.5.1, given in [70], we make use of the continuity of the mapping

$$\mathcal{D}^{(M_p)} \ni \eta \mapsto (S * \varphi)(\tilde{T} * \psi)\phi \in L^1$$

to conclude that $(S * \varphi)(\tilde{T} * \psi) \in L^1$ for each $\varphi, \psi \in \mathcal{S}^{(M_p)}$.

Thus we have already proved the equivalence of conditions $\left(\overset{1}{\star}\right) - \left(\overset{5}{\star}\right)$, $\left(\overset{4_1}{\star}\right)$, $\left(\overset{4_2}{\star}\right)$ and $\left(\overset{6}{\star}\right)$.

Proof of $\binom{6}{*} \Rightarrow \binom{8_2}{*}$. Since, for given (η_k), $(\tilde{\eta}_k) \in \bar{\mathbf{1}}^{(M_p)}(\mathbb{R}^n)$, we have

$$\langle (\eta_k S) * (\tilde{\eta}_k T), \varphi \rangle = \langle (\eta_k S) \otimes (\tilde{\eta}_k T), \varphi^\triangle \rangle = \langle (S \otimes T)\varphi^\triangle, \vartheta_k \rangle,$$

where $\vartheta_k := \eta_k \otimes \tilde{\eta}_k$ and $(\vartheta_k) \in \mathbf{1}^{(M_p)}(\mathbb{R}^{2n})$, it follows from the equivalence $\binom{4_0}{*} \Leftrightarrow \binom{6}{*}$ that the limit in $\binom{7_2}{*}$ exists for all (η_k), $(\tilde{\eta}_k) \in \bar{\mathbf{1}}^{(M_p)}(\mathbb{R}^n)$ and $S \overset{8_0}{*} T = S \overset{5}{*} T$.

Proof of $\binom{8_2}{*} \Rightarrow \binom{8_1}{*}$; $\binom{8_2}{*} \Rightarrow \binom{8_0}{*}$; $\binom{7_2}{*} \Rightarrow \binom{7_1}{*}$; $\binom{7_2}{*} \Rightarrow \binom{7_0}{*}$. It suffices to show the first implication. It is worth noticing that, since the class $\mathbf{1}^{(M_p)}$ is closed under extracting subsequences, condition $\binom{8_2}{*}$ implies that the double limit $\lim_{p,q\to\infty}(\eta_p S) * (\tilde{\eta}_q T)$ exists and equals $\langle S \overset{8_0}{*} T, \varphi \rangle$ for arbitrary (η_k), $(\tilde{\eta}_k) \in \mathbf{1}^{(M_p)}$ and $\varphi \in \mathcal{D}^{(M_p)}$. Consequently,

$$\langle S \overset{8_0}{*} T, \varphi \rangle = \lim_{p\to\infty}\lim_{q\to\infty}\langle (\eta_p S) * (\tilde{\eta}_q T), \varphi \rangle = \lim_{p\to\infty}\langle (\eta_p S) * T, \varphi \rangle.$$

The implications and the respective identities of the convolutions follow.

Proof of $S \overset{8_1}{*} T = S \overset{8_0}{*} T$; $\binom{8_1}{*} \Rightarrow \binom{7_1}{*}$; $\binom{8_0}{*} \Rightarrow \binom{7_0}{*}$. The implications are obvious.

Proof of $\binom{7_1}{*} \Rightarrow \binom{3}{*}$; $\binom{7_0}{*} \Rightarrow \binom{2}{*}$. Since $\langle (\eta_k S) * T, \varphi \rangle = \langle S(\tilde{T} * \varphi), \eta_k \rangle$, we deduce from $\binom{7_1}{*}$ that $S(\tilde{T} * \varphi) \in \mathcal{D}'^{(M_p)}_{L^1}$ for $\varphi \in \mathcal{D}^{(M_p)}$, i.e., $\binom{3}{*}$ holds and $S \overset{7_0}{*} T = S \overset{3}{*} T$. The second implication follows by the symmetry argument. Since $\binom{2}{*} \Leftrightarrow \binom{3}{*} \Leftrightarrow \binom{6}{*}$, the proof is complete. $\square$

6.6 Existence of $\mathcal{D}'^{(M_p)}-$ and $\mathcal{S}'^{(M_p)}-$ convolution

Our purpose now is to give various sufficient conditions for the existence of $\mathcal{D}'^{(M_p)}-$convolution and $\mathcal{S}'^{(M_p)}-$convolution of two ultradistributions. Two types of sufficient conditions are considered in the next two sections: a) conditions expressed in terms of the supports of the ultradistributions involved; b) conditions which rely on distinguishing subspaces of ultradistributions on which convolution can be defined as a bilinear mapping. We emphasize that, for simplicity, we shall consider in Sections 6.6, 6.7, and

6.8 only the one-dimensional case of functions, distributions, and ultradistributions defined on the real line; but all results can be obtained similarly in the n-dimensional case.

In Section 6.7 we are going to prove that, under appropriate compatibility conditions on the supports of two ultradistributions belonging to the space $\mathcal{D}'^{(M_p)}$ or to the space $\mathcal{S}'^{(M_p)}$, their $\mathcal{D}'^{(M_p)}$–convolution (introduced in several equivalent ways in Section 6.2 and denoted by $*$) or, respectively, $\mathcal{S}'^{(M_p)}$–convolution (introduced in several equivalent ways in Section 6.4 and denoted by $\star$) exist. In the case of the space $\mathcal{D}'^{(M_p)}$, the compatibility condition which guarantees the existence of the $\mathcal{D}'^{(M_p)}$–convolution coincides with the well known compatibility condition for distributions in $\mathcal{D}'$. In the case of the space $\mathcal{S}'^{(M_p)}$, we define the notion of M–compatible supports of tempered ultradistributions which corresponds to the concepts of polynomial compatibility introduced for the supports of tempered distributions in [66] and of M_p–compatibility introduced for the supports of generalized functions of Gelfand and Shilov in [144]. We will prove that the condition of M–compatibility of the supports of two ultradistributions in $\mathcal{S}'^{(M_p)}$ implies the existence of the $\mathcal{S}'^{(M_p)}$–convolution. These results are obtained under the assumption that the given numerical sequence (M_p), which defines the respective spaces of test functions and of ultradistributions, satisfies the conditions $(M.1)$ and $(M.3')$.

In Section 6.8 we examine the weighted $\mathcal{D}'_{L^q}$ and $\mathcal{D}'^{(M_p)}_{L^q}$ spaces (distributions and ultradistributions), denoted by $\mathcal{D}'_{L^q,\mu}$ and $\mathcal{D}'^{(M_p)}_{L^q,\mu}$ (for $q \in [1,\infty]$), respectively. Let us mention here two of the known results obtained for convolution in these spaces. Toward this aim fix $\mu, \nu \in \mathbb{R}$ with $\mu + \nu \geq 0$ and $q, r \in [1,\infty]$ with $1/q + 1/r \in [1,2]$; and put $\rho = \min(\mu,\nu)$; $s = (1/q + 1/r - 1)^{-1}$.

$1°$ if the given sequence (M_p) satisfies conditions $(M.1)$, $(M.2')$ and $(M.3')$, then $f \in \mathcal{D}'_{L^q,\mu}$ and $g \in \mathcal{D}'^{(M_p)}_{L^r,\nu}$, imply $f * g = f \star g \in \mathcal{D}'^{(M_p)}_{L^s,\rho}$ and the mapping

$$\mathcal{D}'_{L^q,\mu} \times \mathcal{D}'^{(M_p)}_{L^r,\nu} \ni (f,g) \mapsto f * g \in \mathcal{D}'^{(M_p)}_{L^s,\rho}$$

is continuous;

$2°$ if the sequence (M_p) satisfies conditions $(M.1)$, $(M.2)$ and $(M.3')$, then $f \in \mathcal{D}'^{(M_p)}_{L^q,\mu}$ and $g \in \mathcal{D}'^{(M_p)}_{L^r,\nu}$ imply $f * g = f \star g \in \mathcal{D}'^{(M_p)}_{L^s,\rho}$ and the mapping

$$\mathcal{D}'^{(M_p)}_{L^q,\mu} \times \mathcal{D}'^{(M_p)}_{L^r,\nu} \ni (f,g) \mapsto f * g \in \mathcal{D}'^{(M_p)}_{L^s,\rho}$$

is continuous.

Again we note that the one-dimensional case will be considered for the remainder of this chapter; we assume that $n = 1$ for the remainder of this chapter.

We shall need the following two assertions concerning representations of convergent sequences of ultradistributions and tempered ultradistributions.

Theorem 6.6.1. *Let* $f \in \mathcal{D}'^{(M_p)}$, $f_k \in \mathcal{D}'^{(M_p)}$ *for* $k \in \mathbb{N}$ *and suppose that* $f_k \to f$ *in* $\mathcal{D}'$, *as* $k \to \infty$. *Then, for each open, relatively compact set* G *in* $\mathbb{R}$, *there are measures* f_α, $f_{k,\alpha} \in \mathcal{C}'(\bar{G})$ *for* $k \in \mathbb{N}$, $\alpha \in \mathbb{N}_0$ *and positive constants* L *and* B *such that*

$$f|_G = \sum_{\alpha \in \mathbb{N}_0} f_\alpha^{(\alpha)}, \qquad f_k|_G = \sum_{\alpha \in \mathbb{N}_0} f_{k,\alpha}^{(\alpha)}, \qquad (6.18)$$

$$\|f_\alpha\|_{\mathcal{C}'(\bar{G})} \leq BL^\alpha/M_\alpha, \quad \|f_{k,\alpha}\|_{\mathcal{C}'(\bar{G})} \leq BL^\alpha/M_\alpha, \qquad (6.19)$$

for arbitrary $\alpha, k \in \mathbb{N}_0$ *and*

$$\lim_{k \to \infty} \|f_{k,\alpha} - f_\alpha\|_{\mathcal{C}'(\bar{G})} = 0. \qquad (6.20)$$

for every $\alpha \in \mathbb{N}_0$.

Proof. The above proposition is an extension of Theorem 8.1 of [82] to the case of a sequence of ultradistributions. Note that the existence of the representations (6.18) follows directly from that theorem, but we cannot deduce from it that the constants B and L in the second of the inequalities (6.19) do not depend on k. This and (6.20) follow, however, from the proof of the mentioned theorem with small modifications. For completeness, we present here the whole proof.

Let K be the closure of G. We shall prove that the restrictions of f and f_k to $\mathcal{D}_K^{(M_p)}$ have representations (6.18) and the series in (6.18) are convergent in the strong topology of $(\mathcal{D}_K^{(M_p)})'$. Since the inclusion mapping $\mathcal{D}^{(M_p)}(G) \to \mathcal{D}_K^{(M_p)}$ is continuous, both series in (6.18) converge also in the strong topology of the space $\mathcal{D}^{(M_p)}(G)$. Note that

$$\mathcal{D}_K^{(M_p)} := \mathop{\text{proj lim}}_{j \to \infty} X_j,$$

where X_j is the Banach space of all $\phi \in \mathcal{D}_K$ such that

$$\lim_{\alpha \to \infty} \left(M_\alpha^{-1} j^\alpha \|\phi^{(\alpha)}\|_{C(K)} \right) = 0$$

with the norm

$$\|\phi\|_{X_j} := M_\alpha^{-1} \sup_{\alpha \in \mathbb{N}_0} j^\alpha \|\phi^{(\alpha)}\|_{C(K)}.$$

Since $\mathcal{D}_K^{(M_p)}$ is a strict projective limit of the spaces X_j, it follows from a theorem in [50] (2.6 Satz, p.147) that the restrictions of f and f_k to $\mathcal{D}_K^{(M_p)}$ can be extended to elements g and g_k of the same space X_j' for $k \in \mathbb{N}$, so that $g_k \to g$ in X_j' as $k \to \infty$.

On the other hand, using an appropriate mapping, one can identify X_j with a closed subspace of the space

$$Y_j := \{\phi = (\phi_\alpha) : \phi_\alpha \in \mathcal{C}(K), \lim_{\alpha \to \infty} \left(M_\alpha^{-1} j^\alpha \|\phi_\alpha\|_{\mathcal{C}(K)}\right) = 0\}$$

and Y_j with a closed subspace of $\mathcal{C}_0(K^\mathbb{N})$, where $\mathcal{C}(K)$ is the space of all continuous functions with supports in K, the set $K^\mathbb{N}$ is the disjoint union of countably many copies of K regarded as a locally compact space, and $\mathcal{C}_0(K^\mathbb{N})$ is the space of all continuous functions vanishing at the boundary of $K^\mathbb{N}$. By the Hahn-Banach theorem, we extend g and g_k to measures $\tilde{g}$ and $\tilde{g}_k$ on $K^\mathbb{N}$ (for $k \in \mathbb{N}$), so that $\tilde{g}_k \to \tilde{g}$ in $\mathcal{C}_0'(K^\mathbb{N})$ as $k \to \infty$. This implies the existence of measures f_α, $f_{k,\alpha} \in \mathcal{C}'(K)$ for $k \in \mathbb{N}, \alpha \in \mathbb{N}_0$, which satisfy (6.19) and (6.20). For $\phi \in X_j$, we have

$$\langle \tilde{g}_k, \phi \rangle = \langle f_k, \phi \rangle = \sum_{\alpha \in \mathbb{N}_0} (-1)^\alpha \langle f_{k,\alpha}, \phi^{(\alpha)} \rangle = \sum_{\alpha \in \mathbb{N}_0} \langle f_{k,\alpha}^{(\alpha)}, \phi \rangle$$

and, similarly, $\langle \tilde{g}, \phi \rangle = \langle f, \phi \rangle = \sum_{\alpha \in \mathbb{N}_0} \langle f_\alpha^{(\alpha)}, \phi \rangle$. $\square$

Theorem 6.6.2. *Suppose that $f \in \mathcal{S}'^{(M_p)}$ and $f_k \in \mathcal{S}'^{(M_p)}$ for $k \in \mathbb{N}_0$, and $f_k \to f$ in $\mathcal{S}'^{(M_p)}$ as $k \to \infty$. Then there are functions F_α, $F_{k,\alpha} \in L^2$ for $k, \alpha \in \mathbb{N}_0$ and positive constants λ, L and B such that*

$$f|_G = \sum_{\alpha \in \mathbb{N}_0} (e_\lambda^M F_\alpha)^{(\alpha)}, \quad f_k|_G = \sum_{\alpha \in \mathbb{N}_0} (e_\lambda^M F_{k,\alpha})^{(\alpha)}$$
$$\|F_\alpha\|_{L^2} \leq BL^\alpha/M_\alpha, \quad \|F_{k,\alpha}\|_{L^2} \leq BL^\alpha/M_\alpha,$$

for arbitrary $\alpha, n \in \mathbb{N}_0$, and

$$\lim_{k \to \infty} \|F_{k,\alpha} - F_\alpha\|_{L^2} = 0$$

for every $\alpha \in \mathbb{N}_0$, where $e_\lambda^M(x) := e^{M(\lambda|x|)}$ for $x \in \mathbb{R}$.

Proof. One can prove the proposition following the idea of the proof of the structural theorem for $\mathcal{S}'^{(M_p)}$ by first noticing that $\mathcal{S}^{(M_p)}$ is a strict projective limit of the spaces $\mathcal{S}_2^{(M_p),m}$ (see [92]) and then, similarly to the proof of Theorem 6.6.1, applying a theorem proved in [50] (2.6 Satz, p.147). $\square$

6.7 Compatibility conditions on supports

Let us recall the notion of compatibility of supports used in the theory of distributions: two subsets X and Y of $\mathbb{R}$ are called *compatible* if one of the following equivalent conditions holds (cf. [2], [59] and [76]):

(a) for every bounded interval I in $\mathbb{R}$ the set $(X \times Y) \cap I^{\triangle}$ is bounded in $\mathbb{R}^2$, where $I^{\triangle} := \{(x, y) : x + y \in I\}$;

(b) for every bounded interval I in $\mathbb{R}$ the set $X \cap (I - Y)$ is bounded in $\mathbb{R}$;

(c) for every bounded interval I in $\mathbb{R}$ the set $(X - I) \cap Y$ is bounded in $\mathbb{R}$;

(d) $x_k \in X, y_k \in Y$ $(k \in \mathbb{N})$, $|x_k| + |y_k| \to \infty$ implies $|x_k + y_k| \to \infty$;

(e) for every $R > 0$ the set $T_R := \{(x, y) : x \in X, y \in Y, |x + y| \leq R\}$ is bounded in $\mathbb{R}^2$.

In [70] (see also [72] and [73]), it is proved that several general conditions for the existence of the convolution $f * g$ of given ultradistributions f and g in $\mathcal{D}'$ are equivalent. In the theorem below we shall prove that if the supports of ultradistributions f and g are compatible, then these conditions are satisfied .

Theorem 6.7.1. *Suppose that a given sequence (M_p) satisfies conditions $(M.1)$ and $(M.3')$.*

(i) If $f, g \in \mathcal{D}'^{(M_p)}$ are ultradistributions whose supports supp f and supp g are compatible, then the $\mathcal{D}'^{(M_p)}$–convolution $f * g$ exists.

(ii) If $f_k, g_k \in \mathcal{D}'^{(M_p)}$, supp $f_k \subseteq X$, supp $g_k \subseteq Y$ for $k \in \mathbb{N}$, where X, Y are compatible sets in $\mathbb{R}$, and if $f_k \to f$, $g_k \to g$ in $\mathcal{D}'^{(M_p)}$, then $f_k * g_k \to f * g$ in $\mathcal{D}'^{(M_p)}$ as $k \to \infty$.

Proof. It suffices to show that one of the equivalent conditions for the existence of $\mathcal{D}'^{(M_p)}$–convolution in $\mathcal{D}'$ is satisfied. We shall prove that the limit

$$\lim_{k \to \infty} \langle f \otimes g, \eta_k \phi^{\triangle} \rangle$$

exists for every $\phi \in \mathcal{D}^{(M_p)}(\mathbb{R})$ and for an arbitrary strong approximate unit (η_k) in $\mathcal{D}^{(M_p)}(\mathbb{R}^2)$ (cf. [70]).

Assume that supp ϕ is contained in the ball of radius $R > 0$. Since the set T_R is bounded, there exist open bounded sets Ω_1 and Ω_2 in $\mathbb{R}$ such that the closure K of T_R is contained in $\Omega_1 \times \Omega_2$. By Theorem 8.1 of [82], there are measures $f_\alpha \in \mathcal{C}'(\overline{\Omega}_1)$, $g_\beta \in \mathcal{C}'(\overline{\Omega}_2)$ on Ω_1 and Ω_2 for α, $\beta \in \mathbb{N}_0$ and positive constants B and L such that

$$f|_{\Omega_1} = \sum_{\alpha \in \mathbb{N}_0} f_\alpha^{(\alpha)}, \qquad g|_{\Omega_2} = \sum_{\beta \in \mathbb{N}_0} g_\beta^{(\beta)},$$

$$\|f_\alpha\|_{\mathcal{C}'(\overline{\Omega}_1)} \le BL^\alpha/M_\alpha, \quad \|g_\beta\|_{\mathcal{C}'(\overline{\Omega}_2)} \le BL^\beta/M_\beta$$

for α, $\beta \in \mathbb{N}_0$. This and $(M.1)$ imply that

$$\|f_\alpha \otimes g_\beta\|_{\mathcal{C}'(\overline{\Omega}_1 \times \overline{\Omega}_2)} \le AB^2 \frac{(HL)^{\alpha+\beta}}{M_{\alpha+\beta}} \qquad (6.21)$$

for α, $\beta \in \mathbb{N}_0$.

Let $\eta \in \mathcal{D}((M_p), \Omega_1 \times \Omega_2)$ be equal to 1 on $\overline{T}_R$. There exists an index $k_0 \in \mathbb{N}$ such that

$$\langle f \otimes g, \eta_k \phi^\triangle \rangle = \langle f \otimes g, \eta_k \phi^\triangle \eta \rangle = \sum_{\alpha \in \mathbb{N}_0} \sum_{\beta \in \mathbb{N}_0} \langle f_\alpha^{(\alpha)} \otimes g_\beta^{(\beta)}, \phi^\triangle \eta \rangle$$

$$= \sum_{\alpha \in \mathbb{N}_0} \sum_{\beta \in \mathbb{N}_0} \langle f_\alpha^{(\alpha)} \otimes g_\beta^{(\beta)}, \phi^\triangle \rangle$$

for $k > k_0$. Put

$$a_{p,q} := \sum_{\alpha \le p} \sum_{\beta \le q} \langle f_\alpha^{(\alpha)} \otimes g_\beta^{(\beta)}, \phi^\triangle \eta \rangle$$

for $p, q \in \mathbb{N}_0$ and notice that

$$\lim_{q \to \infty} a_{p,q} = \sum_{\alpha \le p} \langle f_\alpha^{(\alpha)} \otimes g, \phi^\triangle \eta \rangle, \qquad p \in \mathbb{N};$$

$$\lim_{p \to \infty} a_{p,q} = \sum_{\beta \le q} \langle f \otimes g_\beta^{(\beta)}, \phi^\triangle \eta \rangle, \qquad q \in \mathbb{N}.$$

Let (p_j) and (q_j) be arbitrary strictly increasing sequences of positive integers and set $\tilde{a}_j := a_{p_j, q_j}$ for $j \in \mathbb{N}$. It follows from (6.21) that

$$|\tilde{a}_m - \tilde{a}_l| \le \sum_{p_l < \alpha \le p_m} \sum_{q_l < \beta \le q_m} |\langle f_\alpha \otimes g_\beta, (\phi^\triangle)^{(\alpha+\beta)} \rangle|$$

$$\le C \sum_{p_l < \alpha \le p_m} \sum_{q_l < \beta \le q_m} \frac{(LH)^{\alpha+\beta}}{M_{\alpha+\beta}} \sup_{|x+y| \le R} |\phi^{(\alpha+\beta)}(x+y)|$$

$$\le 2^{-2k} C \sup_{r \in \mathbb{N}_0} \sup_{|t| \le R} \frac{(4LH)^r}{M_p} |\phi^{(r)}(t)| \sum_{r=1}^{\infty} 2^{-r} \qquad (6.22)$$

for $m > l > k > k_0$, where $C = \pi R^2 AB$. This implies that $(\tilde{a}_j)$ is a Cauchy sequence, i.e., it converges to a certain number a. It is easy to see that a

does not depend on the choice of sequences (p_j) and (q_j). Consequently, the double limit $\lim\limits_{p,q\to\infty} a_{p,q}$ exists and

$$\lim_{p,q\to\infty} a_{p,q} = \lim_{p\to\infty} \lim_{q\to\infty} a_{p,q} = \lim_{q\to\infty} \lim_{p\to\infty} a_{p,q} = a. \qquad (6.23)$$

Now, by (6.23), we have

$$\langle f * g, \phi \rangle = \langle f \otimes g, \eta_k \phi^\triangle \eta \rangle = \sum_{\alpha,\beta\in\mathbb{N}_0} |\langle f_\alpha \otimes g_\beta, (\phi^\triangle)^{(\alpha+\beta)} \rangle|$$

$$= \sum_{\alpha,\beta\in\mathbb{N}_0} \langle f_\alpha * g_\beta, \phi^{(\alpha+\beta)} \rangle, \qquad (6.24)$$

which proves (i).

Let us prove now the second assertion of the theorem. By Theorem 6.6.1, there exist measures $f_{k,\alpha}$ and $g_{k,\beta}$ and constants $B, L > 0$ such that

$$f_k|_{\Omega_1} = \sum_{\alpha\in\mathbb{N}_0} f_{k,\alpha}^{(\alpha)}, \qquad g_k|_{\Omega_2} = \sum_{\alpha\in\mathbb{N}_0} g_{k,\alpha}^{(\alpha)}, \qquad (6.25)$$

$$\|f_{k,\alpha}\|_{\mathcal{C}'(\bar{\Omega}_1)} \le BL^\alpha/M_\alpha, \qquad \|g_{k,\alpha}\|_{\mathcal{C}'(\bar{\Omega}_2)} \le BL^\alpha/M_\alpha, \qquad (6.26)$$

for $\alpha, k \in \mathbb{N}_0$ and

$$\|f_{k,\alpha} - f_k\|_{\mathcal{C}'(\bar{\Omega}_1)} \to 0, \qquad \|g_{k,\alpha} - g_k\|_{\mathcal{C}'(\bar{\Omega}_2)} \to 0, \qquad (6.27)$$

for $\alpha \in \mathbb{N}_0$ as $k \to \infty$. By the first part of the theorem, the $\mathcal{D}'^{(M_p)}$-convolutions $f * g$ and $f_k * g_k$ exist for $k \in \mathbb{N}$ and, due to (6.24) and the estimation used in (6.22), we have

$$|\langle f_k * g_k, \phi \rangle| \le \sum_{\alpha,\beta\in\mathbb{N}_0} |\langle f_{k,\alpha} * g_{k,\beta}, \phi^{(\alpha+\beta)} \rangle|$$

$$= \sum_{\alpha,\beta\in\mathbb{N}_0} |\langle f_{k,\alpha} \otimes g_{k,\beta}, (\phi^\triangle)^{(\alpha+\beta)} \rangle|$$

$$\le C \sup_{\gamma\in\mathbb{N}_0} \sup_{|t|\le R} \frac{(2LH)^\gamma}{M_\gamma} |\phi^{(\gamma)}(t)| < \infty \qquad (6.28)$$

for each fixed $\phi \in \mathcal{D}^{(M_p)}$. Hence

$$|\langle f_k * g_k - f * g, \phi \rangle| \le \sum_{\alpha,\beta\in\mathbb{N}_0} |\langle f_{k,\alpha} * g_{k,\beta} - f_\alpha * g_\beta, \phi^{(\alpha+\beta)} \rangle|$$

$$\le \left(\sum_{\substack{\alpha\le m \\ \beta\le m}} + \sum_{\substack{\alpha>m \\ \beta\in\mathbb{N}_0}} + \sum_{\substack{\alpha\in\mathbb{N}_0 \\ \beta>m}} \right) |\langle f_{k,\alpha} * g_{k,\beta} - f_\alpha * g_\beta, \phi^{(\alpha+\beta)} \rangle| \qquad (6.29)$$

where $m \in \mathbb{N}_0$.

Let $\varepsilon > 0$ be fixed. Notice that the series

$$\sum_{\alpha,\beta \in \mathbb{N}_0} \langle f_\alpha * g_\beta, \phi^{(\alpha+\beta)} \rangle$$

is convergent and all the series

$$\sum_{\alpha,\beta \in \mathbb{N}_0} \langle f_{k,\alpha} * g_{k,\beta}, \phi^{(\alpha+\beta)} \rangle$$

for $k \in \mathbb{N}$ are commonly bounded by a convergent series (see (6.28)). Therefore each of the last two sums in (6.29) is less than $\varepsilon/3$ for m large enough. On the other hand, in view of (6.26), we have

$$|\langle f_{k,\alpha} * g_{k,\beta} - f_\alpha * g_\beta, \phi^{(\alpha+\beta)} \rangle|$$
$$\leq \|\phi^{(\alpha+\beta)}\|_{C(K)} \, (\|f_{k,\alpha} * (g_{k,\beta} - g_\beta)\|_{C'(K)}$$
$$+ \|(f_{k,\alpha} - f_\alpha) * g_\beta\|_{C'(K)})$$
$$\leq C\|\phi^{(\alpha+\beta)}\|_{C(K)} \, (L^\alpha M_\alpha^{-1}\|g_{k,\beta} - g_\beta\|_{C'(K)}$$
$$+ L^\beta M_\beta^{-1}\|f_{k,\alpha} - f_\alpha\|_{C'(K)})$$

for some constant $C > 0$ and all α, $\beta \in \mathbb{N}_0$ and $k \in \mathbb{N}$. Consequently, by (6.27), it follows that

$$\sum_{\substack{\alpha \leq m \\ \beta \leq m}} |\langle f_{k,\alpha} * g_{k,\beta} - f_\alpha * g_\beta, \phi \rangle| \leq \varepsilon/3$$

for k large enough, which completes the proof of (ii) as well as the whole theorem. $\square$

The following notion is a modification of polynomial compatibility of supports, introduced for tempered distributions in [65] (see also [66]) and then generalized to the case of the space $K\{M_p\}'$ of Gelfand and Shilov in [144] (see also [145] and [76]). Coincidentally, the sequence of functions defining a Gelfand-Shilov space is denoted also by (M_p) as the sequence defining the space of ultradistributions and the above mentioned condition in the Gelfand-Shilov spaces is called M_p−compatibility. To avoid misunderstanding, we shall call the notion introduced below M−*compatibility*, which is justified by the use of the associated function M in its definition.

Definition 6.7.1. Sets X, $Y \subseteq \mathbb{R}$ are said to be M−compatible if

$$M(|x|) + M(|y|) \leq M(d|x+y|), \qquad x \in X, \ y \in Y. \tag{6.30}$$

for some $d > 0$ (or, equivalently, $d \geq 1$).

The following assertion is an analogue of Theorem 6.7.1 for the $\mathcal{S}'^{(M_p)}$-convolution $f \star g$ of tempered ultradistributions f and g.

Theorem 6.7.2. *Suppose that the sequence (M_p) satisfies conditions $(M.1)$, $(M.2)$ and $(M.3)'$.*

(i) If $f, g \in \mathcal{S}'^{(M_p)}$ and supp f and supp g are $M-$compatible, then the $\mathcal{S}'^{(M_p)}-$convolution $f \star g$ exists.

(ii) If $f_k, g_k \in \mathcal{S}'^{(M_p)}$ and supp $f_k \subseteq X$, supp $g_k \subseteq Y$ for $k \in \mathbb{N}$, where X, Y are $M-$compatible sets in $\mathbb{R}$, and if $f_k \to f$, $g_k \to g$ in $\mathcal{S}'^{(M_p)}$, then $f_k \star g_k \to f \star g$ in $\mathcal{S}'^{(M_p)}$ as $k \to \infty$.

Proof. By [75], there exist constants $\lambda, L, B > 1$ and functions f_α, $g_\beta \in L^2$ such that

$$f = \sum_{\alpha \in \mathbb{N}_0} (e_\lambda^M f_\alpha)^{(\alpha)}, \quad g = \sum_{\beta \in \mathbb{N}_0} (e_\lambda^M g_\beta)^{(\beta)}, \tag{6.31}$$

and

$$\|f_\alpha\|_{L^2} \le BL^\alpha / M_\alpha, \quad \|g_\beta\|_{L^2} \le BL^\beta / M_\beta, \tag{6.32}$$

for $\alpha, \beta \in \mathbb{N}$, where $e_\lambda^M(x) = e^{M(\lambda|x|)}$ for $x \in \mathbb{R}$.

Let $X := \operatorname{supp} f$ and $Y := \operatorname{supp} g$. Fix a function $\phi \in \mathcal{S}^{(M_p)}$, a strong approximate unit (η_k) in $\mathcal{D}^{(M_p)}(\mathbb{R}^2)$ and indices $l, m \in \mathbb{N}$, $m > l$, denoting $\eta_{l,m} := \eta_m - \eta_l$, $K_{l,m} := \operatorname{supp}(\eta_m - \eta_l)$ and $D_{l,m} := |\langle f \otimes g, \eta_{l,m} \phi^\triangle \rangle|$.

By (6.31), we have

$$D_{l,m} = |\langle f \otimes g, \eta_{l,m} \phi^\triangle \rangle| \le \sum_{\alpha, \beta \in \mathbb{N}_0} |\langle (e_\lambda^M f_\alpha)^{(\alpha)} \otimes (e_\lambda^M g_\beta)^{(\beta)}, \eta_{l,m} \phi^\triangle \rangle|$$

$$\le \sum_{\alpha, \beta \in \mathbb{N}_0} \sum_{\substack{\gamma \le \alpha \\ \delta \le \beta}} \binom{\alpha}{\gamma} \binom{\beta}{\delta} C_{\gamma,\delta}^{\alpha,\beta}, \tag{6.33}$$

where

$$C_{\gamma,\delta}^{\alpha,\beta} := \int_X \int_Y |[(e_\lambda^M f_\alpha) \otimes (e_\lambda^M g_\beta)] \eta_{l,m}^{(\alpha-\gamma,\beta-\delta)} (\phi^{(\gamma+\delta)})^\triangle|.$$

In view of (6.30), we have

$$C_{\gamma,\delta}^{\alpha,\beta} \le \int_X \int_Y |(f_\alpha \otimes g_\beta) \eta_{l,m}^{(\alpha-\gamma,\beta-\delta)} (e_{d\lambda}^M \phi^{(\gamma+\delta)})^\triangle|. \tag{6.34}$$

Notice that $(M.2)$ implies the inequality

$$2M(t) \le M(c_0 t) + c_0, \quad t > 0,$$

for some constant $c_0 \geq 1$. Hence there exists a constant $c \geq 1$ such that

$$M(d\lambda t) \leq -2M(t) + M(ct) + c, \qquad t > 0.$$

Denote

$$A_{\alpha,\beta} := \sup_{k \in \mathbb{N}} \sup_{(x,y) \in \mathbb{R}^2} |\eta^{(\alpha,\beta)}(x,y)|, \qquad B_\alpha := \sup_{t \in \mathbb{R}} e^{M(b|t|)} |\phi^{(\alpha)}(t)|,$$

$$\kappa_{l,m} := \sup\{e^{-M(|x|)} : x \in K_{l,m}\}, \qquad \kappa_0 := \int_{\mathbb{R}} e^{-M(|t|)} \, dt.$$

By (6.34), the Schwarz inequality and inequalities (6.32), (2.1) and (2.3), we obtain

$$
\begin{aligned}
C^{\alpha,\beta}_{\gamma,\delta} &\leq A_{\alpha-\gamma,\beta-\delta} \, B_{\gamma+\delta} \int_{\mathbb{R}} e^{-2M(|t|)} \Big[\int_{\mathbb{R}} |f_\alpha(x) g_\beta(t-x)| \, dx \Big] dt \\
&\leq \kappa_0 \, \kappa_{l,m} \, A_{\alpha-\gamma,\beta-\delta} \|f_\alpha\|_{L^2} \, \|g_\beta\|_{L^2} \\
&\leq \kappa_0 \, \kappa_{l,m} A \, B^2 \, (LH)^{\alpha+\beta} \frac{A_{\alpha-\gamma,\beta-\delta}}{M_{\alpha+\beta-\gamma-\delta}} \frac{B_{\gamma+\delta}}{M_{\gamma+\delta}} \\
&\leq \frac{\kappa_0 \kappa_{l,m} A \, B^2}{4^{\alpha+\beta}} \cdot \frac{(4LH)^{\alpha+\beta-\gamma-\delta}}{M_{\alpha+\beta-\gamma-\delta}} A_{\alpha-\gamma,\beta-\delta} \cdot \frac{(4LH)^{\gamma+\delta}}{M_{\gamma+\delta}} B_{\gamma+\delta}.
\end{aligned}
$$

By (6.33) and (6.34), this yields

$$D_{l,m} \leq \sum_{\alpha,\beta \in \mathbb{N}_0} \sum_{\substack{\gamma \leq \alpha \\ \delta \leq \beta}} \binom{\alpha}{\gamma} \binom{\beta}{\delta} C^{\alpha,\beta}_{\gamma,\delta} \leq \sum_{\alpha,\beta \in \mathbb{N}_0} \frac{\kappa_{l,m} C}{2^{\alpha+\beta}} = \kappa_{l,m} C$$

for some constant $C > 0$, in view of (2.54) and (6.2), i.e., for arbitrary $\varepsilon > 0$, we have

$$D_{l,m} = |\langle f \otimes g, (\eta_m - \eta_l)\phi^\triangle \rangle| < \varepsilon \qquad (6.35)$$

for sufficiently large $l, m \in \mathbb{N}$. Hence we conclude that the sequence (D_k), where $D_k := |\langle f \otimes g, \eta_k \phi^\triangle \rangle|$, is a Cauchy sequence. It is easy to see that its limit does not depend on the choice of a strong approximate unit (η_k) and this proves the first assertion (cf. [70]).

The second assertion can be proved in a similar way as the second statement of Theorem 6.7.1. $\square$

6.8   Convolution in weighted spaces

The weighted ultradistributional spaces $\mathcal{D}'^{(M_p)}_{L^s,\mu}$ are defined in [90], analogously to the distributional spaces $\mathcal{D}'_{L^s,\mu}$ introduced in [107]. We recall some definitions and assertions from [90].

Let $s \in [1, \infty]$, $h > 0$, $\mu \in \mathbb{R}$. $\mathcal{D}_{L^s,\mu}^{M_p,h}$ is defined to be the space of all functions $\phi \in \mathcal{E}$ such that $((\langle \cdot \rangle^\mu \phi)^{(\alpha)} \in L^s$ (see 6.15) for each $\alpha \in \mathbb{N}_0$ and

$$\|\phi\|_{\mathcal{D}_{L^s,\mu}^{M_p,h}} := \sum_{\alpha \in \mathbb{N}_0} \frac{h^\alpha}{M_\alpha} \|(\langle \cdot \rangle^\mu \phi)^{(\alpha)}\|_{L^s} < \infty$$

and which is equipped with the topology induced by the norm $\| \cdot \|_{\mathcal{D}_{L^s,\mu}^{M_p,h}}$. (Recall Section 1.1 for the notation $\langle \cdot \rangle^\mu$.) Further we define

$$\mathcal{D}_{L^s,\mu}^{(M_p)} := \operatorname*{proj\,lim}_{h \to \infty} \mathcal{D}_{L^s,\mu}^{M_p,h}.$$

The space $\dot{\mathcal{B}}_\mu^{(M_p)}$ is defined to be the subspace of $\mathcal{D}_{L^\infty,\mu}^{(M_p)}$ which is the completion of the space $\mathcal{D}^{(M_p)}$ in the topology of the family of the norms $\| \cdot \|_{\mathcal{D}_{L^s,\mu}^{M_p,h}}$.

One can easily prove that the mappings:

$$\mathcal{D}_{L^s}^{(M_p)} \ni \phi \mapsto \langle \cdot \rangle^{-\mu} \phi \in \mathcal{D}_{L^s,\mu}^{(M_p)},$$
$$\dot{\mathcal{B}}^{(M_p)} \ni \phi \mapsto \langle \cdot \rangle^{-\mu} \phi \in \dot{\mathcal{B}}_\mu^{(M_p)}$$

are homeomorphisms. The spaces $\mathcal{D}_{L^s,\mu}^{(M_p)}$ and $\dot{\mathcal{B}}_\mu^{(M_p)}$ are (FG)–spaces and the spaces $\mathcal{D}_{L^s,\mu}^{(M_p)}$ are (FS)–spaces for $s > 1$ (for the definitions see [50]). The proof of this assertion may be done in a way similar to the proof given in [107] for $\mathcal{D}_{L^s}^{(M_p)}$, $s > 1$.

Let us recall some properties of the spaces defined above which will be needed in the sequel; for their proofs we refer to [90]. Let $\mu, \nu \in \mathbb{R}$ and $q, r, s \in [1, \infty]$. If $1/q + 1/r \geq 1/s$, then the pointwise multiplications

$$\mathcal{D}_{L^q,\mu}^{(M_p)} \times \mathcal{D}_{L^r,\nu}^{(M_p)} \ni (\phi, \psi) \mapsto \phi\psi \in \mathcal{D}_{L^s,\mu+\nu}^{(M_p)},$$
$$\mathcal{D}_{L^s,\mu}^{(M_p)} \times \dot{\mathcal{B}}_\nu^{(M_p)} \ni (\phi, \psi) \mapsto \phi\psi \in \mathcal{D}_{L^s,\mu+\nu}^{(M_p)},$$
$$\mathcal{D}_{L^\infty,\mu}^{(M_p)} \times \dot{\mathcal{B}}_\nu^{(M_p)} \ni (\phi, \psi) \mapsto \phi\psi \in \dot{\mathcal{B}}_{\mu+\nu}^{(M_p)},$$
$$\dot{\mathcal{B}}_\mu^{(M_p)} \times \dot{\mathcal{B}}_\nu^{(M_p)} \ni (\phi, \psi) \mapsto \phi\psi \in \dot{\mathcal{B}}_{\mu+\nu}^{(M_p)}$$

are continuous mappings. If (a) $q < r$ and $\nu < \mu$ or (b) $q > r$ and $\nu < \mu + 1/q - 1/r$, then $\mathcal{D}_{L^q,\mu}^{(M_p)} \hookrightarrow \mathcal{D}_{L^r,\nu}^{(M_p)}$, where the symbol $E \hookrightarrow F$ means that the space E is continuously embedded into the space F and E is dense in F.

Given a $q \in (1, \infty]$ define r to satisfy the equation $1/q + 1/r = 1$. We denote by $\mathcal{D}'^{(M_p)}_{L^q,\mu}$ and $\mathcal{D}'^{(M_p)}_{L^1,\mu}$ the strong duals of the spaces $\mathcal{D}_{L^r,-\mu}^{(M_p)}$ and $\dot{\mathcal{B}}_{-\mu}^{(M_p)}$, respectively.

An ultradistribution $g \in \mathcal{D}'^{(M_p)}$ belongs to $\mathcal{D}'^{(M_p)}_{L^q,\mu}$ if there exist $h > 0$ and a sequence $(G_\alpha)_{\alpha \in \mathbb{N}}$ of elements of L^q such that

$$g = \sum_{\alpha \in \mathbb{N}_0} G_\alpha^{(\alpha)} \qquad (6.36)$$

and

$$\sum_{\alpha \in \mathbb{N}_0} \frac{M_\alpha}{h^\alpha} \| \langle \cdot \rangle^\mu G_\alpha \|_{L^q} < \infty. \qquad (6.37)$$

Conversely, if (6.37) holds for a sequence $(G_\alpha)_{\alpha \in \mathbb{N}_0}$ in L^q, then g defined by (6.36) belongs to $\mathcal{D}'^{(M_p)}_{L^q,\mu}$.

Let $q, r \in [1, \infty]$ and $\mu, \nu \in \mathbb{R}$. If (a) $q \leq r$ and $\nu \leq \mu$; or (b) $q > r$ and $\nu < \mu + 1/q - 1/r$, then we have the following embeddings:

$$\mathcal{E}' \hookrightarrow \mathcal{D}'_{L^r,\nu} \hookrightarrow \mathcal{D}'_{L^q,\mu} \hookrightarrow \mathcal{S}' \hookrightarrow \mathcal{D}'$$

$$\Big\uparrow \qquad \Big\uparrow \qquad \Big\uparrow \qquad \Big\uparrow \qquad \Big\uparrow$$

$$\mathcal{E}'^{(M_p)} \hookrightarrow \mathcal{D}'^{(M_p)}_{L^r,\nu} \hookrightarrow \mathcal{D}'^{(M_p)}_{L^q,\mu} \hookrightarrow \mathcal{S}'^{(M_p)} \hookrightarrow \mathcal{D}'^{(M_p)}.$$

The space $\mathcal{D}'_{L^q,\mu}$ is a proper subset of $\mathcal{D}'^{(M_p)}_{L^q,\mu} \bigcap \mathcal{D}'$.

Now we shall prove some results (Theorems 6.8.1 and 6.8.2) on the convolution in the weighted distributional and ultradistributional spaces.

Theorem 6.8.1. *Fix* $\phi \in \mathcal{S}^{(M_p)}$ *and* $q \in (1, \infty]$. *If the sequence* (M_p) *satisfies conditions* (M.1) *and* (M.3'), *we have*

(i) $f \in \mathcal{D}'_{L^q,\mu}$ *implies* $f * \phi \in \mathcal{D}^{(M_p)}_{L^q,\mu}$;

(ii) $f \in \mathcal{D}'_{L^1,\mu}$ *implies* $f * \phi \in \dot{\mathcal{B}}^{(M_p)}_\mu$.

If (M_p) *satisfies conditions* (M.1), (M.2') *and* (M.3'), *we have*

(iii) $f \in \mathcal{D}'^{(M_p)}_{L^q,\mu}$ *implies* $f * \phi \in \mathcal{D}_{L^q,\mu}$;

(iv) $f \in \mathcal{D}'^{(M_p)}_{L^1,\mu}$ *implies* $f * \phi \in \dot{\mathcal{B}}_\mu$.

If (M_p) *satisfies conditions* (M.1), (M.2) *and* (M.3'), *we have*

(v) $f \in \mathcal{D}'^{(M_p)}_{L^q,\mu}$ *implies* $f * \phi \in \mathcal{D}^{(M_p)}_{L^q,\mu}$;

(vi) $f \in \mathcal{D}'_{L^1,\mu}$ *implies* $f * \phi \in \dot{\mathcal{B}}^{(M_p)}_\mu$.

Proof. To prove (i) suppose that $f \in \mathcal{D}'_{L^q,\mu}$. In view of a result of [107], it follows that $f * \phi \in \mathcal{D}_{L^q,\mu}$ for each $\phi \in \mathcal{S}$. Therefore

$$\langle \cdot \rangle^{-\mu} (f * \phi)^{(\alpha)} = \langle \cdot \rangle^{-\mu} (f * \phi^{(\alpha)}) \in L^q$$

for $\alpha \in \mathbb{N}$. Since $f \in \mathcal{D}'_{L^q,\mu}$, there exist $\gamma \in \mathbb{N}_0$ and functions F_β from L^q such that

$$f = \sum_{\beta \leq \gamma} (\langle \cdot \rangle^{-\mu} F_\beta)^{(\beta)}$$

in $\mathcal{D}'^{(M_p)}$. Applying suitable properties of convolution (see [70]) and the following elementary inequality

$$\langle a + b \rangle^\mu \leq 4^{|\mu|} \langle a \rangle^\mu \langle b \rangle^{|\mu|}, \qquad a, b, \mu \in \mathbb{R},$$

(called *Peetre's inequality*), we get

$$\sum_{\alpha \in \mathbb{N}_0} \frac{h^\alpha}{M_\alpha} \| \langle \cdot \rangle^\mu (f * \phi)^{(\alpha)} \|_{L^q} = \sum_{\alpha \in \mathbb{N}_0} \frac{h^\alpha}{M_\alpha} \| \langle \cdot \rangle^\mu \sum_{\beta \leq \gamma} (\langle \cdot \rangle^{-\mu} F_\beta) * \phi^{(\alpha+\beta)} \|_{L^q}$$

$$\leq \sum_{\alpha \in \mathbb{N}_0} \sum_{\beta \leq \gamma} \frac{h^\alpha}{M_\alpha} \left(\int_{\mathbb{R}} \langle x \rangle^{\mu q} \left(\int_{\mathbb{R}} \langle x - t \rangle^{-\mu} |\phi^{(\alpha+\beta)}(x - t) F_\beta(t)| \, dt \right)^q dx \right)^{1/q}$$

$$\leq 4^{|\mu|} \sum_{\alpha \in \mathbb{N}_0} \sum_{\beta \leq \gamma} \frac{h^\alpha}{M_\alpha} \| \langle \cdot \rangle^{|\mu|} |\phi^{(\alpha+\beta)}| * |F_\beta| \|_{L^q}.$$

Hence, by Young's inequality,

$$\sum_{\alpha \in \mathbb{N}_0} \frac{h^\alpha}{M_\alpha} \| \langle \cdot \rangle^\mu (f * \phi)^{(\alpha)} \|_{L^q} \leq 4^{|\mu|} \sum_{\beta \leq \gamma} \sum_{\alpha \in \mathbb{N}_0} \frac{h^\alpha}{M_\alpha} \| F_\beta \|_{L^q} \| \langle \cdot \rangle^{|\mu|} \phi^{(\alpha+\beta)} \|_{L^1}$$

$$\leq 4^{|\mu|} \max_{\beta \leq \gamma} \| F_\beta \|_{L^q} \max_{\beta \leq \gamma} (M_{\alpha+\beta} M_\alpha^{-1} h^{-\beta}) \sum_{\gamma \in \mathbb{N}_0} \frac{h^\gamma}{M_\gamma} \| \langle \cdot \rangle^{|\mu|} \phi^{(\gamma)} \|_{L^1} < \infty.$$

This yields $f * \phi \in \mathcal{D}_{L^q,\mu}^{(M_p)}$, as desired.

The proof of (ii) is similar.

To prove (iii) suppose that $f \in \mathcal{D}'^{(M_p)}_{L^q,\mu}$. There are functions F_β in L^q and $h > 0$ such that

$$f = \sum_{\beta \in \mathbb{N}_0} (\langle \cdot \rangle^{-\mu} F_\beta)^{(\beta)}; \qquad \sum_{\beta \in \mathbb{N}_0} \frac{M_\beta}{h^\beta} \| F_\beta \|_{L^q} < \infty \qquad (6.38)$$

(cf. (6.36) and (6.37)). From the proof of Theorem 6.10 in [107], p. 71, it follows that $f * \phi \in \mathcal{E}$. From condition $(M.2')$ we obtain constants $A \geq 1$ and $H \geq 1$ such that

$$M_{\alpha+\beta} \leq A^\alpha H^{\alpha(\alpha+\beta)} M_\beta, \ \alpha, \beta \in \mathbb{N}_0.$$

Using this fact and Young's inequality, we obtain

$$\|\langle\cdot\rangle^\mu (f*\phi)^{(\alpha)}\|_{L^q} = \|\langle\cdot\rangle^\mu \sum_{\beta\in\mathbb{N}_0}(\langle\cdot\rangle^{-\mu}F_\beta)*\phi^{(\alpha+\beta)}\|_{L^q}$$

$$\leq 4^{|\mu|}\sum_{\beta\in\mathbb{N}_0}\|F_\beta*(\langle\cdot\rangle^{|\mu|}\phi^{(\alpha+\beta)})\|_{L^q} \leq C\sum_{\beta\in\mathbb{N}_0}\frac{h^\beta}{M_\beta}\|\langle\cdot\rangle^{|\mu|}\phi^{(\alpha+\beta)}\|_{L^1}$$

$$\leq C_\alpha \sum_{\beta\in\mathbb{N}_0}\frac{(H^\alpha h)^{\alpha+\beta}}{M_{\alpha+\beta}}\|\langle\cdot\rangle^{|\mu|}\phi^{(\alpha+\beta)}\|_{L^1}$$

$$\leq C_\alpha \sum_{\beta\in\mathbb{N}_0}\frac{(H^\alpha h)^\gamma}{M_\gamma}\|\langle\cdot\rangle^{|\mu|}\phi^{(\gamma)}\|_{L^1} < \infty$$

for each $\alpha\in\mathbb{N}_0$, where

$$C := 4^{|\mu|}\sup_{\beta\in\mathbb{N}_0}\frac{M_\beta}{\alpha^\beta}\|F_\beta\|; \qquad C_\alpha := \frac{CA^\alpha}{M_\alpha}. \tag{6.39}$$

(C and C_α are finite, in view of (6.38).) Hence $f*\phi\in\mathcal{D}_{L^{q,\mu}}$.

The proof of (iv) is similar.

To prove (v) suppose as before that $f\in\mathcal{D}'^{(M_p)}_{L^{q,\mu}}$. By (6.38), condition (2.1) and Young's inequality, we obtain

$$\sum_{\alpha\in\mathbb{N}_0}\frac{h^\alpha}{M_\alpha}\|\langle\cdot\rangle^\mu(f*\phi)^{(\alpha)}\|_{L^q}$$

$$\leq \sum_{\beta\in\mathbb{N}_0}\sum_{\alpha\in\mathbb{N}_0}\frac{h^\alpha}{M_\alpha}\|\langle\cdot\rangle^\mu|(\langle\cdot\rangle^{-\mu}F_\beta)*\phi^{(\alpha+\beta)}|\,\|_{L^q}$$

$$\leq 4^{|\mu|}\sum_{\alpha,\beta\in\mathbb{N}_0}\frac{h^\alpha}{M_\alpha}\|F_\beta*(\langle\cdot\rangle^{|\mu|}\phi^{(\alpha+\beta)})\|_{L^q}$$

$$\leq C\sum_{\alpha,\beta\in\mathbb{N}_0}\frac{h^{\alpha+\beta}}{M_\alpha M_\beta}\|\langle\cdot\rangle^{|\mu|}\phi^{(\alpha+\beta)}\|_{L^1}$$

$$\leq AC\sum_{\alpha,\beta\in\mathbb{N}_0}\frac{(Hh)^{\alpha+\beta}}{M_{\alpha+\beta}}\|\langle\cdot\rangle^{|\mu|}\phi^{(\alpha+\beta)}\|_{L^1}$$

$$\leq AC\sum_{\gamma\in\mathbb{N}_0}\frac{(Hh)^\gamma}{M_\gamma}\|\langle\cdot\rangle^{|\mu|}\phi^\gamma\|_{L^1} < \infty$$

where C is the constant defined in (6.39). This means that $f * \phi \in \mathcal{D}_{L^q, \mu}$.

The proof of (vi) is similar.□

Theorem 6.8.2. *Suppose that the sequence (M_p) satisfies conditions $(M.1)$ and $(M.3')$. Fix $\mu, \nu \in \mathbb{R}$ such that $\mu + \nu \geq 0$ and $q, r \in [1, \infty]$ such that $1 \leq 1/q + 1/r \leq 2$.*

(i) *If (M_p) satisfies additionally condition $(M.2')$ and if $f \in \mathcal{D}'_{L^q, \mu}$ and $g \in \mathcal{D}'^{(M_p)}_{L^r, \nu}$, then $f * g = f \star g \in \mathcal{D}'^{(M_p)}_{L^s, \rho}$ and the mapping*

$$\mathcal{D}'_{L^q, \mu} \times \mathcal{D}'^{(M_p)}_{L^r, \nu} \ni (f, g) \mapsto f * g \in \mathcal{D}'^{(M_p)}_{L^s, \rho} \tag{6.40}$$

is continuous, where $\rho := \min(\mu, \nu)$ and $s := (1/q + 1/r - 1)^{-1}$ (then $s \in [1, \infty]$).

(ii) *If (M_p) satisfies additionally condition $(M.2)$ and if $f \in \mathcal{D}'^{(M_p)}_{L^q, \mu}$ and $g \in \mathcal{D}'^{(M_p)}_{L^r, \nu}$, then $f * g = f \star g \in \mathcal{D}'^{(M_p)}_{L^s, \rho}$ and the mapping*

$$\mathcal{D}'^{(M_p)}_{L^q, \mu} \times \mathcal{D}'^{(M_p)}_{L^r, \nu} \ni (f, g) \mapsto f * g \in \mathcal{D}'^{(M_p)}_{L^s, \rho}, \tag{6.41}$$

is continuous, where s and ρ are defined above.

Proof. The proof is divided into five steps.

1° We shall show that, for every strong approximate unit (η_k) in $\mathcal{D}^{(M_p)}(\mathbb{R}^2)$ and $\psi \in \mathcal{D}^{(M_p)}_{L^{s/(s-1)}, -\rho}$, the limit

$$\lim_{k \to \infty} \langle f \otimes g, \eta_k \psi^\triangle \rangle$$

exists and the mapping

$$\mathcal{D}^{(M_p)}_{L^{s/(s-1)}, -\rho} \ni \psi \mapsto \lim_{k \to \infty} \langle f \otimes g, \eta_k \psi^\triangle \rangle \in \mathbb{R}$$

defines an element of $\mathcal{D}'_{L^s, \rho}{}^{(M_p)}$. This will imply that

$$f * g = f \star g \in \mathcal{D}'^{(M_p)}_{L^s, \rho}.$$

By (6.36) and the representation

$$f = \sum_{\alpha \leq \alpha_0} (\langle \cdot \rangle^{-\mu} F_\alpha)^{(\alpha)}, \quad (F_\alpha)_\alpha \in L^q, \quad \alpha \leq \alpha_0,$$

we have

$$f \otimes g = \sum_{\alpha \leq \alpha_0} \sum_{\beta \in \mathbb{N}_0} F^{(\alpha)}_{\mu, \alpha} \otimes G^{(\beta)}_{\nu, \beta},$$

where, for simplicity of notation, we adopt

$$F_{\mu, \alpha} := \langle \cdot \rangle^{-\mu} F_\alpha, \quad G_{\nu, \beta} := \langle \cdot \rangle^{-\nu} G_\beta;$$

the sum on the right side converges weakly in $\mathcal{D}'^{(M_p)}(\mathbb{R}^2)$.

Consider an arbitrary subsequence of a given strong approximate unit (η_i) and select from it a subsequence (η_{m_i}) for which there exist increasing sequences of positive numbers (a_i) and (b_i) divergent to ∞ such that, for all $i \in \mathbb{N}$, we have $a_i < b_i < a_{i+1}$, $\eta_{m_i} \equiv 1$ on $[-a_i, a_i] \times [-a_i, a_i]$ and supp $\eta_{m_i} \subseteq [-b_i, b_i] \times [-b_i, b_i]$.

Let (p_i) be another increasing sequence of positive integers. We shall prove that $((A(m_i, p_i))_{i \in \mathbb{N}}$ is a Cauchy sequence, where

$$A(m_i, p_i) := \sum_{\alpha \leq \alpha_0} \sum_{\beta \leq p_i} \langle F_{\mu,\alpha}^{(\alpha)} \otimes G_{\nu,\beta}^{(\beta)}, \bar{\eta}_{m_i} \psi^\triangle \rangle.$$

It will also be seen that the limit does not depend on the sequences (m_i) and (p_i) and this will imply that the limit

$$\lim_{i \to \infty} \sum_{\alpha \leq \alpha_0} \sum_{\beta \in \mathbb{N}_0} \langle F_{\mu,\alpha}^{(\alpha)} \otimes G_{\nu,\beta}^{(\beta)}, \eta_i \psi^\triangle \rangle$$

exists and is equal to

$$\sum_{\alpha \leq \alpha_0} \sum_{\beta \in \mathbb{N}_0} \lim_{i \to \infty} \langle F_{\mu,\alpha}^{(\alpha)} \otimes G_{\nu,\beta}^{(\beta)}, \eta_i \psi^\triangle \rangle.$$

First note that, for each $\varepsilon > 0$, there exists a $\beta_0 \in \mathbb{N}$ such that

$$\sum_{\beta \geq \beta_0} \frac{M_\beta}{h^\beta} \|G_\beta\|_{L^r} < \varepsilon. \tag{6.42}$$

Next, notice that the inequality succeeding (6.38), which follows from $(M.2')$, and (2.3) imply that, for each $d \geq 1$ and $\alpha, \beta, \gamma, \delta \in \mathbb{N}$ such that $\gamma \leq \alpha$ and $\delta \leq \beta$, we have

$$\frac{1}{M_\beta} \leq \frac{A^\beta H^{(\alpha+\beta)\beta}}{M_{\alpha+\beta}} \leq \frac{(dAH^\beta)^{\alpha+\beta}}{d^\beta M_{\alpha+\beta}} \leq \frac{(dAH^\beta)^{\alpha+\beta}}{d^\beta M_{\alpha-\gamma+\beta-\delta} M_{\gamma+\delta}}. \tag{6.43}$$

Now fix the sequence (m_i) and denote $\bar{\eta}_i = \eta_{m_i}$. Next fix indices $i, k \in \mathbb{N}$ so that $i < k$ and denote for short $a := a_i$, $b := b_\alpha$ (hence we have $0 < a < b$). It is easy to see that

$$\mathrm{supp}\,(\bar{\eta}_k - \bar{\eta}_i) \subseteq K_1 \cup K_2,$$

where

$$K_1 := (J_1 \cup J_2) \times J; \qquad K_2 := J \times (J_1 \cup J_2),$$

and

$$J_1 := [-2b, -a]; \quad J_2 := [a, 2b]; \quad J := [-2b, 2b].$$

Putting

$$c_{\gamma,\delta}^{\alpha,\beta} := \binom{\alpha}{\gamma}\binom{\beta}{\delta}, \quad d_{\gamma,\delta} := \|(\bar{\eta}_\alpha - \bar{\eta}_i)^{(\gamma,\delta)}\|_{L^\infty},$$

for $0 \le \gamma \le \alpha$, $0 \le \delta \le \beta$, we have

$$\sum_{\substack{\alpha \le \alpha_0 \\ p_i \le \beta \le p_k}} |\langle F_{\mu,\alpha}^{(\alpha)} \otimes G_{\nu,\beta}^{(\beta)}, (\bar{\eta}_k - \bar{\eta}_i)\psi^\triangle \rangle| = \sum_{\substack{\alpha \le \alpha_0 \\ p_i \le \beta \le p_k}} \sum_{\substack{\gamma \le \alpha \\ \delta \le \beta}} c_{\gamma,\delta}^{\alpha,\beta} d_{\gamma,\delta}(I_1 + I_2) \quad (6.44)$$

where

$$I_i = \int_{K_i} |(F_{\mu,\alpha} \otimes G_{\nu,\beta})(\psi^{(\alpha-\gamma,\beta-\delta)})^\triangle| \qquad (6.45)$$

for $i = 1, 2$.

Assume first that $\rho = \nu$, i.e., $\nu \le \mu$, $\mu \ge 0$. In this case, Peetre's inequality yields

$$\langle x \rangle^{-\mu} \langle t - x \rangle^{-\nu} \le 4^{|\nu|} \langle x \rangle^{-\mu+|\nu|} \langle t \rangle^{-\nu} \le 4^{|\nu|} \langle t \rangle^{-\nu}$$

for arbitrary $x, t \in \mathbb{R}$ and thus

$$I_1 = \int_{J_1 \cup J_2} \left(\int_{J_x} |F_{\mu,\alpha}(x) G_{\nu,\beta}(t-x) \psi^{(\alpha+\beta-\gamma-\delta)}(t)| \, dt \right) dx$$

$$\le 4^{|\nu|} \int_{J_1 \cup J_2} \int_{J'} |F_\alpha(x) G_\beta(t-x) \langle t \rangle^{-\nu} \psi^{(\alpha+\beta-\gamma-\delta)}(t)| \, dx \, dt$$

where

$$J_x := \{t \in \mathbb{R} : t - x \in J\}$$

and

$$J' := (J_1 \cup J_2) + J = [-4b, 2b-a] \cup [-2b+a, 4b].$$

Now, assume that $\rho = \mu$, i.e., $\mu \le \nu$, $\nu \ge 0$. We have, by Peetre's inequality,

$$\langle y - t \rangle^{-\mu} \langle y \rangle^{-\nu} \le 4^{|\mu|} \langle y \rangle^{|\mu|-\nu} \langle t \rangle^\mu \le 4^{|\mu|} \langle t \rangle^{-\mu}$$

for arbitrary $y, t \in \mathbb{R}$; and, since $J \times J^y = (J_1 \cup J_2) \times J_x$, we have

$$I_1 = \int_J \left(\int_{J^y} |F_{\mu,\alpha}(y-t) G_{\nu,\beta}(y) \psi^{(\alpha+\beta-\gamma-\delta)}(t)| \, dt \right) dy$$

$$\le 4^{|\mu|} \int_{J_1 \cup J_2} \left(\int_{J_x} |F_\alpha(x) G_\beta(t-x) \langle t \rangle^{-\mu} \psi^{(\alpha+\beta-\gamma-\delta)}(t)| \, dt \right) dx$$

$$\le 4^{|\mu|} \int_{J_1 \cup J_2} \int_{J'} |F_\alpha(x) G_\beta(t-x) \langle t \rangle^{-\mu} \psi^{(\alpha+\beta-\gamma-\delta)}(t)| \, dx \, dt,$$

where $J^y := \{t \in \mathbb{R} : y - t \in J_1 \cup J_2\}$.

Hence, putting $C := 4^{|\rho|}$, $D_{\gamma,\delta}^{\alpha,\beta} := \|\langle\cdot\rangle^{-\rho}\psi^{(\alpha+\beta-\gamma-\delta)}\|_{L^{s/(s-1)}}$ and denoting by F'_α the function which equals to 0 on the interval $[-a,a]$ and coincides with F_α otherwise, we get in both cases

$$
I_1 \le C \int_{-\infty}^{\infty} |\langle t\rangle^{-\rho}\psi^{(\alpha+\beta-\gamma-\delta)}(t)| \left(\int_{\mathbb{R}\setminus[-a,a]} |F_\alpha(x)G_\beta(t-x)|\, dx \right) dt
$$

$$
\le C \int_{-\infty}^{\infty} |\langle t\rangle^{-\rho}\psi^{(\alpha+\beta-\gamma-\delta)}(t)| \, (|F'_\alpha| * |G_\beta|)(t)\, dt
$$

$$
\le C\, D_{\gamma,\delta}^{\alpha,\beta} \, \| \, |F'_\alpha| * |G_\beta| \, \|_{L^s} \le C\, D_{\gamma,\delta}^{\alpha,\beta} \, \|F'_\alpha\|_{L^q} \|G_\beta\|_{L^r}, \tag{6.46}
$$

in view of Hölder's and Young's inequalities. Analogously, we prove that

$$
I_2 \le C D_{\gamma,\delta}^{\alpha,\beta} \, \|F_\alpha\|_{L^q} \|G'_\beta\|_{L^r}, \tag{6.47}
$$

where G'_β are functions equal to 0 on $[-a,a]$ and to G_β otherwise. Combining (6.44) - (6.47) and taking into account that

$$
\sup_{\beta,\gamma\in\mathbb{N}_0} \frac{(dAH^\beta)^\gamma}{M_\gamma} < \infty, \qquad \sup_{\alpha,\beta\in\mathbb{N}_0} \frac{(dAH^\beta)^\alpha}{M_\alpha} < \infty,
$$

we get

$$
\sum_{\substack{\alpha\le\alpha_0 \\ p_i\le\beta\le p_k}} |\langle F_{\mu,\alpha}^{(\alpha)} \otimes G_{\nu,\beta}^{(\beta)}, (\tilde{\eta}_k - \tilde{\eta}_i)\psi^\triangle\rangle|
$$

$$
\le C \sum_{\substack{\alpha\le\alpha_0 \\ p_i\le\beta\le p_k}} \sum_{\substack{\gamma\le\alpha \\ \delta\le\beta}} c_{\gamma,\delta}^{\alpha,\beta} d_{\gamma,\delta} D_{\gamma,\delta}^{\alpha,\beta}
$$

$$
\cdot \frac{M_\beta}{h^\beta} \left(\|F_\alpha\|_{L^q(\mathbb{R}\setminus[-a,a])} \|G_\beta\|_{L^r} + \|F_\alpha\|_{L^q} \|G_\beta\|_{L^r(\mathbb{R}\setminus[-a,a])} \right).
$$

If now i,j with $j > i$ vary so that $i \to \infty$, then $a = a_i \to \infty$ and the above estimate, in view of (6.42), shows that $((A(m_i,p_i))_{i\in\mathbb{N}}$ is a Cauchy sequence. It is easy to see that its limit does not depend on subsequences $(\tilde{\eta}_i)$. Consequently,

$$
\langle f * g, \psi\rangle = \lim_{i\to\infty} \langle f \otimes g, \eta_i \psi^\triangle\rangle
$$

$$
= \lim_{i\to\infty} \sum_{\substack{\alpha\le\alpha_0 \\ \beta\in\mathbb{N}_0}} \langle F_{\mu,\alpha}^{(\alpha)} \otimes G_{\nu,\beta}^{(\beta)}, \eta_i \psi^\triangle\rangle = \sum_{\substack{\alpha\le\alpha_0 \\ \beta\in\mathbb{N}_0}} \langle F_{\mu,\alpha}^{(\alpha)} * G_{\nu,\beta}^{(\beta)}, \psi\rangle.
$$

Hence (see [119]) f and g are convolvable, $f * g = f \star g$ and

$$
\langle f*g, \psi\rangle = \sum_{\substack{\alpha\in\mathbb{N}_0 \\ \beta\in\mathbb{N}_0}} \langle F_{\mu,\alpha} * G_{\nu,\beta}^{(\alpha+\beta)}, \psi\rangle
$$

for each $\psi \in \mathcal{D}_{L^{s/(s-1)},-\rho}^{(M_p)}$, which finishes the proof of the first step.

$2°$ Let us prove that under the assumptions of (ii), we have $f * g = f \star g$. According to (6.36), there exist a sequence $(F_\alpha)_\alpha$ of elements of L^q, a sequence $(G_\beta)_\beta$ of elements of $L^r(\mathbb{R})$ and $h > 0$ such that

$$f = \sum_{\alpha \in \mathbb{N}_0} F_{\mu,\alpha}^{(\alpha)}, \qquad g = \sum_{\beta \in \mathbb{N}_0} G_{\nu,\beta}^{(\beta)},$$

where the two series converge in the weak sense, and that

$$\sum_{\alpha \in \mathbb{N}_0} \frac{M_\alpha}{h^\alpha} \|F_\alpha\|_{L^r} < \infty, \qquad \sum_{\beta \in \mathbb{N}_0} \frac{M_\beta}{h^\beta} \|G_\beta\|_{L^q} < \infty.$$

It follows that

$$f \otimes g = \sum_{\alpha \in \mathbb{N}_0} \sum_{\beta \in \mathbb{N}_0} F_{\mu,\alpha}^{(\alpha)} \otimes G_{\nu,\beta}^{(\beta)},$$

where the series on the right side converges weakly in $\mathcal{D}'^{(M_p)}(\mathbb{R}^2)$.

Conditions (2.1) and (2.3) imply that

$$1 \leq \frac{A H^{\alpha+\beta} M_\alpha M_\beta}{M_{\alpha+\beta}} \leq \frac{A(dH)^{\alpha+\beta}}{M_{\alpha-\gamma+\beta-\delta} M_{\gamma+\delta}} \frac{M_\alpha M_\beta}{d^\alpha d^\beta}$$

for each $d > 0$, $\alpha, \beta \in \mathbb{N}$ and $\gamma, \delta \in \mathbb{N}$ such that $\gamma \leq \alpha$, $\delta \leq \beta$. Applying the above estimate instead of (6.43) and repeating the arguments used in $1°$, one can conclude that f and g are convolvable and

$$\langle f * g, \psi \rangle = \sum_{\alpha \in \mathbb{N}_0} \sum_{\beta \in \mathbb{N}_0} \langle F_{\mu,\alpha} * G_{\nu,\beta}^{(\alpha+\beta)}, \psi \rangle$$

for $\psi \in \mathcal{D}_{L^{s/(s-1)},-\rho}^{(M_p)}$.

The next part of the proof is similar to the proof of Proposition 9 in [107].

$3°$ Suppose that condition $(M.2')$ holds. If $f \in \mathcal{D}'_{L^{q,\mu}}$ is fixed, then the mapping

$$\mathcal{D}_{L^{q,\nu}}'^{(M_p)} \ni g \mapsto f * g \in \mathcal{D}_{L^{s,\rho}}'^{(M_p)},$$

has the closed graph. Indeed, if (g_i) converges to zero in $\mathcal{D}_{L^{r,\nu}}'^{(M_p)}$ and $(f * g_i)$ converges to h in $\mathcal{D}_{L^{s,\rho}}'^{(M_p)}$, then

$$\langle h, \phi \rangle = \lim_{i \to \infty} \langle f * g_i, \phi \rangle = \lim_{i \to \infty} \langle f_i(\tilde{u} * \phi), 1 \rangle$$

for each $\phi \in \mathcal{D}^{(M_p)}$ (see [107], Proposition 6). Note that $\tilde{f} * \phi \in \mathcal{D}_{L^{q,\mu}}^{(M_p)}$. Moreover, the multiplication, as a mapping from $\mathcal{D}_{L^{q,\mu}}^{(M_p)} \times \mathcal{D}_{L^{r,\nu}}'^{(M_p)}$ into

$\mathcal{D}'^{(M_p)}_{L^1,\mu+\nu}$, is separately continuous, and $\mathcal{D}'^{(M_p)}_{L^1,\mu+\nu}$ is continuously embedded into $\mathcal{D}'^{(M_p)}_{L^1}$. This implies that $g_i(\tilde{f}*\phi)$ converges to zero in $\mathcal{D}'^{(M_p)}_{L^1}$ as $i \to \infty$. Therefore $\langle h, \phi \rangle = 0$ for each $\phi \in \mathcal{D}^{(M_p)}$. Consequently, the linear mapping in question has the closed graph.

If $g \in \mathcal{D}'^{(M_p)}_{L^q,\nu}$ is fixed, then the mapping

$$\mathcal{D}'_{L^q,\mu} \ni f \mapsto g * f \in \mathcal{D}'^{(M_p)}_{L^s,\rho}$$

has the closed graph. Indeed, if (f_i) converges to zero in $\mathcal{D}'_{L^q,\mu}$ and (f_i*g) converges to h in $\mathcal{D}'^{(M_p)}_{L^s,\rho}$, then

$$\langle h, \phi \rangle = \lim_{i\to\infty} \langle f_i*g, \phi \rangle = \lim_{i\to\infty} \langle f_i(\tilde{g}*\phi), 1 \rangle$$

for each $\phi \in \mathcal{D}^{(M_p)}$ (see [119], Proposition 6, or [70]). We have $\tilde{g}*\phi \in \mathcal{D}_{L^r,\nu}$ and, moreover, the multiplication, as a mapping from $\mathcal{D}_{L^q,\mu} \times \mathcal{D}'_{L^r,\nu}$ to $\mathcal{D}'_{L^1,\mu+\nu}$, is separately continuous, and $\mathcal{D}'_{L^1,\mu+\nu}$ is continuously embedded into $\mathcal{D}'_{L^1}$. Therefore $f_i(\tilde{g}*\phi)$ converges to zero in $\mathcal{D}'_{L^1}$ as $i \to \infty$ and thus $\langle h, \phi \rangle = 0$ for each $\phi \in \mathcal{D}^{(M_p)}$. Consequently, the linear mapping in question has a closed graph.

4° Similarly, one can prove that if condition $(M.2)$ is satisfied and $g \in \mathcal{D}'^{(M_p)}_{L^r,\nu}$ is fixed, then the mapping

$$\mathcal{D}'^{(M_p)}_{L^q,\mu} \ni f \mapsto f * g \in \mathcal{D}'^{(M_p)}_{L^s,\rho},$$

has the closed graph.

5° Note that

(a) the space $\mathcal{D}'^{(M_p)}_{L^q,\mu}$ is the strong dual of a Fréchet space;

(b) the space $\mathcal{D}'_{L^q,\nu}, q \in [1,\infty]$, is an inductive limit of Banach spaces (see [107], Theorem 9).

As we have noted, since $\mathcal{D}'^{(M_p)}_{L^1} = (\mathcal{B}_C^{(M_p)})'$ and the space $\mathcal{B}_C^{(M_p)}$ is semireflexive (see [119]), it follows that $\mathcal{D}'^{(M_p)}_{L^1}$ is barrelled. Therefore the space $\mathcal{D}'^{(M_p)}_{L^1,\nu}$, which is isomorphic to $\mathcal{D}'^{(M_p)}_{L^1}$, is barrelled. It follows that the spaces $\mathcal{D}^{(M_p)}_{L^1,\nu}$ are distinguished Fréchet spaces (see [59], p. 228, Proposition 1). Thus $\mathcal{D}'^{(M_p)}_{L^1,\nu}$ is a bornological space (see [59], p. 289, Theorem 1) and hence, being complete, is an inductive limit of Banach spaces.

It follows from statements given in [123] (Appendix, p. 159, Theorem 3, and p. 164) that if E is an inductive limit of Banach spaces (which implies that E is ultrabornological), F is a strong dual of a Fréchet space and if $T : E \to F$ is a sequentially closed linear mapping, then T is continuous. Consequently, the convolution mappings (6.40) and (6.41) are partially continuous by the closed graph theorem. But then they are continuous (see [59], p. 364, Exercise 10). □

Chapter 7

Integral Transforms of Tempered Ultradistributions

7.1 Introductory remarks

Various integral transforms on the spaces $\mathcal{S}'^{(M_p)}$ and $\mathcal{S}'^{\{M_p\}}$ of tempered Beurling and Roumieu type ultradistributions will be defined and analyzed in this chapter (see [74] and [75]). Also the Hermite expansions of elements of the basic spaces and their duals will be considered; the Hermite expansions can be regarded as generalized integral transforms in the sense of [157], Chapter IX. The use of Hermite expansions will enable us to obtain in Sections 7.3-7.4 , in a similar manner as was done in [118], results about the Wigner distribution, the Fourier, Bargmann and Laplace transforms and the boundary value representation of elements of $\mathcal{S}'^{(M_p)}$ and $\mathcal{S}'^{\{M_p\}}$ (see [75]). In particular, in Sections 7.3 and 7.4 we characterize basic spaces by Fourier and Laplace transforms.

Janssen and van Eijndhoven (see [64]) studied the Gel'fand-Shilov inductive limit type spaces $W_M^{M^\times}$ (see [52]), where $M^\times$ is the Young conjugate of a suitable function M. They characterized elements of these spaces by expansions in Hermite series, Fourier transform, Wigner distribution and Bargmann transform. In the special case where $M(x) = \alpha x^{1/\alpha}$, $x > 0$, $1/2 \le \alpha < 1$, and $M_p = p^{\alpha p}$, $p \in \mathbb{N}_0$, both of the spaces $W_M^{M^\times}$ and $\mathcal{S}^{\{M_p\}}$ are equal to the Gel'fand-Shilov space $\mathcal{S}_\alpha^\alpha$. But, in the general case, the spaces $W_M^{M^\times}$ and the space $\mathcal{S}^*$ are different. In the case of $W_M^{M^\times}$, the function M tends to infinity faster than x and slower than x^2. For $\mathcal{S}^*$ the role of M is taken by the function associated with the sequence (M_p), which is increasing and tends to infinity slower than x. For example, if $M_p = p!^\alpha$, $\alpha > 1$, $p \in \mathbb{N}_0$, then $M(x) \sim \mathcal{C}x^{1/\alpha}$, and Young's conjugate for such a function does not exist at all. Using a technique which is quite different from Janssen and van Eijndhoven's method, we prove that results analogous to theirs hold also for the spaces $\mathcal{S}^*$.

For the sake of simplicity we give most of the definitions, theorems and their proofs in the one-dimensional case in this chapter, although all of the obtained results can be generalized to the multi-dimensional case. In particular, the study of the Hilbert transform in Section 7.6 is presented in both the one-dimensional case and the multi-dimensional case. Further, Section 7.7 is obtained in the multi-dimensional case.

In Section 7.2 the basic definitions of test function spaces are recalled and structural theorems are stated. The main assertion of the section is Theorem 7.2.2, which essentially describes the properties of $\mathcal{S}^{(M_p)}$ and $\mathcal{S}^{\{M_p\}}$ via Hermite expansions and which is the basis for the results of Section 7.3. The proof of Theorem 7.2.2 is given in Section 7.5.

In Sections 7.6 and 7.7 we will study the Hilbert transform and, more generally, singular integral operators on the spaces of tempered ultradistributions of Beurling and Roumieu type.

Note that in this chapter we will use other versions, which we denote $\mathcal{F}_0$ and $\mathcal{F}_0^{-1}$, of the Fourier and inverse Fourier transforms instead of the versions $\mathcal{F}$ and $\mathcal{F}^{-1}$ previously defined in (1.7) and (1.8), respectively. The Fourier transform $\mathcal{F}_0$ and inverse Fourier transform $\mathcal{F}_0^{-1}$ are defined in the n-dimensional case by

$$\mathcal{F}_0[\varphi](x) := \int_{\mathbb{R}^n} \varphi(t) e^{-i\langle x, t \rangle} \, dt, \qquad \varphi \in L^1, \tag{7.1}$$

and

$$\mathcal{F}_0^{-1}[\varphi](x) := \frac{1}{(2\pi)^n} \int_{\mathbb{R}^n} \varphi(t) e^{i\langle x, t \rangle} \, dt, \qquad \varphi \in L^1. \tag{7.2}$$

This does not cause any misinterpretation of results that have been obtained previously in this book. We do this in order to coordinate our notation in this chapter of this book with the notation of the existing literature concerning the results of this chapter. Consequently, the Laplace transform considered in Section 7.4 will be defined using the Fourier transform $\mathcal{F}_0$ and will be denoted by $\mathcal{L}_0$.

7.2   Definitions

In the one-dimensional case, Hermite functions h_α are defined by

$$h_\alpha(x) := (-1)^\alpha \left(\sqrt[4]{\pi} \sqrt{2^\alpha \alpha!} \right)^{-1} e^{x^2/2} \left(e^{-x^2} \right)^{(\alpha)}, \qquad \alpha \in \mathbb{N}, \ x \in \mathbb{R}. \tag{7.3}$$

In the n-dimensional case, we have a multi-indexed sequence of Hermite functions:

$$h_\alpha(x) := h_{\alpha_1}(x_1) h_{\alpha_2}(x_2) \cdot \ldots \cdot h_{\alpha_n}(x_n),$$

where $\alpha := (\alpha_1, \ldots, \alpha_n) \in \mathbb{N}^n$, $x := (x_1, \ldots, x_n) \in \mathbb{R}^n$ and $h_{\alpha_i}(x_i)$ are defined by (7.3). We will use the fact that the set of Hermite functions forms an orthonormal base of the space $L^2(\mathbb{R}^n)$.

The Wigner distribution and the Bargmann transform are defined, respectively, by the formulas:

$$\mathbf{W}(x, y; f) := \frac{1}{\sqrt{2\pi}} \int_{\mathbb{R}} \exp(-iyt) f(x + t/2) \overline{f(y - t/2)} \, dt,$$

$$x, y \in \mathbb{R}, f \in L^2,$$

$$(\mathbf{A}f)(\zeta) := \pi^{-1/4} \int_{\mathbb{R}} \exp(-1/2(\zeta^2 + x^2) + \sqrt{2}\zeta x) f(x) \, dx,$$

$$\zeta \in \mathbb{C}, f \in L^2.$$

For the properties of the Wigner distribution and the Bargmann transform we refer to [13], [62] and [63]. We will assume in this section that conditions $(M.1)$ and $(M.3')$ are satisfied and $M_0 = 1$.

The basic spaces and ultrapolynomials are defined (see the cited papers and [110]) in the n-dimensional case via the multi-indexed sequence M_α, $\alpha = (\alpha_1, \ldots, \alpha_n) \in \mathbb{N}_0^n$. Notice that under $(M.2)$ this multi-indexed sequence and the sequence $\tilde{M}_\alpha = \prod_{i=1}^n M_{\alpha_i}$, $\alpha \in \mathbb{N}_0^n$, define the same spaces of ultradistributions and ultrapolynomials (see [110]).

The Wigner distribution and the Bargmann transform are investigated only in the one-dimensional case in [13], [62], and [63]. Their n-dimensional analogues may be similarly examined.

Let $m > 0$ and $r \in [1, \infty]$ be given. In Section 2.5, the definitions of the norms $\sigma_{m,r}$, $\sigma_{m,\infty}$, $\sigma'_{m,r}, \tau_{m,r}$ for $m > 0, r \in [1, \infty]$ and $\sigma_{(a_p),(b_p),r}(\varphi)$, $\sigma'_{(a_p),(b_p),r}(\varphi)$, $\tau_{(a_p),(b_p)}$ for $(a_p), (b_p) \in \mathcal{R}$ and $r \in [1, \infty]$ have been given for the spaces $\mathcal{S}^{(M_p)}$ and $\mathcal{S}^{\{M_p\}}$ (see Definitions 2.5.1 - 2.5.3). We will consider now additional norms in these spaces.

Definition 7.2.1. Fix a $\varphi \in \mathcal{S}^{(M_p)}$ with $\varphi \overset{L^2}{=} \sum_{k \in \mathbb{N}_0} c_k h_k$. We define the following norms:

$$\sigma''_{m,r}(\varphi) := \sum_{\alpha, \beta \in \mathbb{N}_0} \frac{m^{\alpha+\beta}}{M_\alpha M_\beta} \| [x^\beta \varphi]^{(\alpha)} \|_{L^r};$$

$$\tau_m(\varphi) := \sum_{k \in \mathbb{N}_0} |c_k|^2 \exp[2M(m\sqrt{2k+1})]$$

for $m > 0$, $r \in [1, \infty)$.

Fix $\varphi \in \mathcal{S}^{\{M_p\}}$ with $\varphi \overset{L^2}{=} \sum_{k \in \mathbb{N}_0} c_k h_k$. We define the following norms:

$$\sigma''_{(a_p),(b_p),r}(\varphi) := \sum_{\alpha, \beta \in \mathbb{N}} \frac{\|[x^\beta \varphi]^{(\alpha)}\|_{L^r}}{M_\alpha A_\alpha M_\beta B_\beta};$$

$$\tau_{(a_p)}(\varphi) := \sum_{k \in \mathbb{N}_0} |c_k|^2 \exp[2N_{(a_p)}(\sqrt{2k+1})]$$

for fixed $(a_p), (b_p) \in \mathcal{R}$ and $r \in [1, \infty)$, where

$$A_\alpha := \prod_{p=1}^{\alpha} a_p, \quad B_\beta := \prod_{p=1}^{\beta} b_p$$

if $\alpha, \beta \in \mathbb{N}$ and $A_0 := 1 =: B_0$, according to the notation introduced in (2.57).

Definition 7.2.2. Denote

$$\mathbf{S}''_r := \{\sigma''_{m,r} : m > 0\}, \quad \tilde{\mathbf{S}}''_r := \{\sigma''_{(a_p),(b_p),r} : (a_p), (b_p) \in \mathcal{R}\},$$

$$\mathbf{T} := \{\tau_m : m > 0\}, \quad \tilde{\mathbf{T}} := \{\tau_{(a_p)} : (a_p) \in \mathcal{R}\}$$

for $r \in [1, \infty]$.

Recall the following structural theorem (Theorem 2.5.1 in Section 2.5) for $\mathcal{S}^{\{M_p\}}$:

Theorem 7.2.1. *Let $(a_p), (b_p) \in \mathcal{R}$ and let $\mathcal{S}^{(M_p)}_{(a_p),(b_p)}$ be the space of smooth functions φ on $\mathbb{R}$ such that $\sigma_{(a_p),(b_p),\infty}(\varphi) < \infty$, equipped with the topology induced by the norm $\sigma_{(a_p),(b_p),\infty}$. We have*

$$\mathcal{S}^{\{M_p\}} = \operatorname*{proj\,lim}_{(a_p),\,(b_p)\,\in\,\mathcal{R}} \mathcal{S}^{(M_p)}_{(a_p),(b_p)}.$$

Various norms defined above (recall Definitions 2.5.4 and 7.2.2) lead to different characterizations of test spaces with ultrapolynomial decrease. We prove the following in Section 7.5 below.

Theorem 7.2.2. *The above defined families of norms have the following properties:*

(1) *the families $\mathbf{S}_\infty$ and $\mathbf{S}'_\infty$ (respectively, $\tilde{\mathbf{S}}_\infty$ and $\tilde{\mathbf{S}}'_\infty$) of norms in the space $\mathcal{S}^{(M_p)}$ (respectively, $\mathcal{S}^{\{M_p\}}$) are equivalent;*

(2) *if condition* $(M.2')$ *holds, then for every* $r \in [1, \infty]$ *the families* $\mathbf{S}_r$, $\mathbf{S}'_r$ *and* $\mathbf{S}$ *(respectively,* $\tilde{\mathbf{S}}_r$, $\tilde{\mathbf{S}}'_r$ *and* $\tilde{\mathbf{S}}$*) of norms in the space* $\mathcal{S}^{(M_p)}$ *(respectively,* $\mathcal{S}^{\{M_p\}}$*) are equivalent;*

(3) *if condition* $(M.2)$ *holds, the families* $\mathbf{S}_2$, $\mathbf{S}'_2$ *and* $\mathbf{T}$ *(respectively,* $\tilde{\mathbf{S}}_2$, $\tilde{\mathbf{S}}'_2$ *and* $\tilde{\mathbf{T}}$*) of norms in the space* $\mathcal{S}^{(M_p)}$ *(respectively,* $\mathcal{S}^{\{M_p\}}$*) are equivalent;*

(4) *if condition* $(M.2)$ *holds and a given smooth function* φ *on* $\mathbb{R}$ *satisfies for every (respectively, for some)* $\ell > 0$ *the inequalities:*

$$p_{\ell,\beta}(\varphi) := \sup_{\alpha \in \mathbb{N}_0} \frac{\|x^\beta \varphi^{(\alpha)}\|_{L^2}}{\ell^\alpha M_\alpha} < \infty \qquad (7.4)$$

for all $\alpha \in \mathbb{N}_0$ *and*

$$q_{\ell,\alpha}(\varphi) := \sup_{\beta \in \mathbb{N}_0} \frac{\|x^\beta \varphi^{(\alpha)}\|_{L^2}}{\ell^\beta M_\beta}, \qquad (7.5)$$

for all $\beta \in \mathbb{N}_0$, *then for every (respectively, for some)* $\ell > 0$,

$$a_\ell(\varphi) := \sum_{\alpha, \beta \in \mathbb{N}_0} \frac{\|x^\beta \varphi^{(\alpha)}\|_{L^2}}{\ell^{\alpha+\beta} M_\alpha M_\beta} < \infty. \qquad (7.6)$$

Remark 7.2.1. Notice that a) if condition $(M.2')$ is satisfied, the space $\mathcal{S}_2^{(M_p),m}$ in the definition of $\mathcal{S}^*$ can be replaced by $\mathcal{S}_r^{(M_p),m}$, $r \in [1, \infty]$;
b) part (3) of Theorem 7.2.2 is a characterization of Hermite expansions of elements of test function spaces for the space of tempered distributions;
c) part (4) of Theorem 7.2.2 is an analogue of the following Kashpirovski's result: $\mathcal{S}_\alpha^\alpha = \mathcal{S}^\alpha \cap \mathcal{S}_\alpha$ (see [78] and [46]).

Let us recall from [92] that $\mathcal{S}^{(M_p)}$ and $\mathcal{S}'^{\{M_p\}}$ are $(F\bar{S})$ spaces, that $\mathcal{S}^{\{M_p\}}$ and $\mathcal{S}'^{(M_p)}$ are (LS) spaces and if $(M.2')$ is satisfied then

$$\mathcal{D}^* \hookrightarrow \mathcal{S}^* \hookrightarrow \mathcal{E}^*; \quad \mathcal{S}^* \hookrightarrow \mathcal{S}; \quad \mathcal{E}'^* \hookrightarrow \mathcal{S}'^* \hookrightarrow \mathcal{D}'^*; \quad \mathcal{S}' \hookrightarrow \mathcal{S}'^*.$$

The following theorem is a characterization of Hermite expansions of tempered ultradistributions.

Theorem 7.2.3. *If condition* $(M.2)$ *is satisfied, then the spaces* $\mathcal{S}^{(M_p)}$ *and* $\mathcal{S}'^{(M_p)}$ *are* $(FN)-$*spaces; and the spaces* $\mathcal{S}^{\{M_p\}}$ *and* $\mathcal{S}'^{(M_p)}$ *are* $(LN)-$*spaces, respectively.*

Proof. If condition $(M.2)$ is satisfied, the spaces $\mathcal{S}^{(M_p)}$ and $\mathcal{S}^{\{M_p\}}$ are isomorphic to the spaces of projective and inductive limits of Köthe spaces $\ell^2(b_k)$ and $\ell^2(c_k)$ (see [50]), respectively, where

$$b_k = (b_{1,k}, b_{2,k}, \ldots), \quad b_{k,k} = \exp[M(k\sqrt{2k+1})],$$

$$c_k = (c_{1,k}, c_{2,k}, \ldots), \quad c_{k,k} = \exp[M((1/k)\sqrt{2k+1})],$$

respectively, for $n, k \in \mathbb{N}$. The isomorphism is given by the mapping: $\varphi \mapsto (a_k)$, where $\varphi = \sum_{k=0}^{\infty} a_k h_k$ (see Theorem 7.2.2). In order to prove the assertion it is enough to show that the inequalities:

$$\sum_{k \in \mathbb{N}_0} b_{k,k}/b_{k,\ell} < \infty, \qquad \sum_{k \in \mathbb{N}_0} c_{k,k}/c_{k,\ell} < \infty \qquad (7.7)$$

are satisfied for some $\ell > k$ (see for example [50], p. 112, 4.3). It follows from the inequalities

$$M(k\rho) + M(\rho) \le 2M((k+1)\rho), \qquad 2M(\rho) \le M(H\rho) + \log_+ A,$$

which are true for $k \in \mathbb{N}_0$ and $\rho > 0$, that

$$\sum_{k \in \mathbb{N}_0} \frac{b_{k,k}}{b_{k,\ell}} \le \sum_{k \in \mathbb{N}_0} \exp(-M(\sqrt{2k+1})) < \infty$$

for $\ell > H(k+1)$; thus the first of the inequalities in (7.7) holds true. The proof of the second inequality is similar. $\square$

We now present another structural theorem for the spaces $\mathcal{S}'^*$.

Theorem 7.2.4. *Assume that condition $(M.2)$ is satisfied and let $f \in \mathcal{D}'^{(M_p)}$ (respectively, $f \in \mathcal{D}'^{\{M_p\}}$). Then $f \in \mathcal{S}'^{(M_p)}$ (respectively, $f \in \mathcal{S}'^{\{M_p\}}$) if and only if*

$$f = \sum_{k \in \mathbb{N}_0} a_k h_k \qquad in \ \mathcal{S}'^*$$

and, for some (respectively, for every) $\delta > 0$,

$$\sum_{k \in \mathbb{N}_0} \mid a_k \mid^2 \exp\left[-2M(\delta\sqrt{2k+1})\right] < \infty.$$

Remark 7.2.2. All of the definitions given in this section can be easily generalized to the n-dimensional case. For example $\mathcal{S}_r^{M_p,m}(\mathbb{R}^n)$, $r \ge 0$, and $m > 0$, is the space of smooth functions φ on $\mathbb{R}^n$ which satisfy the inequality

$$\sigma_{m,r}(\varphi) = \left(\sum_{\alpha,\beta \in \mathbb{N}_0^n} \int_{\mathbb{R}^n} \mid \frac{m^{\alpha+\beta}}{M_\alpha M_\beta} < x >^\beta \varphi^{(\alpha)}(x) \mid^r dx \right)^{1/r} < \infty,$$

equipped with the topology induced by the norm $\sigma_{m,r}$. Similarly, it is easy to verify that the proofs of the theorems of this section still hold in the n-dimensional case. The proofs can be written in the same way if we use the fact that the multi-indexed sequences M_α and $\tilde{M}_\alpha$, $\alpha \in \mathbb{N}_0^n$, where $\tilde{M}_\alpha$ was defined in the third paragraph preceding Definition 7.2.1, determine the same spaces of ultradistributions under the assumption that $(M.2)$ holds, and that the operator $\Re^\alpha$, $\alpha \in \mathbb{N}_0^n$, is defined by

$$\Re^\alpha = (x_1^2 - \frac{\partial^2}{\partial x_1^2})^{\alpha_1}(x_2^2 - \frac{\partial^2}{\partial x_2^2})^{\alpha_2} \ldots (x_n^2 - \frac{\partial^2}{\partial x_n^2})^{\alpha_n}, \ \alpha = (\alpha_1, .., \alpha_n) \in \mathbb{N}_0^n.$$

Notice that α^β denotes $\alpha_1^{\beta_1} \ldots \alpha_n^{\beta_n}$, where $\alpha = (\alpha_1, \ldots, \alpha_n) \in \mathbb{N}_0^n$ and $\beta = (\beta_1, \ldots, \beta_n) \in \mathbb{N}_0^n$ as defined in Section 1.1.

7.3 Characterizations of some integral transforms

Suppose that conditions $(M.1)$, $(M.2)$ and $(M.3')$ are satisfied . The Fourier transform is an isomorphism of $\mathcal{S}^*$ onto itself. In the next theorem we give the characterizations which are similar to the ones given in [64].

Theorem 7.3.1. *Functions of the class $\mathcal{S}^{(M_p)}$ (respectively, $\mathcal{S}^{\{M_p\}}$) can be characterized in terms of certain integral transforms in the following way:*

1. [CHARACTERIZATION VIA THE FOURIER TRANSFORM] *A function φ belongs to $\mathcal{S}^{(M_p)}$ (respectively, to $\mathcal{S}^{\{M_p\}}$) if and only if it is square integrable and for every (respectively, for some) $h > 0$,*

$$\varphi(x) = \mathcal{O}\left(\exp\left[-M(h \mid x \mid)\right]\right) \ and \ (\mathcal{F}_0\varphi)(x) = \mathcal{O}\left(\exp\left[-M(h \mid x \mid)\right]\right).$$

2. [CHARACTERIZATION VIA THE WIGNER DISTRIBUTION] *A function $\varphi \in \mathcal{S}^{(M_p)}$ (respectively, $\varphi \in \mathcal{S}^{\{M_p\}}$) if and only if for every (respectively, for some) $\lambda > 0$*

$$\mathbf{W}(x, y; \varphi) = \mathcal{O}(\exp\left[-M(\lambda(x^2 + y^2)^{1/2})\right]).$$

3. [CHARACTERIZATION VIA THE BARGMANN TRANSFORM] *A function $\varphi \in \mathcal{S}^{(M_p)}$ (respectively, $\varphi \in \mathcal{S}^{\{M_p\}}$) if and only if for every (respectively, for some) $\lambda > 0$ there exists a $C > 0$ such that*

$$|(\mathbf{A}\varphi)(\zeta)| \leq C \exp\left[\frac{1}{2}|\zeta|^2 - M(\lambda|\zeta|)\right], \qquad \zeta \in \mathbb{C}.$$

Proof. Parts (1) and (3) of Theorem 7.2.2 imply that our assertion 1 holds. Hence, according to parts (2) and (4) of Theorem 7.2.2 and a calculation based on the properties of the function M, parts 2 and 3 of our assertion follow. $\square$

7.4   Laplace transform

Assume that $(M.1)$, $(M.2)$ and $(M.3)$ are satisfied.

Denote by $\mathcal{S}'^{*}_{+}$ the subspace of $\mathcal{S}'^{*}$ consisting of elements supported by $[0, \infty)$. Let $g \in \mathcal{S}'^{*}_{+}$. For fixed $y < 0$ we define the ultradistribution $g \exp[y \cdot]$ as an element of $\mathcal{S}'^{*}$ such that

$$\langle g \exp[y \cdot], \varphi \rangle := \langle g, \varrho \exp[y \cdot]\varphi \rangle, \quad \varphi \in \mathcal{S}^{*},$$

where ϱ is a function in $\mathcal{E}^{*}$ such that, for some $\varepsilon > 0$, $\varrho(x) = 1$ if $x \in (-\varepsilon, \infty)$ and $\varrho(x) = 0$ if $x \in (-\infty, -2\varepsilon)$. As an example of such a ϱ we may take $\varrho := f * \omega$, where $\omega \in \mathcal{D}^{*}$, $\int \omega = 1$, $\mathrm{supp}\, \omega \subseteq [-\varepsilon/2, \varepsilon/2]$ and $f(x) = 1$ if $x \geq -3\varepsilon/2$, $f(x) = 0$ if $x < -3\varepsilon/2$. (For the existence of a function ω with the mentioned properties we refer to [82], Theorem 4.2.) It is easy to see that the definition does not depend on the choice of ϱ. As in the case of $\mathcal{S}'_{+}$ (see for example [149]) we define the Laplace transform of $g \in \mathcal{S}'^{*}_{+}$ by

$$(\mathcal{L}_0 g)(\zeta) := \mathcal{F}_0(g \exp[y \cdot])(x), \quad \zeta = x + iy \in \mathbb{C}_{-}.$$

Clearly, $\mathcal{L}_0 g$ is an element of $\mathcal{S}'^{*}$ for every fixed $y < 0$.

Let

$$G(\zeta) := \langle g, \eta \exp[-i\zeta \cdot] \rangle, \quad \zeta = x + iy \in \mathbb{C}_{-}, \tag{7.8}$$

where η is as above. The function G is analytic on $\mathbb{C}_{-}$ and does not depend on η.

The next two theorems were proved in [118] for the spaces of Beurling-Gevrey tempered ultradistributions $\Sigma'_{\alpha} = \mathcal{S}^{(p^{\alpha p})}$, $\alpha > 1/2$, on the real line, supported by $[0, \infty)$. Their proofs in the general case are quite analogous.

Theorem 7.4.1. *Let* $g \in \mathcal{S}'^{(M_p)}_{+}$ *(respectively, $g \in \mathcal{S}'^{\{M_p\}}_{+}$) and let G be defined by (7.8). Then*

1. for every $\varepsilon > 0$ there are a $k > 0$ and a $C > 0$ (respectively, for every $\varepsilon > 0$ and $k > 0$ there exists a $C > 0$) such that

$$|G(\zeta)| \leq C \exp\left(\varepsilon y + \left(M(k|x|) + \tilde{M}\left(k|y|^{-1}\right)\right)\right), \quad \zeta = x + iy \in \mathbb{C}_{-};$$

2. $(\mathcal{L}_0 g)(x + iy) = G(x + iy)$, for every fixed $y < 0$ and for all $x \in \mathbb{R}$;

3. there is a $G(\cdot + i0)$ in $\mathcal{S}'^{(M_p)}_{+}$ (respectively, in $\mathcal{S}'^{\{M_p\}}_{+}$) such that

$$G(x + iy) \to G(x + i0) \quad \text{as } y \to 0-$$

in the sense of convergence in $\mathcal{S}'^{(M_p)}$ (respectively, in $\mathcal{S}'^{\{M_p\}}$) and

$$G(x + i0) = (\mathcal{F}_0 g)(x), \quad x \in \mathbb{R};$$

4. if $G_k(\zeta) = (\mathcal{L}g_k)(\zeta)$ for $\zeta \in \mathbb{C}_{-}$ and $k = 1, 2$ and $G_1(x + i0) = G_2(x + i0)$ for $x \in \mathbb{R}$, then $g_1 = g_2$.

Applying the above theorem, as in [118], one can obtain a representation of elements of the space $\mathcal{S}'^{(M_p)}$, $M_p = p^{\alpha p}$, of tempered ultradistributions as boundary values of appropriate harmonic functions in the lower half-plane, which are expanded into series of Hermite functions of second type.

Let $b > 0$ (respectively, $(b_p) \in \mathcal{R}$) be given and let P_b (respectively, $P_{(b_p)}$) be an entire function such that, for some constants $L > 0$ and $\mathcal{C}$,

$$|P_b(\zeta)| \le \mathcal{C} \exp[M(L|\zeta|)] \left(\text{respectively, } |P_{(b_p)}(\zeta)| \le \mathcal{C} \exp[N_{(b_p)}(L|\zeta|)] \right) \tag{7.9}$$

for all $\zeta = x + iy \in \mathbb{C}$ and

$$\exp[M(b|\zeta|)] \le P_b(\zeta) \left(\text{respectively, } \exp[N_{(b_p)}(|\zeta|)] \le P_{(b_p)}(\zeta) \right), \tag{7.10}$$

for $\zeta = x + iy \in \mathbb{C}$ such that $|x| \ge |y|$. If conditions $(M.1)$, $(M.2)$ and $(M.3)$ are satisfied, an example of such an entire function is

$$P_b(\zeta) = \prod_{\alpha=1}^{\infty} (1 + \frac{\zeta^2}{b^2 m_\alpha^2})$$

$$\left(\text{respectively, } P_{(b_p)}(\zeta) = \prod_{\alpha=1}^{\infty} (1 + \frac{\zeta^2}{b_\alpha^2 m_\alpha^2}) \right), \zeta \in \mathbb{C}.$$

From [82], p. 91, it follows that this entire function satisfies (7.9). Notice that (7.10) holds, because for $\zeta = x + iy \in \mathbb{C}$ and $|x| \ge |y|$, we have

$$\left| \prod_{\alpha=1}^{\infty} (1 + \frac{\zeta^2}{b_\alpha^2 m_\alpha^2}) \right| \ge \sup_{\beta \in \mathbb{N}} \prod_{\alpha=1}^{\beta} \left| 1 + \frac{\zeta^2}{b_\alpha^2 m_\alpha^2} \right|$$

$$\ge \sup_{\beta \in \mathbb{N}} \prod_{\alpha=1}^{\beta} \left| \frac{\zeta^2}{b_\alpha^2 m_\alpha^2} \right| = \exp[2N_{b_p}(|\zeta|)].$$

Theorem 7.4.2. [CHARACTERIZATION VIA THE LAPLACE TRANSFORM] *Let G be an analytic function on $\mathbb{C}_-$. Then G is the Laplace transform of some $g \in \mathcal{S}'^{(M_p)}_+$ (respectively, $\mathcal{S}'^{\{M_p\}}_+$) if and only if for every $\varepsilon > 0$ there are a constant $k > 0$, an ultradifferential operator P_b and constant $\mathcal{C}$ (respectively, for every $\varepsilon > 0$ there is an ultradifferential operator P_{b_p} such that for every $k > 0$ there exists a constant $\mathcal{C}$) for which*

$$\left\| \frac{G(\cdot + iy)}{P_b(\cdot + iy)} \right\|_{L^2} \le \mathcal{C} \exp(\varepsilon y + \tilde{M}(k|y|^{-1})), \qquad y < 0,$$

$$\left(\text{respectively, } \left\| \frac{G(\cdot + iy)}{P_{b_p}(\cdot + iy)} \right\|_{L^2} \le \mathcal{C} \exp(\varepsilon y + \tilde{M}(k|y|^{-1})), \ y < 0 \right).$$

Moreover,

$$(\mathcal{F}_0 g)(x) = \lim_{y \to 0-} G(x + iy), \qquad x \in \mathbb{R},$$

in $\mathcal{S}'^{(M_p)}$ *(respectively, in* $\mathcal{S}'^{\{M_p\}}$*).*

Remark 7.4.1. The proofs of the theorems of this section are valid in the n-dimensional case; they can be written in the same way if we use the conventions mentioned in Remark 7.2.2.

7.5   Proof of equivalence of families of norms

In order to prove the assertion of Theorem 7.2.2 we need the following two lemmas concerning Hermite functions proved in [78] (see also [3]).

Lemma 7.5.1. *Let* $m, k \in \mathbb{N}$. *Then*

$$x^m h_k(x) = 2^{-m/2} \sum_{j=0}^{m} c_{j,m}^{(k)} \, h_{k-m+2j}(x), \qquad x \in \mathbb{R}, \qquad (7.11)$$

where

$$|c_{j,m}^{(k)}| \leq \binom{m}{j} [(2k+1)^{m/2} + m^{m/2}], \qquad (7.12)$$

with the convention $h_{k-m+2j} = 0$ *in case* $k - m + 2j < 0$.

Let $\Re^0$ be the identity operator, $\Re^j = (x^2 - d^2/dx^2)^j$ for $j \in \mathbb{N}$ and denote $\Re := \Re^1$.

Lemma 7.5.2. *The operator* $\Re$ *is self-adjoint in* $\mathcal{S}$, $\Re \, h_k = (2k+1)h_k$ *for* $k \in \mathbb{N}$ *and* $\Re^j$ *is of the form*

$$\Re^j \varphi(x) = \sum_{\substack{p+q=2k \\ 0 \leq k \leq j}} C_{p,q}^{(j)} x^p \varphi^{(q)}(x), \qquad (7.13)$$

where the coefficients $C_{p,q}$ *satisfy the estimate*

$$|C_{p,q}^{(j)}| \leq 10^j \, j^{j - \frac{p+q}{2}} \qquad (7.14)$$

for $\varphi \in \mathcal{S}$, $x \in \mathbb{R}$ *and* $j \in \mathbb{N}$.

We now give the

Proof of Theorem 7.2.2.
Recall that Parts (1) and (2) already have been proved in Theorem
2.5.2. We give here the proofs of Parts (3) and (4).

Proof of Part (3). Let us prove the equivalence of systems $\{b_{m,2}\colon m >
0\}$ and $\{\tau_m\colon m > 0\}$ of norms, which together with the fact that

$$s'_{m,2}(\varphi) = \frac{1}{\sqrt{2\pi}} b_{m,2}(\mathcal{F}_0\varphi), \qquad \varphi \in \mathcal{S}^*,$$

and

$$\tau_m(\mathcal{F}_0\varphi) = \tau_m(\varphi), \qquad \varphi \in \mathcal{S}^*,$$

implies the equivalence of the systems $\{\bar{s}_{m,2},\, m > 0\}$ and $\{\theta_\delta,\, \delta > 0\}$ of
norms. It is enough to prove that if (7.6) holds then the estimation

$$\sum_{k \in \mathbb{N}_0} |a_k|^2 \exp[2M(\delta\sqrt{2k+1})] < \infty, \tag{7.15}$$

holds for $\delta = (\sqrt{20e}\,(1+H)^4 \ell)^{-1}$, and conversely that (7.15) implies (7.6)
with $\ell = H\sqrt{8/\delta}$.

Proof of Part (4). Suppose that (7.6) holds. In the estimates which are
to follow we shall use the fact that for every $k \in \mathbb{N}_0$ and $L > 0$,

$$L^{-(j-k)} \frac{j!}{k!} \frac{M_k}{M_j} \to 0 \qquad \text{as } j \to \infty, \tag{7.16}$$

which follows from $(M.3')$ since, for $j > k$,

$$L^{-(j-k)} \frac{j!}{k!} \frac{M_j}{M_k} = \frac{j}{Lm_j} \cdot \frac{j+1}{Lm_{j+1}} \cdots \frac{j-k+1}{Lm_{j-k+1}} \to 0 \qquad \text{as } j \to \infty,$$

where $m_k := M_k/M_{k-1}$, $k = 1, 2, \ldots$ (see [82], (4.5)).

By Theorem 7.2.4 and Lemma 7.5.2, we have, for every $j \in \mathbb{N}_0$,

$$\Re^j \varphi = \sum_{k \in \mathbb{N}_0} a_k \Re^j h_k = \sum_{k \in \mathbb{N}_0} a_k (2k+1)^{2j} h_k,$$

so, by (7.13), (7.14), (7.4) and (7.5), there exists a constant $\mathcal{C} > 0$ such
that

$$\left(\sum_{k \in \mathbb{N}_0} |a_k|^2 (2k+1)^{2j} \right)^{1/2} = \|\Re^j \varphi\|_{L^2} \sum_{\substack{p+q=2k \\ 0 \le k \le j}} C_{p,q}^{(j)} \|x^p \varphi^{(q)}\|_{L^2}$$

$$\le \mathcal{C}(10j)^j \sum_{\substack{p+q=2k \\ 0 \le k \le j}} \left(\frac{p+q}{2} \right)^{-\frac{p+q}{2}} \ell^{p+q} M_p M_q.$$

Hence, by formula (2.3) obtained due to condition $(M.1)$, by the two inequalities following from Stirling's formula and by condition $(M.2)$, we get

$$\left(\sum_{k \in \mathbb{N}_0} |a_k|^2 (2k+1)^{2j}\right)^{1/2} \leq \mathcal{C}(10j)^j \sum_{0 \leq k \leq j} k^{-k} \ell^k M_{2k}$$

$$\leq \mathcal{C}(20eH^2\ell^2)^j \sum_{0 \leq k \leq j} \frac{j!}{k!2^k} \frac{M_k M_k}{(H\ell)^{2(j-k)} M_j M_j} M_{2j}$$

$$\leq \mathcal{C}\,(20eH^2\ell^2)^j M_{2j}.$$

Moreover, from the above and condition $(M.2')$, it follows, for some $\mathcal{C}$, that

$$\left(\sum_{k \in \mathbb{N}_0} |a_k|^2 (2k+1)^{2j-1}\right)^{1/2} \leq \mathcal{C}\,(20eH^4\ell^2)^j\, M_{2j-1}.$$

Thus there exists a constant $\mathcal{C}$ such that, for every j and $k \in \mathbb{N}_0$,

$$|a_k|^2 (2k+1)^j \leq \mathcal{C}\,(\sqrt{20e}\,(1+H)^2\,\ell)^{2j}\, M_j^2.$$

By putting $j+2$ instead of j in the above inequality and by using $(M.2')$ it follows that for every $j, k \in \mathbb{N}_0$ and $\delta = (\sqrt{20e}\,(1+H)^4\,\ell)^{-1}$ there exists a $\mathcal{C} > 0$ such that

$$\frac{|a_k|^2 \delta^{2j}(2k+1)^j}{M_j^2} \leq \frac{\mathcal{C}}{(2k+1)^2},$$

which implies

$$|a_k|^2 \exp[2M(\delta\sqrt{2k+1}\,)] = |a_k|^2 \sup_{j \in \mathbb{N}_0} \frac{\delta^{2j}(2k+1)^j}{M_j^2} \leq \frac{\mathcal{C}}{(2k+1)^2}.$$

Therefore,

$$\sum_{k \in \mathbb{N}_0} |a_k|^2 \exp[2M(\delta\sqrt{2k+1}\,)] < \infty.$$

Suppose that (7.15) holds for some $\delta > 0$ and fix an $m \in \mathbb{N}_0$. Applying inequalities (7.11) and (7.12) in Lemma 7.5.1 and the Cauchy-Schwarz inequality we have

$$\|x^m \varphi\|_{L^2} = \|\sum_{k \in \mathbb{N}_0} a_k x^m h_k\|_{L^2} \leq 2^{-m/2} \sum_{k \in \mathbb{N}_0} |a_k| \left(\sum_{j \leq m} \binom{m}{j} \bar{C}_k\right)$$

$$\leq 2^{m/2} \left(\sum_{k \in \mathbb{N}_0} |a_k|^2 C_k^2\right)^{1/2} \left(\sum_{k \in \mathbb{N}_0} C_k^2 \bar{C}_k^2\right)^{1/2},$$

where
$$C_k := \exp[M(\delta(2k+1))], \qquad \bar{C}_k := (2k+1)^{m/2} + m^{m/2}.$$

Hence there exists a constant $\mathcal{C}$ such that

$$\|x^m\varphi\|_{L^2} \le \mathcal{C}\, 2^{m/2} \left[\sum_{k\in\mathbb{N}_0} C_k^2 \bar{C}_k^2\right]^{1/2}$$

$$\le \mathcal{C}\, 2^{m/2}\, \bar{C} \left[\sum_{k\in\mathbb{N}_0} \exp[-M(\delta(2k+1))]\right]^{1/2}$$

$$\le \mathcal{C} \left(\frac{2}{\delta}\right)^{m/2} M_m \delta^{m/2}\tilde{C} + \frac{\delta^{m/2}m^{m/2}}{M_m},$$

where
$$\bar{C} := \sup_{k\in\mathbb{N}_0}\,[(2k+1)^{m/2} + m^{m/2}]\, \exp[-\tfrac{1}{2}M(\delta(2k+1))]$$

and
$$\tilde{C} := \sup_{k\in\mathbb{N}_0}\, \frac{(2k+1)^{m/2}\exp[-\tfrac{1}{2}M(\delta(2k+1))]}{M_m}.$$

Notice that
$$\frac{\delta^{m/2}m^{m/2}}{M_m} \le \frac{m!e^m\delta^{m/2}}{M_m}$$

and
$$\delta^{m/2}\tilde{C} = \frac{1}{M_m}\sup_{k\in\mathbb{N}_0}\left(\frac{\delta^m(2k+1)^m}{\exp[M(\delta(2k+1))]}\right)^{1/2}$$

for all $m \in \mathbb{N}$. Since
$$\lim_{m\to\infty} \frac{m!e^m\delta^{m/2}}{M_m} = \lim_{m\to\infty} \frac{\sqrt{M_m}}{M_m} = 0,$$

which follows from [82], (3.3), we conclude that
$$\|x^m\varphi\|_{L^2} \le \mathcal{C}\left(\sqrt{2/\delta}\right)^m M_m. \tag{7.17}$$

Applying the Fourier transform, we get

$$\|\varphi^{(k)}\|_{L^2} = \frac{1}{\sqrt{2\pi}}\|\mathcal{F}_0[\varphi^{(k)}]\|_{L^2} = \frac{1}{\sqrt{2\pi}}\|x^k\mathcal{F}_0[\varphi]\|_{L^2}$$

$$= \frac{1}{\sqrt{2\pi}}\|x^k\sum_{j\in\mathbb{N}_0} a_j h_j\|_{L^2} = \frac{1}{\sqrt{2\pi}}\|x^k\varphi\|_{L^2}$$

$$\le \mathcal{C}\left(\sqrt{\frac{2}{\delta}}\right)^k M_k \tag{7.18}$$

for every $k \in \mathbb{N}_0$.

Fix $\alpha, \beta \in \mathbb{N}_0$ and denote $\gamma := \min\{\alpha, 2\beta\}$. We have

$$\left(\|x^\beta \varphi^{(\alpha)}\|_{L^2}\right)^2 = (x^\beta \varphi^{(\alpha)}, x^\beta \varphi^{(\alpha)})_{L^2} = |((x^{2\beta}\varphi^{(\alpha)})^{(\alpha)}, \varphi)_{L^2}|$$

$$\leq \left| \sum_{\kappa=0}^{\gamma} \binom{\alpha}{\kappa} \frac{(2\beta)!}{(2\beta-\kappa)!} (x^{2\beta-\kappa}\varphi^{(2\alpha-\kappa)}, \varphi)_{L^2} \right|$$

$$\leq \sum_{\kappa=0}^{\gamma} \binom{\alpha}{\kappa}\binom{2\beta}{\kappa} \kappa! \, \|x^{2\beta-\kappa}\varphi\|_{L^2} \, \|\varphi^{(2\alpha-\kappa)}\|_{L^2}.$$

Hence, by (7.17), (7.18) and conditions $(M.1)$, $(M.3')$ and $(M.2)$, we obtain

$$(\|x^\beta \varphi^{(\alpha)}\|_{L^2})^2 \leq \mathcal{C} \sum_{\kappa=0}^{\gamma} \binom{\alpha}{\kappa}\binom{2\beta}{\kappa} \kappa! \, (2/\delta)^{(\alpha+\beta-\kappa)} \frac{M_\kappa^2}{M_\kappa^2} M_{2\alpha-\kappa} M_{2\beta-\kappa}$$

$$\leq \mathcal{C} \sum_{\kappa=0}^{\gamma} \binom{\alpha}{\kappa}\binom{2\beta}{\kappa} (2/\delta)^{(\alpha+\beta-\kappa)} \frac{\kappa!}{M_\kappa^2} M_{2\alpha} M_{2\beta}$$

$$\leq \mathcal{C} H^{2(\alpha+\beta)} M_\alpha^2 M_\beta^2 \sum_{\kappa=0}^{\gamma} \binom{\alpha}{\kappa}\binom{2\beta}{\kappa} (2/\delta)^{\alpha+\beta}$$

$$\leq \mathcal{C} 8^{\alpha+\beta} H^{2(\alpha+\beta)} \delta^{-(\alpha+\beta)} M_\alpha^2 M_\beta^2, \qquad (7.19)$$

which imply that (7.6) holds for $\ell = 4H/\sqrt{\delta}$.

Applying analogous reasoning as in (7.19) one can prove the last part of the theorem. $\square$

7.6 Hilbert transform

The Hilbert transform on distribution and ultradistribution spaces has been studied by many mathematicians, see e.g. Tillmann [142], Beltrami and Wohlers [4], Vladimirov [149], Singh and Pandey [136], Ishikawa [61], Ziemian [159] and Pilipović [112]. In all of these papers the Hilbert transform is defined by one of two methods: by the method of adjoints or by considering a generalized function on the kernel which belongs to the corresponding test function space.

Ziemian [159] defined the right and left Hilbert transform of a tempered distribution $T \in \mathcal{S}'(\mathbb{R})$ as an element of $\mathcal{S}'(\mathbb{R})$, using the kernel

$$G(z) := \int_0^\infty \vartheta(x) x^{-z-1} \, dx, \qquad z \in \mathbb{C} \setminus \{0\},$$

where $\vartheta \in C_0^\infty(\mathbb{R})$ and $\vartheta = 1$ in a neighborhood of zero.

Koizumi in [80] and [81] considered the generalized Hilbert transform H for $f \in W^2(\mathbb{R})$ defined by

$$Hf(x) := \lim_{\varepsilon \to 0+} \frac{x+i}{\pi} \int_{|t|>\varepsilon} \frac{f(x-t)}{t(x-t+i)} \, dt,$$

where $W^2(\mathbb{R})$ is the space of functions f such that $g \in L^2(\mathbb{R})$, where $g(x) := f(x)/(1+|x|)$ for $x \in \mathbb{R}$. A generalization of the same type was given by Ishikawa [61] who extended the Hilbert transform to the space of tempered distributions.

We follow this latter method and extend Ishikawa's generalized Hilbert transform to the spaces of tempered ultradistributions first in the one-dimensional and then in the n-dimensional case. Our generalization of the Hilbert transform is defined as the composition of the classical Hilbert transform with the operators $\varphi \mapsto P\varphi$ and $\varphi \mapsto (1/P)\varphi$, where P is an elliptic ultradifferential operator.

Subsection 7.6.1 below is devoted to the generalized Hilbert transform defined on the spaces $\mathcal{S}'^{(M_p)}(\mathbb{R})$ and $\mathcal{S}'^{\{M_p\}}(\mathbb{R})$. Structural properties of the basic spaces imply that the Hilbert transform of a tempered ultradistribution is defined uniquely in the sense of hyperfunctions.

In Subsection 7.6.2 we define the Hilbert transform on the spaces $\mathcal{S}'^*(\mathbb{R}^n)$. The simple structure of the kernel enables us to define this transform as the iterations of the one-dimensional Hilbert transform.

7.6.1   *One-dimensional case*

In order to define the Hilbert transform on $\mathcal{S}'^*$ we follow Ishikawa's ideas of introducing this transform for tempered distributions (see [61]) and represent $\mathcal{S}^{(M_p)}$ (respectively, $\mathcal{S}^{\{M_p\}}$) as the projective limit of the spaces $\mathcal{D}_a^{(M_p)}$, $a > 0$ (respectively, $\mathcal{D}_{(a_p)}^{\{M_p\}}$, $(a_p) \in \mathbb{R}^{\mathbb{N}}$). We follow here Ishikawa's method of notation using the symbols $\mathcal{D}_a^{(M_p)}$, $a > 0$ (respectively, $\mathcal{D}_{(a_p)}^{\{M_p\}}$, $(a_p) \in \mathbb{R}^{\mathbb{N}}$) instead of the symbol $\mathcal{S}$ with respective indices to avoid too many spaces denoted by the symbol S in the book. Consequently, it should be remembered that members of the space $\mathcal{D}_a^{(M_p)}$ (respectively, $\mathcal{D}_{(a_p)}^{\{M_p\}}$) do not have compact support in general.

In contrast to the case of tempered distributions, the space $\mathcal{S}^{(M_p)}$ (respectively, $\mathcal{S}^{\{M_p\}}$) is not dense in the space $\mathcal{D}_a^{(M_p)}$ (respectively, $\mathcal{D}_{(a_p)}^{\{M_p\}}$) in general. We overcome this difficulty because of parts 4 and 5 of Theorem 7.6.3 below.

Let us now give definitions and a structural characterization of basic spaces which will be used in our investigation of the Hilbert transform.

As in Section 2.2, $\mathcal{R}$ will denote the family of all sequences of positive numbers increasing to ∞. Recall (see (7.9)-(7.10)) that, for a given $b > 0$ (respectively, $(b_p) \in \mathcal{R}$), we denote by P_b (respectively, by $P_{(b_p)}$) an arbitrary entire function such that there exist positive constants $\mathcal{C}$ and L for which the following inequalities are satisfied:

$$|P_b(\zeta)| \leq \mathcal{C} \exp\left[M(L|\zeta|)\right]$$
$$\left(\text{respectively, } |P_{(b_p)}(\zeta)| \leq \mathcal{C} \exp\left[N_{(b_p)}(L|\zeta|)\right]\right)$$

for all $\zeta \in \mathbb{C}$ and

$$\exp\left[M(b|\zeta|)\right] \leq P_b(\zeta) \left(\text{respectively, } \exp\left[N_{(b_p)}(|\zeta|)\right] \leq P_{(b_p)}(\zeta)\right)$$

for all $\zeta \in \mathbb{C}$ such that $|\text{Re } \zeta| \geq |\text{Im } \zeta|$. We recall that M means here the so-called associated function for the sequence (M_p) and $N_{(b_p)}$ is the associated function for the corresponding sequence (N_p), given by $N_p := M_p \prod_{j=1}^p b_j$ for $p \in \mathbb{N}$; that is,

$$M(t) := \sup_{p \in \mathbb{N}_0} \log_+ (t^p/M_p), \quad N_{(b_p)}(t) := \sup_{p \in \mathbb{N}_0} \log_+ (t^p/N_p)$$

for $t > 0$, where $\log_+ t := \max\left(\log t, 0\right)$. Since we assume that the sequence (M_p) satisfies conditions $(M.1)$ and $(M.3')$ (clearly, (N_p) also fulfills these conditions), it is easy to see that the associated function M (and so $N_{(b_p)}$) is a nondecreasing function on $[0, \infty)$, equal to 0 in a right neighborhood of 0.

In case conditions $(M.1)$, $(M.2)$ and $(M.3)$ are satisfied, an example of such an entire function is given by

$$P_b(\zeta) := \prod_{p=1}^{\infty}\left(1 + \frac{\zeta^2}{b^2 m_p^2}\right) \left(\text{respectively, } P_{(b_p)}(\zeta) := \prod_{p=1}^{\infty}\left(1 + \frac{\zeta^2}{b_p^2 m_p^2}\right)\right)$$

for $\zeta \in \mathbb{C}$, where $m_p := M_p/M_{p-1}$ for $p \in \mathbb{N}$. It follows from (7.9) (see Proposition 4.5 in [82]) that $P_b(D)$ (respectively, $P_{(b_p)}(D)$) is an ultradifferential operator of the class (M_p) (respectively, $\{M_p\}$). It is easy to see that

$$|P_b(\zeta)| \leq P_b(|\zeta|) \quad \left(\text{respectively, } |P_{(b_p)}(\zeta)| \leq P_{(b_p)}(|\zeta|)\right) \qquad (7.20)$$

and the functions P_b and $P_{(b_p)}$ are non-decreasing on the positive half-line of the real axis.

In the sequel, we shall need some estimates of the derivatives of both P_b (respectively, $P_{(b_p)}$) and $1/P_b$ (respectively, $1/P_{(b_p)}$). We will formulate and prove these estimates in the lemma below for the functions $P_{(b_p)}$ and $1/P_{(b_p)}$, but the corresponding inequalities hold for P_b and $1/P_b$ and can be proved in a similar way.

Lemma 7.6.1. *If $P_{(b_p)}$ satisfies (7.9) and (7.10), we have*

(a) *for every $r > 0$ there is a positive constant C (depending on r) such that*

$$\left| \left(P_{(b_p)}(\zeta) \right)^{(\gamma)} \right| \le C \frac{\gamma!}{r^\gamma} P_{(b_p/2)}(|\zeta|), \qquad \zeta \in \mathbb{C}, \ \gamma \in \mathbb{N}_0, \tag{7.21}$$

and in particular,

$$\left| \left(P_{(b_p)}(\xi) \right)^{(\gamma)} \right| \le C \frac{\gamma!}{r^\gamma} |P_{(b_p/2)}(\xi))|, \qquad \xi \in \mathbb{R}, \ \gamma \in \mathbb{N}_0; \tag{7.22}$$

(b) *there exist an $r > 0$ and a $C > 0$ such that*

$$\left| \left(\frac{1}{P_{(b_p)}(\xi)} \right)^{(\gamma)} \right| \le C \frac{\gamma!}{r^\gamma} \exp\left[-N_{(2b_p)}(|\xi|) \right], \qquad \xi \in \mathbb{R}, \ \gamma \in \mathbb{N}_0. \tag{7.23}$$

The corresponding inequalities hold for P_b.

Proof. (a) Fix $r > 0$. Applying Cauchy's formula and (7.20), we get

$$\left| \left(P_{(b_p)}(\zeta) \right)^{(\gamma)} \right| = \left| \frac{\gamma!}{2\pi i} \int_{|z-\zeta|=r} \frac{P_{(b_p)}(z) dz}{(z-\zeta)^{\gamma+1}} \right|$$

$$\le \frac{\gamma!}{r^\gamma} P_{(b_p)}(|\zeta| + r) \le C_r \frac{\gamma!}{r^\gamma} P_{(b_p/2)}(|\zeta|)$$

for $\zeta \in \mathbb{C}$, where $C_r := P_{(b_p/2)}(r)$; and (7.21) is proved. To obtain (7.22) it is enough to notice that

$$0 \le P_{(b_p/2)}(|\xi|) = P_{(b_p/2)}(\xi)$$

for every $\xi \in \mathbb{R}$.

(b) Since $P_{(b_p)}(0) \ne 0$, there exist positive r and C_1 such that $|P_{(b_p)}(\zeta)| \ge C_1$ for $|\zeta| \le (1 + \sqrt{2})r$. By the Cauchy formula,

$$\left| \left(\frac{1}{P_{(b_p)}(\xi)} \right)^{(\gamma)} \right| \le \frac{\gamma!}{2\pi} \int_{|\zeta-\xi|=r} \frac{d\zeta}{|P_{(b_p)}(\zeta)| \cdot |\zeta - \xi|^{\gamma+1}}$$

$$\le \frac{\gamma!}{C_1 r^\gamma} \le C_2 \frac{\gamma!}{r^\gamma} \exp\left[-N_{(2b_p)}(|\xi|) \right] \tag{7.24}$$

for $\xi \in \mathbb{R}$ such that $|\xi| \le \sqrt{2}r$, where $C_2 := C_1^{-1} \exp\left[N_{(2b_p)}(\sqrt{2}r) \right]$. Now let $\xi \in \mathbb{R}$, $|\xi| > \sqrt{2}r$ and let K_ξ be the circle with radius $|\xi|/\sqrt{2}$ and center at ξ. Evidently, every point ζ of K_ξ satisfies the inequality $|\text{Re } \zeta| \ge |\text{Im } \zeta|$.

Applying the Cauchy formula for the circle K_ξ and the estimate (7.10), we obtain

$$
\left| \left(\frac{1}{P_{(b_p)}(\xi)} \right)^{(\gamma)} \right| \leq \frac{\gamma!}{2\pi} \int_{K_\xi} \frac{d\zeta}{|P_{(b_p)}(\zeta)| \cdot |\zeta - \xi|^{\gamma+1}}
$$

$$
\leq \gamma! 2^{\gamma/2} |\xi|^{-\gamma} \sup_{\Theta \in [0,2\pi]} \exp\left[-N_{(b_p)}(|\xi| + |\xi| e^{\Theta i}/\sqrt{2}) \right]
$$

$$
\leq \gamma! r^{-\gamma} \exp\left[-N_{(b_p)}(|\xi|/2) \right] \leq \gamma! r^{-\gamma} \exp\left[-N_{(2b_p)}(|\xi|) \right]
$$

for every $\gamma \in \mathbb{N}_0$ and $\xi \in \mathbb{R}$, $|\xi| > \sqrt{2} r$. This and (7.24) imply (7.23) and the proof is finished. $\square$

For the sake of convenience, we will use in the sequel the following notation for given $(a_p), (b_p) \in \mathcal{R}$:

$$
A_\alpha := \prod_{p=1}^{\alpha} a_p , \quad B_\beta := \prod_{p=1}^{\beta} b_p , \quad \alpha, \beta \in \mathbb{N} , \quad A_0 := B_0 := 1. \quad (7.25)
$$

Definition 7.6.1. Let $a, b > 0$ and let $(a_p), (b_p) \in \mathcal{R}$. The spaces $\mathcal{D}_a^{(M_p),b}$, $\mathcal{D}_{a,b}^{(M_p)}$, $\mathcal{D}_{(a_p)}^{\{M_p\},(b_p)}$, and $\mathcal{D}_{(a_p),(b_p)}^{\{M_p\}}$ are defined to be the sets of all smooth functions φ on $\mathbb{R}$ such that

$$
p_{a,b}(\varphi) := \sup_{\alpha \in \mathbb{N}_0} \frac{a^\alpha}{M_\alpha} \| (P_b \varphi)^{(\alpha)} \|_{L^\infty} < \infty,
$$

$$
q_{a,b}(\varphi) := \sup_{\alpha \in \mathbb{N}_0} \frac{a^\alpha}{M_\alpha} \| P_b \varphi^{(\alpha)} \|_{L^\infty} < \infty,
$$

$$
p_{(a_p),(b_p)}(\varphi) := \sup_{\alpha \in \mathbb{N}_0} \frac{\| (P_{(b_p)} \varphi)^{(\alpha)} \|_{L^\infty}}{M_\alpha A_\alpha} < \infty,
$$

$$
q_{(a_p),(b_p)}(\varphi) := \sup_{\alpha \in \mathbb{N}_0} \frac{\| P_{(b_p)} \varphi^{(\alpha)} \|_{L^\infty}}{M_\alpha A_\alpha} < \infty,
$$

respectively, where A_α, B_α are defined in (7.25), equipped with the topologies induced by the norms $p_{a,b}$, $q_{a,b}$, $p_{(a_p),(b_p)}$ and $q_{(a_p),(b_p)}$, respectively. Further, we define

$$
\mathcal{D}_a^{(M_p)} := \operatorname*{proj\,lim}_{b > 0} \mathcal{D}_a^{(M_p),b}, \quad \mathcal{D}_{(a_p)}^{\{M_p\}} := \operatorname*{proj\,lim}_{(b_p) \in \mathcal{R}} \mathcal{D}_{(a_p)}^{\{M_p\},(b_p)},
$$

$$
\mathcal{D}^{(M_p),b} := \operatorname*{proj\,lim}_{a > 0} \mathcal{D}_a^{(M_p),b}, \quad \mathcal{D}^{\{M_p\},(b_p)} := \operatorname*{proj\,lim}_{(a_p) \in \mathcal{R}} \mathcal{D}_{(a_p)}^{\{M_p\},(b_p)}.
$$

Theorem 7.6.1. *If condition $(M.2')$ is satisfied, we have*

$$\mathcal{S}^{(M_p)} = \underset{a > 0}{\text{proj lim}} \ \mathcal{D}_a^{(M_p)} = \underset{b > 0}{\text{proj lim}} \ \mathcal{D}^{(M_p),b}$$

and

$$\mathcal{S}^{\{M_p\}} = \underset{(a_p) \in \mathcal{R}}{\text{proj lim}} \ \mathcal{D}_{(a_p)}^{\{M_p\}} = \underset{(b_p) \in \mathcal{R}}{\text{proj lim}} \ \mathcal{D}^{(M_p),(b_p)}.$$

Proof. We shall prove the assertion only in the case $* = \{M_p\}$, which is more complicated than for $* = (M_p)$, but the ideas of the proof in both cases are similar.

First recall that every sequence (M_p) satisfying conditions $(M.1)$ and $(M.3')$ tends very quickly to infinity. More precisely, it follows from these conditions that $p M_{p-1}/M_p \to 0$ as $p \to \infty$ (see [82], (4.6)) and this implies that

$$\frac{a^p p!}{M_p} \to 0 \qquad \text{as } p \to \infty \tag{7.26}$$

for an arbitrary $a > 0$.

Define

$$\gamma_{(a_p),(b_p)}(\varphi) := \sup_{\alpha,\beta \in \mathbb{N}_0} \sup_{x \in \mathbb{R}} \frac{\langle t \rangle^{\beta} |\varphi^{(\alpha)}(t)|}{M_\alpha A_\alpha M_\beta B_\beta},$$

where $\langle t \rangle := (1 + t^2)^{1/2}$ for $t \in \mathbb{R}$. Since $\langle t \rangle^\beta \le 2^{\beta/2}(1 + |t|^\beta)$ for $t \in \mathbb{R}$, we get from (7.26) and the definition of the functions $N_{(b_p)}$ the following estimate:

$$\gamma_{(a_p),(b_p)}(\varphi) \le \sup_{\alpha,\beta \in \mathbb{N}_0} \frac{2^{\beta/2}\|\varphi^{(\alpha)}\|_{L^\infty}}{M_\alpha A_\alpha M_\beta B_\beta} + \sup_{\alpha,\beta \in \mathbb{N}_0} \sup_{t \in \mathbb{R}} \frac{2^{\beta/2}|t|^\beta |\varphi^{(\alpha)}(t)|}{M_\alpha A_\alpha M_\beta B_\beta}$$

$$\le \mathcal{C} \left(\sup_{\alpha \in \mathbb{N}_0} \frac{\|\varphi^{(\alpha)}\|_{L^\infty}}{M_\alpha A_\alpha} + \sup_{\alpha \in \mathbb{N}_0} \frac{\|\exp [N_{(b_p/\sqrt{2})}]\varphi^{(\alpha)}\|_{L^\infty}}{M_\alpha A_\alpha} \right)$$

$$\le \mathcal{C} \sup_{\alpha \in \mathbb{N}_0} \frac{\|P_{(b_p/\sqrt{2})}\varphi^{(\alpha)}\|_{L^\infty}}{M_\alpha A_\alpha} = \mathcal{C} q_{(a_p),(b_p/\sqrt{2})}(\varphi)$$

for each $\varphi \in C^\infty$, where the constants A_α, B_α are defined in (7.25). On the other hand, inequality (7.9) yields

$$q_{(a_p),(b_p)}(\varphi) \le \mathcal{C} \sup_{\alpha \in \mathbb{N}_0} \sup_{t \in \mathbb{R}} \frac{\exp[N_{(b_p)}(L|t|)]|\varphi^{(\alpha)}(t)|}{M_\alpha A_\alpha}$$

$$\le \mathcal{C} \sup_{\alpha \in \mathbb{N}_0} \sup_{\beta \in \mathbb{N}_0} \sup_{t \in \mathbb{R}} \frac{|\langle t \rangle^\beta \varphi^{(\alpha)}(t)|}{M_\alpha A_\alpha M_\beta (B_\beta/L^\beta)} \le \mathcal{C} \gamma_{(a_p),(b_p/L)}(\varphi)$$

for every $\varphi \in C^\infty$. The two estimates just proved show that the families $\{q_{(a_p),(b_p)} : (a_p), (b_p) \in \mathcal{R}\}$ and $\{\gamma_{(a_p),(b_p)} : (a_p), (b_p) \in \mathcal{R}\}$ of seminorms are equivalent.

Let us prove now the equivalence of the families $\{p_{(a_p),(b_p)} : (a_p), (b_p) \in \mathcal{R}\}$ and $\{q_{(a_p),(b_p)} : (a_p), (b_p) \in \mathcal{R}\}$. First notice that for an arbitrary $(a_p) \in \mathcal{R}$ we have $A_p A_q \le A_{p+q}$ and, because of condition $(M.1)$, $M_p M_q \le M_{p+q}$ for $p, q \in \mathbb{N}$. Therefore, applying (7.22) for a fixed $r > 0$ and (7.26), we have

$$
\begin{aligned}
p_{(a_p),(b_p)}(\varphi) &= \sup_{\alpha \in \mathbb{N}_0} \frac{\| (P_{(b_p)}\varphi)^{(\alpha)} \|_{L^\infty}}{M_\alpha A_\alpha} \\
&\le \sup_{\alpha \in \mathbb{N}_0} \frac{1}{M_\alpha A_\alpha} \sum_{\gamma \le \alpha} \binom{\alpha}{\gamma} \| P_{(b_p)}^{(\gamma)} \varphi^{(\alpha-\gamma)} \|_{L^\infty} \\
&\le \mathcal{C} \sup_{\alpha \in \mathbb{N}_0} \sum_{\gamma \le \alpha} \binom{\alpha}{\gamma} \frac{\gamma!}{M_{\alpha-\gamma} A_{\alpha-\gamma} M_\gamma A_\gamma r^\gamma} \| P_{(b_p/2)} \varphi^{(\alpha-\gamma)} \|_{L^\infty} \\
&\le \mathcal{C} \sup_{\alpha \in \mathbb{N}_0} \frac{1}{2^\alpha} \sup_{\gamma \le \alpha} \frac{\gamma!}{M_\gamma A_\gamma'} \frac{\| P_{(b_p/2)} \varphi^{(\alpha-\gamma)} \|_{L^\infty}}{M_{\alpha-\gamma} A_{\alpha-\gamma}'} \sum_{\gamma \le \alpha} \binom{\alpha}{\gamma} \\
&\le \mathcal{C} q_{(a_p/2),(b_p/2)}(\varphi),
\end{aligned}
$$

where A_α, B_α are defined in (7.25) and

$$
A_\gamma' := \prod_{p=1}^\gamma a_p/2, \quad \gamma \in \mathbb{N}; \qquad A_0' := 1. \tag{7.27}
$$

Let (a_p) and (b_p) be given sequences in $\mathcal{R}$ and choose $(b_p') \in \mathcal{R}$ such that $b_p' \le b_p/(2L)$ for $p \in \mathbb{N}$, where L is the constant from (7.9). This implies that

$$
\tilde{N}(t) := \exp\left[N_{(b_p)}(L|t|) - N_{(2b_p')}(|t|) \right] \le 1, \qquad t \in \mathbb{R}. \tag{7.28}
$$

Inequalities (7.9), (7.23), (7.28) and the properties of the sequences (M_p) and (A_p) mentioned above imply that there exist constants $\mathcal{C} > 0$ and $r > 0$

such that, for every $\varphi \in C^\infty$, we have

$$q_{(a_p),(b_p)}(\varphi) = \sup_{\alpha \in \mathbb{N}_0} \frac{1}{M_\alpha A_\alpha} \|P_{(b_p)}\varphi^{(\alpha)}\|_{L^\infty}$$

$$= \sup_{\alpha \in \mathbb{N}_0} \frac{1}{M_\alpha A_\alpha} \left\| P_{(b_p)} \sum_{\gamma \leq \alpha} \binom{\alpha}{\gamma} \left(1/P_{(b'_p)}\right)^{(\alpha-\gamma)} (P_{(b'_p)}\varphi)^{(\gamma)} \right\|_{L^\infty}$$

$$\leq C \sup_{\alpha \in \mathbb{N}_0} \sum_{\gamma \leq \alpha} \binom{\alpha}{\gamma} \frac{(\alpha-\gamma)!}{r^{\alpha-\gamma} M_\alpha A_\alpha} \|(P_{(b'_p)}\varphi)^{(\gamma)}\|_{L^\infty}$$

$$\leq C \sup_{\alpha \in \mathbb{N}_0} \frac{1}{2^\alpha} \sum_{\gamma \leq \alpha} \binom{\alpha}{\gamma} \frac{(\alpha-\gamma)!}{M_{\alpha-\gamma} A'_{\alpha-\gamma}} \frac{1}{M_\gamma A'_\gamma} \|(P_{(b'_p)}\varphi)^{(\gamma)}\|_{L^\infty}$$

$$\leq C \sup_{\beta \in \mathbb{N}_0} \frac{1}{M_\beta A'_\beta} \|(P_{(b'_p)}\varphi)^{(\beta)}\|_{L^\infty} = C p_{(a_p/2),(b'_p)}(\varphi),$$

where A_γ and A'_γ are defined in (7.25) and (7.27). $\square$

Remark 7.6.1. From the preceding proof it follows that for every $a > 0$ (respectively, $(a_p) \in \mathcal{R}$) there exists a $b > 0$ (respectively, $(b_p) \in \mathcal{R}$) with $a < b$ (respectively, $(a_p) \preceq (b_p)$, i.e., $a_p \leq b_p$ for sufficiently large $p \in \mathbb{N}$) such that $\mathcal{D}_b^{(M_p)} \subseteq \mathcal{D}_a^{(M_p)}$ (respectively, $\mathcal{D}_{(b_p)}^{\{M_p\}} \subseteq \mathcal{D}_{(a_p)}^{\{M_p\}}$) and the inclusion mapping is continuous.

Definition 7.6.2. For given $a > 0$ and $(a_p) \in \mathcal{R}$, we define the Hilbert transforms $\mathcal{H}_a$ and $\mathcal{H}_{(a_p)}$ on $\mathcal{D}_a^{(M_p)}$ and $\mathcal{D}_{(a_p)}^{\{M_p\}}$, respectively, in the following way:

$$(\mathcal{H}_a \varphi)(x) := \frac{1}{\pi P_a(x)} \,\mathrm{pv} \int_{-\infty}^\infty \frac{P_a(x-t)\varphi(x-t)}{t}\, dt,$$

$$\varphi \in \mathcal{D}_a^{(M_p)}, x \in \mathbb{R},$$

and

$$(\mathcal{H}_{(a_p)}\varphi)(x) := \frac{1}{\pi P_{(a_p)}(x)} \,\mathrm{pv} \int_{-\infty}^\infty \frac{P_{(a_p)}(x-t)\varphi(x-t)}{t}\, dt,$$

$$\varphi \in \mathcal{D}_{(a_p)}^{\{M_p\}}, x \in \mathbb{R}.$$

Theorem 7.6.2. *For given $a > 0$ and $(a_p) \in \mathcal{R}$, the Hilbert transforms $\mathcal{H}_a : \mathcal{D}_a^{(M_p)} \to \mathcal{D}_a^{(M_p)}$ and $\mathcal{H}_{(a_p)} : \mathcal{D}_{(a_p)}^{\{M_p\}} \to \mathcal{D}_{(a_p)}^{\{M_p\}}$ are linear continuous surjections such that*

$$\mathcal{H}_a \mathcal{H}_a \varphi = -\varphi, \qquad \varphi \in \mathcal{D}_a^{(M_p)};$$

$$\mathcal{H}_{(a_p)} \mathcal{H}_{(a_p)} \varphi = -\varphi, \qquad \varphi \in \mathcal{D}_{(a_p)}^{\{M_p\}}.$$

Proof. We give the proof only in the case $* = \{M_p\}$; the proof in the case $* = (M_p)$ is analogous. The linearity and the continuity of $\mathcal{H}_{(a_p)}$ follow immediately from the fact that $\mathcal{H}_{(a_p)}$ is a composition of the three linear and continuous mappings $T_{(a_p)} \colon \mathcal{D}_{(a_p)}^{\{M_p\}} \to \mathcal{D}_{L^2}^{\{M_p\}}$, $\mathcal{H} \colon \mathcal{D}_{L^2}^{\{M_p\}} \to \mathcal{D}_{L^2}^{\{M_p\}}$ and $T_{(a_p)}^{-1} \colon \mathcal{D}_{L^2}^{\{M_p\}} \to \mathcal{D}_{(a_p)}^{\{M_p\}}$, defined by the formulas:

$$T_{(a_p)}(\varphi)(x) := P_{(a_p)}(x)\varphi(x), \qquad \varphi \in \mathcal{D}_{(a_p)}^{\{M_p\}}, \tag{7.29}$$

$$(\mathcal{H}\varphi)(x) := \frac{1}{\pi}\mathrm{pv}\int_{-\infty}^{\infty} \frac{\varphi(t)\,dt}{t-x}, \qquad \varphi \in \mathcal{D}_{L^2}^{\{M_p\}}, \tag{7.30}$$

$$T_{(a_p)}^{-1}(\varphi) := \varphi(x)/P_{(a_p)}(x), \qquad \varphi \in \mathcal{D}_{L^2}^{\{M_p\}}, \tag{7.31}$$

for $x \in \mathbb{R}$. Note that the Hilbert transform is considered in [112] only on $\mathcal{D}_{L^2}^{(M_p)}$, i. e. in the Beurling case, but it can be examined in a similar way in the Roumieu case.

From the definition of $\mathcal{H}_{(a_p)}$ and the properties of the classical Hilbert transform on $\mathcal{D}_{L^2}^{\{M_p\}}$, it follows that

$$\mathcal{H}_{(a_p)}(\mathcal{H}_{(a_p)}\varphi) = T_{(a_p)}^{-1}(\mathcal{H}T_{(a_p)}(T_{(a_p)}^{-1}(\mathcal{H}(T_{(a_p)}\varphi))))$$

$$= T_{(a_p)}^{-1}(\mathcal{H}(\mathcal{H}(T_{(a_p)}\varphi))) = T_{(a_p)}^{-1}(-T_{(a_p)}\varphi) = -\varphi$$

for every $(a_p) \in \mathcal{R}$ and $\varphi \in \mathcal{D}_{(a_p)}^{\{M_p\}}$. This completes the proof. $\square$

Definition 7.6.3. The generalized Hilbert transforms $\mathbf{H}_a$ and $\mathbf{H}_{(a_p)}$ are defined for $f \in \mathcal{D}'^{(M_p)}_a$ and $f \in \mathcal{D}'^{\{M_p\}}_{(a_p)}$ by

$$\langle \mathbf{H}_a f, \varphi \rangle := -\langle f, \mathcal{H}_a \varphi \rangle, \qquad \varphi \in \mathcal{D}_a^{(M_p)},$$

and

$$\langle \mathbf{H}_{(a_p)} f, \varphi \rangle := -\langle f, \mathcal{H}_{(a_p)}\varphi \rangle, \qquad \varphi \in \mathcal{D}_{(a_p)}^{\{M_p\}},$$

respectively.

In the theorem below, we list several properties of the Hilbert transforms $\mathbf{H}_a$ and $\mathbf{H}_{(a_p)}$ defined above. In particular, we shall prove that

$$\langle \mathcal{F}_0(\mathbf{H}_* f), \varphi \rangle = \begin{cases} -i\langle \mathcal{F}_0 f, \varphi \rangle, & \text{if } \mathrm{supp}\ \varphi \subset (0, \infty), \\ i\langle \mathcal{F}_0 f, \varphi \rangle, & \text{if } \mathrm{supp}\ \varphi \subset (-\infty, 0), \end{cases} \tag{7.32}$$

for every $\varphi \in \mathcal{D}^{(M_p)}$ if $f \in \mathcal{D}'^{(M_p)}$ and for every $\varphi \in \mathcal{D}^{\{M_p\}}$ if $f \in \mathcal{D}'^{\{M_p\}}$, where the symbol $\mathbf{H}_*$ means $\mathbf{H}_{(M_p)}$ and $\mathbf{H}_{\{M_p\}}$ in the respective cases.

Theorem 7.6.3. *The above defined Hilbert transforms* $\mathbf{H}_a: \mathcal{D}'^{(M_p)}_a \to \mathcal{D}'^{(M_p)}_a$ *and* $\mathbf{H}_{(a_p)}: \mathcal{D}'^{\{M_p\}}_{(a_p)} \to \mathcal{D}'^{\{M_p\}}_{(a_p)}$ *have the following properties:*

1. $\mathbf{H}_a$ *and* $\mathbf{H}_{(a_p)}$ *are linear continuous surjections;*

2. $\mathbf{H}_a(\mathbf{H}_a f) = -f$ *for* $f \in \mathcal{D}'^{(M_p)}_a$ *and* $\mathbf{H}_{(a_p)}(\mathbf{H}_{(a_p)} f) = -f$ *for* $f \in \mathcal{D}'^{\{M_p\}}_{(a_p)}$;

3. If $f \in \mathcal{D}'^{(M_p)}_a$, *formula (7.32) holds for all* $\varphi \in \mathcal{D}^{(M_p)}$; *and if* $f \in \mathcal{D}'^{\{M_p\}}_{(a_p)}$, *formula (7.32) holds for all* $\varphi \in \mathcal{D}^{\{M_p\}}$;

4. Under conditions $(M.2)$ *and* $(M.3)$, *if* $f \in \mathcal{D}'^{(M_p)}_a$ *(respectively,* $f \in \mathcal{D}'^{\{M_p\}}_{(a_p)}$) *with* $0 < a < b$ *(respectively,* $(a_p) \preceq (b_p)$) *such that* $f|_{\mathcal{D}^{(M_p)}_b} \in \mathcal{D}'^{(M_p)}_b$ *(respectively,* $f|_{\mathcal{D}^{\{M_p\}}_{(b_p)}} \in \mathcal{D}'^{\{M_p\}}_{(b_p)}$) *(see Remark 2.1), we have that* $\mathbf{H}_a f - \mathbf{H}_b f|_{\mathcal{D}^{(M_p)}_b}$ *(respectively,* $\mathbf{H}_{(a_p)} f - \mathbf{H}_{(b_p)} f|_{\mathcal{D}^{\{M_p\}}_{(b_p)}}$) *is an ultrapolynomial of the class* (M_p) *(respectively,* $\{M_p\}$);

5. If $f, g \in \mathcal{D}'^{(M_p)}_a$ *(respectively,* $f, g \in \mathcal{D}'^{\{M_p\}}_{(a_p)}$) *and* $f|_{\mathcal{D}^{(M_p)}} = g|_{\mathcal{D}^{(M_p)}}$ *(respectively,* $f|_{\mathcal{D}^{\{M_p\}}} = g|_{\mathcal{D}^{\{M_p\}}}$), *we have that* $\mathbf{H}_a f - \mathbf{H}_a g$ *(respectively,* $\mathbf{H}_{(a_p)} f - \mathbf{H}_{(a_p)} g$) *is an ultrapolynomial of the class* (M_p) *(respectively,* $\{M_p\}$).

Proof. We shall prove the assertion only in the case $* = \{M_p\}$. Parts 1 and 2 follow immediately from Theorem 7.6.2.

Let us prove part 3. Let $\varphi \in \mathcal{D}^{\{M_p\}}$ and supp $\varphi \subset (0, \infty)$. From the properties of the classical Fourier and Hilbert transforms of functions in L^2, we have

$$\langle \mathcal{F}_0(\mathbf{H}_{(a_p)} f), \varphi \rangle = \langle \mathbf{H}_{(a_p)} f, \mathcal{F}_0 \varphi \rangle = \langle f, T^{-1}_{(a_p)} \mathcal{H} T_{(a_p)} \mathcal{F}_0 \varphi \rangle$$
$$= \langle f, T^{-1}_{(a_p)} \mathcal{H} \mathcal{F}_0(P_{(a_p)}(D)\varphi) \rangle = \langle f, T^{-1}_{(a_p)} \mathcal{F}_0(-i P_{(a_p)}(D)\varphi) \rangle$$
$$= -i\langle f, T^{-1}_{(a_p)} T_{(a_p)} \mathcal{F}_0 \varphi \rangle = -i\langle f, \mathcal{F}_0 \varphi \rangle = -i\langle \mathcal{F}_0 f, \varphi \rangle.$$

In a similar way we can prove part 3 in the case $\varphi \in \mathcal{D}^{\{M_p\}}$ and supp $\varphi \subset (-\infty, 0)$.

Let us now prove part 4. For each $\varphi \in \mathcal{D}^{(M_p)}$ with supp $\varphi \subset (0, \infty)$, we have

$$\langle \mathcal{F}_0(\mathbf{H}_{(a_p)} f - \mathbf{H}_{(b_p)} f), \varphi \rangle = \langle \mathcal{F}_0(\mathbf{H}_{(a_p)} f), \varphi \rangle - \langle \mathcal{F}_0(\mathbf{H}_{(b_p)} f), \varphi \rangle$$
$$= -i\langle \mathcal{F}_0 f, \varphi \rangle - (-i)\langle \mathcal{F}_0 f, \varphi \rangle = 0.$$

Analogously, we have

$$\langle \mathcal{F}_0(\mathbf{H}_{(a_p)} f - \mathbf{H}_{(b_p)} f), \varphi \rangle = 0$$

for $\varphi \in \mathcal{D}^{\{M_p\}}$ with supp $\varphi \subset (-\infty, 0)$. Therefore supp $\mathcal{F}_0(\mathbf{H}_{(a_p)}f - \mathbf{H}_{(b_p)}f) \subseteq \{0\}$. Theorem 3.1 in [82] implies the existence of an ultradifferential operator $P(D)$ such that

$$\mathcal{F}_0(\mathbf{H}_{(a_p)}f - \mathbf{H}_{(b_p)}f) = P(D)\,\delta. \qquad (7.33)$$

Applying the inverse Fourier transform on (7.33), we obtain

$$(\mathbf{H}_{(a_p)}f - \mathbf{H}_{(b_p)}f)(x) = P(x), \qquad x \in \mathbb{R},$$

which proves property 4.

Assertion 5 follows from the fact that

$$\text{supp } \mathcal{F}_0(\mathbf{H}_{(a_p)}f - \mathbf{H}_{(a_p)}g) \subseteq \{0\},$$

which can be proved analogously as in part 4. $\square$

Definition 7.6.4. Since for every $f \in \mathcal{S}'^{(M_p)}$ (respectively, $f \in \mathcal{S}'^{\{M_p\}}$), there is an $a > 0$ (respectively, $(a_p) \in \mathcal{R}$) such that f has a linear and continuous extension F on $\mathcal{D}_a^{(M_p)}$ (respectively, $\mathcal{D}_{(a_p)}^{\{M_p\}}$), we define the Hilbert transform $\mathbf{H}^{(M_a)}f$ (respectively, $\mathbf{H}^{\{M_a\}}f$) of $f \in \mathcal{S}'^{(M_p)}$ (respectively, $f \in \mathcal{S}'^{\{M_p\}}$) by

$$\mathbf{H}^{(M_p)}f := \mathbf{H}_a F \qquad \left(\text{respectively, } \mathbf{H}^{\{M_p\}}f := \mathbf{H}_{(a_p)}F\right).$$

Theorem 7.6.3 shows that the Hilbert transform of an element of the space $\mathcal{S}'^*$ is defined uniquely up to an ultrapolynomial of class $*$.

7.6.2 *Multi-dimensional case*

We now extend the definition of the Hilbert transform given in the preceding subsection to the n-dimensional case. We shall show that all of the results of Subsection 7.6.1, namely Lemma 7.6.1, Theorem 7.6.1, Theorem 7.6.2 and Theorem 7.6.3, remain true in the n-dimensional case.

If $a = (a^1, \ldots, a^n) \in \mathbb{R}^n$ and $\alpha = (\alpha_1, \ldots, \alpha_n) \in \mathbb{N}_0^{\,n}$, we denote

$$a^\alpha := (a^1)^{\alpha_1} \cdot \ldots \cdot (a^n)^{\alpha_n}; \qquad M_\alpha := M_{\alpha_1 + \ldots + \alpha_n}.$$

By $\mathcal{R}^n$ we denote the family of all sequences (a_p) of elements of $\mathbb{R}^n$ of the form

$$a_p = (a_p^1, \ldots, a_p^n), \qquad (a_p^j) \in \mathcal{R}, \qquad j = 1, \ldots, n. \qquad (7.34)$$

For a given sequence $(a_p) \in \mathcal{R}^n$ of the form (7.34) and $\alpha = (\alpha^1, \ldots, \alpha^n) \in \mathbb{N}_0^{\,n}$, we shall use the following extension of the notation (7.25): $A_\alpha := A_{\alpha^1} \cdot \ldots \cdot A_{\alpha^n}$, where $A_0 := 1$ and

$$A_{\alpha^j} := \prod_{p=1}^{\alpha^j} a_p^j \qquad \text{whenever } \alpha^j \in \mathbb{N}$$

for $j = 1, \ldots, n$.

Let $a = (a^1, \ldots, a^n) \in \mathbb{R}^n_+$, i.e., $a^j > 0$ for $j = 1, \ldots, n$, and let $(a_p) \in \mathcal{R}^n$ with elements of the form (7.34). We define

$$P_a(\zeta) := P_{a_1}(\zeta^1) \cdot \ldots \cdot P_{a_n}(\zeta^n); \qquad P_{(a_p)}(\zeta) := P_{(a_p^1)}(\zeta^1) \cdot \ldots \cdot P_{(a_p^n)}(\zeta^n)$$

for $\zeta = (\zeta^1, \ldots, \zeta^n) \in \mathbb{C}^n$.

Remark 7.6.2. The n-dimensional version of Lemma 7.6.1 is valid and the estimates (7.21) and (7.23) in the multi-dimensional case follow easily from the one-dimensional case.

Now the definitions of the seminorms $p_{a,b}$ and $q_{a,b}$, for $a = (a^1, \ldots, a^n) \in \mathbb{R}^n_+$ and $b = (b^1, \ldots, b^n) \in \mathbb{R}^n_+$, and the definitions of the seminorms $p_{(a_p),(b_p)}$ and $q_{(a_p),(b_p)}$, for $(a_p) \in \mathcal{R}^n$ and $(b_p) \in \mathcal{R}^n$, are obtained by modifications of the ones given in Definition 7.6.1 with all of the least upper bounds being taken over $\alpha \in \mathbb{N}_0^n$. The definitions of all spaces given in Definition 7.6.1 are modified accordingly.

Remark 7.6.3. With the above conventions, the n-dimensional analogue of Theorem 7.6.1 is true and its proof is obtained in the same way as in the one-dimensional case.

Definition 7.6.5. For given $a \in \mathbb{R}^n_+$ and $(a_p) \in \mathcal{R}^n$, we define the Hilbert transforms $\mathcal{H}_a$ and $\mathcal{H}_{(a_p)}$ on the spaces $\mathcal{D}_a^{(M_p)}$ and $\mathcal{D}_{a_p}^{\{M_p\}}$, respectively, by the formulas:

$$(\mathcal{H}_a \varphi)(x) := \frac{1}{\pi^n P_a(x)} \, \mathrm{pv} \int_{-\infty}^{\infty} \cdots \int_{-\infty}^{\infty} \frac{P_a(x-t)\varphi(x-t)}{\prod_{j=1}^n t^j} \, dt^1 \ldots dt^n$$

for $\varphi \in \mathcal{D}_a^{(M_p)}$ and

$$(\mathcal{H}_{(a_p)} \varphi)(x)$$

$$:= \frac{1}{\pi^n P_{(a_p)}(x)} \, \mathrm{pv} \int_{-\infty}^{\infty} \cdots \int_{-\infty}^{\infty} \frac{P_{(a_p)}(x-t)\varphi(x-t)}{\prod_{j=1}^n t^j} \, dt^1 \ldots dt^n$$

for $\varphi \in \mathcal{D}_{(a_p)}^{\{M_p\}}$, where $x = (x^1, \ldots, x^n), t = (t^1, \ldots, t^n) \in \mathbb{R}^n$.

Remark 7.6.4. The above defined Hilbert transforms $\mathcal{H}_a$ and $\mathcal{H}_{(a_p)}$ have the same properties as those mentioned in Theorem 7.6.2 and their proofs are analogous. In particular, $\mathcal{H}_{(a_p)}$ is the composition of the mappings given by formulas (7.29), (7.31) and the following extension of formula (7.30):

$$(\mathcal{H}\varphi)(x) = \frac{1}{\pi^n} \mathrm{pv} \int_{-\infty}^{\infty} \cdots \int_{-\infty}^{\infty} \frac{\varphi(t^1, \ldots, t^n) \, dt^1 \ldots dt^n}{\prod_{j=1}^n (t^j - x^j)}$$

for $x = (x^1, \ldots, x^n) \in \mathbb{R}^n$. Though in [112] only the one-dimensional Hilbert transform is considered on $\mathcal{D}_{L^2}^{\{M_p\}}$, we can easily extend it to the n-dimensional case and prove that this is an isomorphism of the space $\mathcal{D}_{L^2}^{\{M_p\}}(\mathbb{R}^n)$, because there is a constant $C > 0$ such that

$$\|\mathcal{H}(\varphi^{(\alpha)})\|_{L^2} \le C\|\varphi^{(\alpha)}\|_{L^2}, \qquad \alpha \in \mathbb{N}^n, \qquad \varphi \in \mathcal{D}_{L^2}^{\{M_p\}}.$$

Definition 7.6.6. For given $a \in \mathbb{R}_+^n$ and $(a_p) \in \mathcal{R}^n$, we define the n-dimensional Hilbert transforms $\mathbf{H}_a$ and $\mathbf{H}_{(a_p)}$ on the dual spaces $\mathcal{D}'_a^{(M_p)}$ and $\mathcal{D}'_{(a_p)}^{\{M_p\}}$ by

$$\langle \mathbf{H}_a f, \varphi \rangle := -\langle f, \mathcal{H}_a \varphi \rangle, \qquad \varphi \in \mathcal{D}_a^{(M_p)},$$

for $f \in \mathcal{D}'_a^{(M_p)}$, and

$$\langle \mathbf{H}_{(a_p)} f, \varphi \rangle := -\langle f, \mathcal{H}_{(a_p)} \varphi \rangle, \qquad \varphi \in \mathcal{D}_{(a_p)}^{\{M_p\}},$$

for $f \in \mathcal{D}'_{(a_p)}^{\{M_p\}}$.

Remark 7.6.5. Theorem 7.6.3 remains valid in the n-dimensional case and the proof of parts 1-3 and 5 can be easily transferred to this case. We shall give below the proof of part 4 for the Hilbert transform $\mathbf{H}_a f$ of an arbitrary $f \in \mathcal{D}'_a^{(M_p)}$.

For given $a = (a^1, \ldots, a^n) \in \mathbb{R}^n$ and $b = (b^1, \ldots, b^n) \in \mathbb{R}^n$, we have

$$\mathbf{H}_a f - \mathbf{H}_b f := \sum_{j=1}^{n} (\mathbf{H}_{a_j} f - \mathbf{H}_{a_{j-1}} f),$$

where $a_0 := b$ and $a_j := (a^1, \ldots, a^j, b^{j+1}, \ldots, b^n)$ for $j = 1, \ldots, n$. Moreover

$$\mathbf{H}_{a_j} f = \mathbf{H}_{\tilde{a}_j}(\mathbf{H}_{a^j} f), \tag{7.35}$$

where $\tilde{a}_j := (a^1, \ldots, a^{j-1}, b^{j+1}, \ldots, b^n) \in \mathbb{R}^{n-1}$ for $j = 1, \ldots, n$. By part 4 of Theorem 7.5.3, it follows from (7.35) that $\mathbf{H}_{a_j} f - \mathbf{H}_{a_{j-1}} f$ is equal to

$$\mathbf{H}_{\tilde{a}_j}(1_{x^1} \otimes \cdots \otimes 1_{x^{j-1}} \otimes P(x^j) \otimes 1_{x^{j+1}} \cdots \otimes 1_{x^n}), \tag{7.36}$$

where $P(x^j)$ is an ultrapolynomial in x^j. Since we have

$$\|(\mathbf{H}\frac{1}{P_b})^{(\alpha)}\|_{L^2} \le C\|(\frac{1}{P_b})^{(\alpha)}\|_{L^2}$$

for some constant $C > 0$, Lemma 7.6.1 implies that (7.36) is an entire function.

7.7 Singular integral operators

In this section we consider general singular integrals with odd and even kernels of tempered ultradistributions. For the L^2 theory of singular integrals we refer to [137]. We use the classical results of Pandey [108] and follow his ideas in our definition of singular integral operators on the spaces of tempered ultradistributions with values in certain spaces of ultradistributions which contain $\mathcal{S}'^{(M_p)}(\mathbb{R}^n)$ and $\mathcal{S}'^{\{M_p\}}(\mathbb{R}^n)$, respectively.

Let Ω be an arbitrary function on $\mathbb{R}^n$, homogeneous of degree zero, such that $\Omega \in C^\infty(\mathbb{R}^n \setminus \{0\})$. Clearly, this function is integrable and square integrable on $\Sigma^{n-1} = \{x : |x| = 1\}$. Put

$$K(t) := \frac{\Omega(t')}{|t|^n}, \qquad t \in \mathbb{R}^n,\ t \neq 0,\ t' = t/|t|. \tag{7.37}$$

If the mapping $t' \mapsto \Omega(t')$ is an odd function on Σ^{n-1}, we say that K is an odd kernel. If $\int_{\Sigma^{n-1}} \Omega(t')\, dt' = 0$, we say that K is an even kernel. Clearly, an odd kernel is also an even kernel (see [137], Chapters IV and VI).

Let the symbol $\mathrm{pv}K$ denote the principal value of K; clearly, it is a tempered distribution. Let

$$\tilde{K} := \mathcal{F}_0(\mathrm{pv}K), \tag{7.38}$$

where $\mathcal{F}_0$ is understood as the Fourier transform of a tempered distribution.

The convolution $\varphi * (\mathrm{pv}K)$, where $\varphi \in L^p$, $1 < p < \infty$, yields the singular integral operator $\mathcal{K}$ with the kernel K. More precisely, the singular integral operator $\mathcal{K}$ on L^p with the kernel K is defined by

$$(\mathcal{K}\varphi)(x) := \lim_{\substack{\varepsilon \to 0 \\ \delta \to \infty}} \int_{\delta \geq |t| \geq \varepsilon > 0} \varphi(x - t) K(t)\, dt, \qquad \varphi \in L^p. \tag{7.39}$$

If the dimension of the space is $n = 1$ and $\Omega(t) = \operatorname{sgn} t/\pi$ then the singular integral operator $\mathcal{K}$ defined by (7.39) is the Hilbert transform.

The operator $\mathcal{K}$ defined on L^p, $1 < p < \infty$, by (7.39), where K is an even kernel, is called an integral operator with even kernel.

We denote by $\hat{\mathcal{S}}^*(\mathbb{R}^n)$ the set of all elements of $\mathcal{S}^*(\mathbb{R}^n)$ whose supports are contained in $\mathbb{R}^n \setminus \{0\}$ with the topology induced by the space $\mathcal{S}^*(\mathbb{R}^n)$ and let

$$\mathcal{S}^*_{\hat{\mathcal{F}}_0}(\mathbb{R}^n) := \mathcal{F}_0\left(\mathcal{S}^*(\overset{\circ}{\mathbb{R}}{}^n)\right),$$

where $\mathcal{F}_0$ denotes the classical Fourier transform of a function as given in (7.1).

Theorem 7.7.1. *Let K be of the form (7.37) with the corresponding $\tilde{K}$ and $\mathcal{K}\varphi$, given by (7.38) and (7.39), respectively. We have*

   1. $\tilde{K}$ is homogeneous of order zero on $\mathbb{R}^n$ and $\tilde{K} \in C^\infty(\mathbb{R}^n \setminus \{0\})$;

   2. $\mathcal{K}\varphi \in \mathcal{S}^*(\mathbb{R}^n)$ for every $\varphi \in \mathcal{S}^*_{\hat{\mathcal{F}}_0}(\mathbb{R}^n)$ and $\tilde{K}\psi \in \hat{\mathcal{S}}^*(\mathbb{R}^n)$ for every $\psi \in \hat{\mathcal{S}}^*(\mathbb{R}^n)$; moreover, the mappings

$$\mathcal{S}^*_{\hat{\mathcal{F}}_0}(\mathbb{R}^n) \ni \psi \mapsto \mathcal{K}\psi \in \mathcal{S}^*(\mathbb{R}^n); \qquad \hat{\mathcal{S}}^*(\mathbb{R}^n) \ni \psi \mapsto \tilde{K}\psi \in \hat{\mathcal{S}}^*(\mathbb{R}^n)$$

are continuous;

   3. The space $\mathcal{S}^*_{\hat{\mathcal{F}}_0}(\mathbb{R}^n)$ is dense in $\mathcal{S}^*(\mathbb{R}^n)$.

Proof.

1. For this assertion we refer to Corollary 9.5 in [109], p. 108.

2. If $\varphi \in \mathcal{S}^*_{\hat{\mathcal{F}}_0}(\mathbb{R}^n)$, we have

$$\mathcal{K}\varphi = (\mathrm{pv}K) * \varphi = \mathcal{F}_0^{-1}(\mathcal{F}_0(\mathrm{pv}K) \cdot \mathcal{F}_0(\varphi)) = \mathcal{F}_0^{-1}(\tilde{K} \cdot \mathcal{F}_0(\varphi));$$

hence $\mathcal{K}\varphi \in \mathcal{S}^*(\mathbb{R}^n)$.

Now, if $\psi \in \hat{\mathcal{S}}^*(\mathbb{R}^n)$, then $\varphi := \mathcal{F}_0^{-1}(\psi) \in \mathcal{S}^*_{\hat{\mathcal{F}}_0}(\mathbb{R}^n)$; and $\mathcal{K}\varphi \in \mathcal{S}^*(\mathbb{R}^n)$, as we have just shown. Hence

$$\mathcal{F}_0(\mathcal{K}\varphi) = \mathcal{F}_0((\mathrm{pv}K) * \varphi) = \tilde{K} \cdot \psi \in \hat{\mathcal{S}}^*(\mathbb{R}^n).$$

The above facts and the continuity of the Fourier transform and its inverse imply the assertion.

3. Let ψ be an arbitrary function in $\mathcal{E}^*$ such that

$$\psi \in \mathcal{C}_0^\infty(\mathbb{R}^n), \ \psi \geq 0, \ \int_{\mathbb{R}^n} \psi(x)dx = 1;$$

and let $\psi_k(x) := \psi(kx)$ for $x \in \mathbb{R}^n$. Fix $\theta \in \mathcal{S}^*(\mathbb{R}^n)$ and put $\kappa := \mathcal{F}_0^{-1}(\theta)$. Then $\kappa\psi_k \in \mathcal{S}^*(\mathring{\mathbb{R}}^n)$ for $k \in \mathbb{N}$ and $(\kappa\psi_k) \to \kappa$ as $k \to \infty$ in $\mathcal{S}^*(\mathring{\mathbb{R}}^n)$. This implies that $\mathcal{F}_0(\kappa\psi_k) \in \mathcal{S}^*_{\hat{\mathcal{F}}_0}(\mathbb{R}^n)$ for $k \in \mathbb{N}$ and $\mathcal{F}_0(\kappa\psi_k) \to \theta$ as $k \to \infty$ in $\mathcal{S}^*(\mathbb{R}^n)$, which completes the proof of Theorem 7.7.1. $\square$

We define now the singular integral with an even kernel of a given tempered ultradistribution $f \in \mathcal{S}'^*$ as an element of $(\mathcal{S}^*_{\hat{\mathcal{F}}_0}(\mathbb{R}^n))'$ by

$$\langle \mathbf{K}f, \theta \rangle := \langle f, \mathcal{K}\theta \rangle, \qquad \theta \in \mathcal{S}^*_{\hat{\mathcal{F}}_0}(\mathbb{R}^n). \tag{7.40}$$

It is easy to prove the following theorem.

Theorem 7.7.2. *Suppose that K is an even kernel of the form (7.37) with the mentioned properties. Then the mapping $\mathbf{K} : \mathcal{S}'^*(\mathbb{R}^n) \to (\mathcal{S}^*_{\hat{\mathcal{F}}_0}(\mathbb{R}^n))'$ defined by (7.40) is linear, continuous and injective.*

Now consider the operator $\mathcal{K}$ on $\mathcal{D}_{L^2}^*(\mathbb{R}^n)$.

Theorem 7.7.3. *If K is an even kernel, then the mapping*

$$\mathcal{D}_{L^2}^* \ni \varphi \mapsto \mathcal{K}\varphi \in \mathcal{D}_{L^2}^*$$

is continuous.

Proof. By [137], Chapter VI, Theorem 3.1, the mapping

$$L^2 \ni \varphi \mapsto \mathcal{K}\varphi \in L^2$$

is well defined and there exists a constant $C > 0$, depending on the dimension n but not on φ, such that

$$\|\mathcal{K}\varphi\|_{L^2} \leq C\|\varphi\|_{L^2}.$$

If $\varphi \in \mathcal{D}_{L^2}^*$, then for each $\alpha \in \mathbb{N}_0^n$ we have

$$(\mathcal{K}\varphi)^{(\alpha)} = ((\mathrm{pv}K) * \varphi)^{(\alpha)} = (\mathrm{pv}K) * \varphi^{(\alpha)} = \mathcal{K}(\varphi^{(\alpha)}),$$

in the sense of distributions. Since the functions $((\mathrm{pv}K) * \varphi)^{(\alpha)}$ and $(\mathrm{pv}K) * \varphi^{(\alpha)}$ are smooth, they are equal in the space L^2. Therefore there exists a constant $C > 0$, which does not depend on α or on φ, such that

$$\|(\mathcal{K}\varphi)^{(\alpha)}\|_{L^2} \leq C\|\varphi^{(\alpha)}\|_{L^2}.$$

This implies that

$$\|\mathcal{K}\varphi\|_{\mathcal{D}_{L^2}^*} \leq C\|\varphi\|_{\mathcal{D}_{L^2}^*}$$

and this completes the proof. $\square$

Using relations (7.29) -(7.31) in the n-dimensional version (see Remark 7.6.4) and Theorem 7.6.2, we define, for $a = (a^1, \ldots, a^n) \in \mathbb{R}_+^n$ and $(a_p) \in \mathcal{R}^n$ with $a_p = (a_p{}^1, \ldots, a_p{}^n)$ for $p \in \mathbb{N}$ (where $(a_p^j) \in \mathcal{R}$ for $j = 1, \ldots, n$), the transforms $\mathbf{K}_a$ and $\mathbf{K}_{(a_p)}$ which act from the spaces $\mathcal{S}'^{(M_p)}(\mathbb{R}^n)$ and $\mathcal{S}'^{\{M_p\}}(\mathbb{R}^n)$ into themselves, respectively, by the formulas:

$$\langle \mathbf{K}_a f, \varphi \rangle = \langle f, \mathcal{K}_a \varphi \rangle, \qquad \mathcal{K}_a \varphi = P_a^{-1} \mathcal{K} P_a \varphi,$$

for $f \in \mathcal{S}'^{(M_p)}(\mathbb{R}^n)$ and $\varphi \in \mathcal{S}^{(M_p)}(\mathbb{R}^n)$, and

$$\langle \mathbf{K}_{(a_p)} f, \varphi \rangle = \langle f, \mathcal{K}_{(a_p)} \varphi \rangle, \qquad \mathcal{K}_{(a_p)} \varphi = P_{(a_p)}^{-1} \mathcal{K} P_{(a_p)} \varphi,$$

for $f \in \mathcal{S}'^{\{M_p\}}(\mathbb{R}^n)$ and $\varphi \in \mathcal{S}^{\{M_p\}}(\mathbb{R}^n)$.

Theorem 7.7.2 implies the following property.

Corollary 7.7.1. *The transform $\mathbf{K}_a$(respectively, the transform $\mathbf{K}_{(a_p)}$) is a continuous linear mapping of the space $\mathcal{S}^{(M_p)}(\mathbb{R}^n)$ (respectively, the space $\mathcal{S}^{\{M_p\}}(\mathbb{R}^n)$) onto itself.*

Let us remark that, in general, we do not have a uniqueness type result (up to ultrapolynomials) as in the case of Hilbert transform for different a, $b \in \mathbb{R}^n_+$ and (a_p), $(b_p) \in \mathcal{R}^n$, respectively.

Remark 7.7.1. If $n = 1$ and $\Omega(t) = \operatorname{sgn} t/t$, then formula (7.40) defines the Hilbert transform on $\mathcal{S}'^*$. Notice that for every $a > 0$ and $(a_p) \in \mathcal{R}$, we have

$$\mathbf{H}_a f\big|_{\mathcal{S}^{(M_p)}_{\mathcal{F}_0}(\overset{\circ}{\mathbb{R}}{}^n)} = \mathbf{K}f, \qquad \mathbf{H}_{(a_p)} f\big|_{\mathcal{S}^{\{M_p\}}_{\mathcal{F}_0}(\overset{\circ}{\mathbb{R}}{}^n)} = \mathbf{K}f,$$

for $f \in \mathcal{S}'^*(\mathbb{R}^n)$, where $\mathbf{K}$ is defined via $K(t) := 1/\pi t$.

Remark 7.7.2.   An important class of singular integral operators with odd kernels is the one consisting of the Riesz transforms. In n dimensions these are n singular integral operators $R_1, R_2, \ldots, R_n$ defined by the kernels

$$K_j(x) := \frac{\Gamma((n+1)/2)}{\pi^{(n+1)/2}} \frac{x_j}{|x|^{n+1}}$$

for $j = 1, 2, \ldots, n$, where $x = (x_1, \ldots, x_n) \in \mathbb{R}^n$.

It is well known that $\sum_{j=1}^n R_j^2 = -I$ on $L^2(\mathbb{R}^n)$. We can apply Corollary 7.7.1 to

$$\mathcal{K}_{a,j} = T_a^{-1} R_j T_a, \qquad \mathcal{K}_{(a_p),j} = T_{(a_p)}^{-1} R_j T_{(a_p)},$$

with $a \in \mathbb{R}^n_+$ and $(a_p) \in \mathcal{R}^n$, respectively, for $j = 1, 2, \ldots, n$. Put

$$\mathbf{K}^2_{a,j} := \mathbf{K}_{a,j} \circ \mathbf{K}_{a,j}, \qquad \mathbf{K}^2_{(a_p),j} := \mathbf{K}_{(a_p),j} \circ \mathbf{K}_{(a_p),j},$$

for $j = 1, 2, \ldots, n$. If $f \in \mathcal{S}'^{(M_p)}$ and $\varphi \in \mathcal{S}^{(M_p)}$ with supp $\varphi \subseteq \mathbb{R}^n \setminus \{0\}$, we have

$$\sum_{j=1}^n \langle \mathcal{F}_0 \mathbf{K}^2_{a,j} f, \varphi \rangle = \sum_{j=1}^n \langle f, T_a^{-1} R_j{}^2 T_a \hat{\varphi} \rangle = -\langle f, \hat{\varphi} \rangle = \langle \hat{f}, \varphi \rangle.$$

Thus if $a, b \in \mathbb{R}^n_+$ and $a \neq b$, then

$$\sum_{j=1}^n (\mathbf{K}^2_{a,j} - \mathbf{K}^2_{b,j}) f = P,$$

where P is an ultrapolynomial of the class (M_p). Analogously if (a_p), $(b_p) \in \mathcal{R}^n$ and $(a_p) \neq (b_p)$, we have

$$\sum_{j=1}^n (\mathbf{K}^2_{(a_p),j} - \mathbf{K}^2_{(b_p),j}) f = P,$$

where P is an ultrapolynomial of the class $\{M_p\}$.

Remark 7.7.3. Let

$$\Omega(x/|x|) := \sum_{\alpha=1}^{m} Y_k(x/|x|),$$

where the Y_k are spherical harmonics of degree k. The kernel $K(x) := \Omega(x')/|x|$ is an example of a singular integral with an even kernel.

Other examples can be deduced on the basis of Theorem 4.7 in [155], Chapter IV.

Bibliography

[1] R. Abraham, J. E. Marsden, T. Ratin, Manifolds, Tensor Analysis and Applications, Second Edition, Springer-Verlag, New York, 1988.

[2] P. Antosik, J. Mikusiński, R. Sikorski, Theory of Distributions. Sequential Approach, PWN-Elsevier, Amsterdam-Warszawa, 1973.

[3] A. Avantaggiati, $\mathcal{S}-$ spaces by means of the behaviour of Hermite-Fourier coefficients, Boll. Un. Mat. Ital. (6) 4-A (1985), 487-495.

[4] E. J. Beltrami, M. R. Wohlers, Distributions and the Boundary Values of Analytic Functions, Academic Press, New York, 1966.

[5] E. J. Beltrami, M. R. Wohlers, Distributional boundary values of functions holomorphic in a half plane, J. Math. Mech. 15 (1966), 137-146.

[6] G. Bengel, Darstellung skalarer und vektorwertiger Distributionen aus $\mathcal{D}'_{L^p}$ durch Randwerte holomorpher Funktionen, Manuscripta Math. 13 (1974), 15-25.

[7] A. Beurling, Quasi-analyticity and General Distributions, Lectures 4 and 5, Amer. Math. Soc. Summer Inst., Stanford, 1961.

[8] G. Björk, Linear partial differential operators and generalized distributions, Ark. Mat. 6 (1966), 351-407.

[9] P. Boggiatto, E. Buzano, L. Rodino, Global hypoellipticity and spectral theory. Mathematical Research, 92, Akademie Verlag, Berlin, 1996.

[10] N. Bourbaki, Intégration, Chap. I-IV, Hermann, Paris, 1952.

[11] R. W. Braun, R. Meise, B. A. Taylor, Ultradifferentiable functions and Fourier analysis, Resultat. Math. 17 (1990), 206-257.

[12] H. J. Bremermann, Distributions, Complex Variables, and Fourier Transforms, Addison-Wesley, New York, 1965.

[13] N. G. de Bruin, A theory of generalized functions, with applications to Wigner distribution and Weyl correspondence, Nieuw Arch. Wisk. (3) 21 (1973), 205-280.

[14] L. De Carli, L^p estimates for the Cauchy transforms of distributions with respect to convex cones, Rend. Sem. Mat. Univ. Padova 88 (1992), 35-53.

[15] B. C. Carlson, Special Functions of Applied Mathematics, Acad. Press, New York, 1977.

[16] R. D. Carmichael, Distributional boundary values in D'_{L^p}. IV, Rend. Sem.

Mat. Univ. Padova 63 (1980), 203-214.

[17] R. D. Carmichael, Generalization of H^p functions in tubes. I, Complex Variables Theory Appl. 2 (1983), 79-101.

[18] R. D. Carmichael, Generalization of H^p functions in tubes. II, Complex Variables Theory Appl. 2 (1984), 243-259.

[19] R. D. Carmichael, Values on the boundary of tubes, Applicable Analysis 20 (1985), 19-22.

[20] R. D. Carmichael, Holomorphic extension of generalizations of H^p functions, Internat. J. Math. Math. Sci. 8 (1985), 417-424.

[21] R. D. Carmichael, Cauchy and Poisson integral representations of generalizations of H^p functions, Complex Variables Theory Appl. 6 (1986), 171-188.

[22] R. D. Carmichael, Holomorphic extension of generalizations of H^p functions, II, Internat. J. Math. Math. Sci. 10 (1987), 1-8.

[23] R. D. Carmichael, Boundary values of generalizations of H^p functions in tubes, Complex Variables Theory Appl. 8 (1987), 83-101.

[24] R. D. Carmichael, Values on the topological boundary of tubes, Generalized Functions, Convergence Structures and Their Applications, B. Stanković, E. Pap, S. Pilipović and V. S. Vladimirov (editors), Plenum Press, New York, 1988, 131-138.

[25] R. D. Carmichael, Distributional and L^2 boundary values on the topological boundary of tubes, Complex Variables Theory Appl. 11 (1989), 135-153.

[26] R. D. Carmichael, Extensions of H^p functions, Progr. Math. 24 (1990), 1-12.

[27] R. D. Carmichael, Generalized Cauchy and Poisson integrals and distributional boundary values, SIAM J. Math. Anal. 4 (1973), 198-219.

[28] R. D. Carmichael, E. K. Hayashi, Analytic functions in tubes which are representable by Fourier-Laplace integrals, Pacific J. Math. 90 (1980), 51-61.

[29] R. D. Carmichael, A. Kamiński, S. Pilipović, Notes on Boundary Values in Ultradistribution Spaces, Lecture Notes Series 49, Seoul National University, Seoul, 1999.

[30] R. D. Carmichael, D. Mitrović, Distributions and Analytic Functions, Longman Scientific and Technical, Harlow, Essex, 1989.

[31] R. D. Carmichael, S. Pilipović, On the convolution and the Laplace transformation in the space of Beurling-Gevrey tempered ultradistributions, Math. Nachr. 158 (1992), 119-132.

[32] R. D. Carmichael, S. Pilipović, Elements of $\mathcal{D}'^{(M_p)}_{L^s}$ and $\mathcal{D}'^{\{M_p\}}_{L^s}$ as boundary values, Trudy Mathematical Institute of V. A. Steklov 203 (1994), 235-248.

[33] R. D. Carmichael, R. S. Pathak, S. Pilipović, Cauchy and Poisson integrals of ultradistributions, Complex Variables Theory Appl. 14 (1990), 85-108.

[34] R. D. Carmichael, R. S. Pathak, S. Pilipović, Holomorphic functions in tubes associated with ultradistributions, Complex Variables Theory Appl. 21 (1993), 49-72.

[35] R. D. Carmichael, R. S. Pathak, S. Pilipović, Ultradistributional boundary values of holomorphic functions, Generalized Functions and Their Applications, R. S. Pathak (editor), Plenum Press, New York, 1993, 11-28.

[36] C. Chevalley, Theory of Distributions, Lecture at Columbia University, 1950-51.

[37] C. C. Chow, Probléme de régularité universelle, C. R. Acad. Sci. Paris 260 (1965), Ser. A, 4397-4399.

[38] J. Chung, S. Y. Chung, D. Kim, Une caractérisation de l'espace de Schwartz, C.R. Acad. Sci. Paris Sér. I Math. 316 (1993), 23-5.

[39] J. Chung, S. Y. Chung, D. Kim, A characterisation for Fourier hyperfunctions, Publ. Res. Inst. Math. Sci. 30 (1994), 203-208.

[40] J. Chung, S. Y. Chung, D. Kim, Positive defined hyperfunctions, Nagoya Math. J. 140 (1995), 139-49.

[41] S. Y. Chung, D. Kim, E. G. Lee, Representation of quasianalytic ultradistributions, Ark. Mat. 31 (1993), 51-60.

[42] S. Y. Chung, D. Kim E. G. Lee, Schwartz kernel theorem for the Fourier hyperfunctions, Tsukuba J. Math. 19 (1995), 377-385.

[43] I. Cioranescu, The characterization of the almost-periodic ultradistributions of Beurling type, Proc. Amer. Math. Soc. 116 (1992), 127-134.

[44] I. Ciorănescu, L. Zsidó, x-ultradistributions and their applications to operator theory, in: W. Żelazko, ed., Spectral Theory, Banach Center Publications 8, PWN- Polish. Sci. Publ., Warsaw 1982, 77–220.

[45] P. Dierolf, J. Voigt, Convolution and $\mathcal{S}'$-convolution of distributions, Collectanea Math. 29 (1978), 185–196.

[46] S. J. L. van Eijndhoven, Functional analytic characterization of the Gelfand-Shilov spaces $\mathcal{S}_\alpha^\beta$, Proc. Nederl. Akad. Wetensch. Proc. Ser. A 90 (1987), 133-144.

[47] S. J. L. van Eijndhoven, J. de Graaf, Trajectory Spaces, Generalized Functions and Unbounded Operators, Lecture Notes in Mathematics 1162, Springer Verlag, Berlin-Heidelberg-New York-Tokyo, 1985.

[48] H. Federer, Geometric Measure Theory, Springer-Verlag, New York, 1969.

[49] H. Flanders, Differential Forms with Applications to the Physical Sciences, Dover Publications, New York, 1989.

[50] K. Floret, J. Wloka, Einführung in die Theorie der lokalkonvexen Räume, Lecture Notes in Math. 56, Springer, Berlin-Heidelberg-New York, 1968.

[51] I. M. Gel'fand, G. E. Shilov, Generalized Functions, Vol. 2, Academic Press, New York-London-Toronto-Sydney-San Francisco, 1968.

[52] I. M. Gel'fand, G. E. Shilov, Generalized Functions, Vol. 3, Academic Press, New York-London-Toronto-Sydney-San Francisco, 1967.

[53] I. M. Gelfand, N. Y. Vilenkin, Generalized functions, Vol. 4, Academic Press, New York, 1964.

[54] T. Gramchev, The stationary phase method in Gevrey classes and Fourier integral operators, Banach Center Publ. 19 1987, 101-112.

[55] T. Gramchev, L. Rodino, Gevrey solvability for semilinear partial differential equations, Indian J. Math. 41 (1999), 1–13.

[56] V. O. Grudzinski, Temperierte Beurling-distributions, Math. Nachr. 91 (1979), 197-220.

[57] Y. Hirata, On the convolutions in the theory of distributions, J. Hiroshima Univ. Ser A 22 (1958), 89–98.

[58] Y. Hirata, H. Ogata, On the exchange formula for distributions, J. Sci. Hiroshima Univ. Ser. A 22 (1958), 147-152.

[59] J. Horváth, Topological Vector Spaces and Distributions, I, Addison-Wesley, Reading, Massachusetts, 1966.

[60] L. Hörmander, The Analysis of Linear Partial Differential Operators I, Springer-Verlag, Berlin-Heidelberg-New York-Tokyo, 1983.

[61] S. Ishikawa, Generalized Hilbert transforms in tempered distributions, Tokyo J. Math. 10 (1987), 119–132.

[62] A. J. E. M. Jansen, Positivity of weighted Wigner distributions, SIAM J. Math. Anal. 12 (1981), 752-758.

[63] A. J. E. M. Jansen, Bargmann transform, Zak transform and coherent states, J. Math. Phys. 23 (1982), 720-731.

[64] A. J. E. M. Jansen, S. J. L. van Eindhoven, Spaces of type W, growth of Hermite coefficients, Wigner distribution and Bargmann transform, J. Math. Anal. Appl. 152 (1990), 368-390.

[65] A. Kamiński, Integration and irregular operations, Ph.D. Thesis, Institute of Mathematics, Polish Academy of Sciences, Warszawa, 1975.

[66] A. Kamiński, On convolutions, products and Fourier transforms of distributions, Bull. Acad. Polon. Sci. Ser. Sci. Math. Astronom. Phys. 25 (1977), 369-374.

[67] A. Kamiński, On the exchange formula, Bull. Acad. Polon. Sci. Sér. Sci. Math. Astronom. Phys. 26 (1978), 19-24.

[68] A. Kamiński, Remarks on delta- and unit-sequences, Bull. Acad. Polon. Sci., Sér. Sci. Math. Astronom Phys. 26 (1978), 25-30.

[69] A. Kamiński, Convolution, product and Fourier transform of distributions, Studia Math. 74 (1982), 83–96.

[70] A. Kamiński, D. Kovačević, S. Pilipović, The equivalence of various definitions of the convolution of ultradistributions, Trudy Mat. Inst. Steklov 203 (1994), 307-322

[71] A. Kamiński, D. Perišić, S. Pilipović, Existence theorems for convolution of ultradistributions, Dissertationes Math. 340 (1995), 79-91.

[72] A. Kamiński, D. Perišić, S. Pilipović, On the convolution of ultradistributions, Dissertationes Math. 340 (1995), 93–114.

[73] A. Kamiński, D. Perišić, S. Pilipović, On the convolution in the space of tempered ultradistributions of Beurling type, Integral Transforms and Special Functions 15 (2003), 323-329.

[74] A. Kamiński, D. Perišić, S. Pilipović, Hilbert transform and singular integrals on the spaces of tempered ultradistributions, Banach Center Publications 53 (2000), 139-153.

[75] A. Kamiński, D. Perišić, S. Pilipović, On various integral transformations of tempered ultradistributions, Demonstratio Math. 33 (2000), 641-655.

[76] A. Kamiński, J. Uryga, Convolution in $K'\{M_p\}$-spaces, Generalized Functions, Convergence Structures and their Applications, B. Stanković, E. Pap, S. Pilipović and V. S. Vladimirov (editors), Plenum Press, New York, 1988, 187-196.

[77] A. Kaneko, Introduction to Hyperfunctions, Kluwer Acad. Publ., Dordrecht,

1988.
[78] O. I. Kashpirovskij, Realization of some spaces of S-type as the spaces of sequences, Vestn. Kiev Univ. Mat. Meh. 21 (1979), 52-58. (in Ukrainian)
[79] K. H. Kim, S. Y. Chung, D. Kim, Fourier hyperfunctions as the boundary values of smooth solutions of heat equations. Publ. Res. Inst. Math. Sci., Kyoto Univ. 29 (1993), 289-300.
[80] S. Koizumi, On the singular integrals I-VI, Proc. Japan Acad. 34 (1958), 193-198; 235-240; 594-598; 653-656; 35 (1959), 1-6; 323-328.
[81] S. Koizumi, On the Hilbert transform I, II, J. Fac. Sci. Hokkaido Univ. Ser. I, 14 (1959), 153-224; 15 (1960), 93-130.
[82] H. Komatsu, Ultradistributions, I: Structure theorems and a characterization, J. Fac. Sci. Univ. Tokyo Sect. IA Math. 20 (1973), 25-105.
[83] H. Komatsu, Ultradistributions, II, J. Fac. Sci. Univ. Tokyo, Sect. IA 24 (1977), 607 - 628.
[84] H. Komatsu, Ultradistributions, III, J. Fac. Sci. Univ. Tokyo, Sect. IA 29 (1982), 653 - 717.
[85] H. Komatsu, Microlocal Analysis in Gevrey Classes and in Complex Domains, Lecture Notes in Math. 1726, Springer, Berlin, 1989, 426 - 493.
[86] H. Komatsu, Ultradistributions, Lecture Notes, Tokyo, 1999.
[87] J. Körner, Roumieu'sche Ultradistributionen als Randverteilung holomorpher Funktionen, Dissertation, Kiel, 1975.
[88] G. Köthe, Die Randverteilungen analytischer Funktionen, Math. Z. 57 (1952), 13-33.
[89] G. Köthe, Dualität in der Funktionentheorie, J. reine angew. Math. 191 (1953), 30-49.
[90] D. Kovačević, The spaces of weighted and tempered ultradistributions I-II, Zb. Rad. Prirod.-Mat. Fak. Ser. Mat. 23 (1993), 77-100; 24 (1994), 171-185.
[91] D. Kovačević, Some operations on the space $S'^{(M_p)}$ of tempered ultradistributions, Zb. Rad. Prirod.-Mat. Fak. Ser. Mat. 23 (1993), 87-106.
[92] D. Kovačević, S. Pilipović, Structural properties of the space of tempered ultradistributions, Proc. Conf. on Complex Analysis and Generalized Functions, Varna 1991, Publ. House of the Bulgar. Acad. Sci., Sofia, 1993, 169-184.
[93] M. Langenbruch, Ultradifferentiable functions on compact intervals, Math. Nachr. 137 (1989), 21-45.
[94] M. Langenbruch, Extension of ultradifferentiable functions, Manuscripta Math. 83 (1994), 123-143.
[95] O. Liess, L. Rodino, Fourier integral operators and inhomogeneous Gevrey classes, Ann. Mat. Pura Appl. 150 (1988), 167-262.
[96] Z. Łuszczki, Z. Zieleźny, Distributionen der Räume D'_{L^p} als Randverteilungen analytischer Funktionen, Colloq. Math. 8 (1961), 125-131.
[97] S. Mandelbrojt, Séries adhérentes, régularisation des suites, applications, Gauthier-Villars, Paris, 1952.
[98] T. Matsuzawa, A calculus approach to hyperfunctions, Nagoya Math J. 108 (1987), 53-66.

[99] T. Matsuzawa, A calculus approach to hyperfunctions II, Trans. Amer. Math. Soc. 313 (1990), 619-654.

[100] T. Matsuzawa, Foundation of a calculus approach to hyperfunctions, Lecture Notes, 1992.

[101] R. Meise, Darstellung temperierter vektorwertiger Distributionen durch holomorphe Funktionen I, Math. Ann. 198 (1972), 147-159.

[102] R. Meise, Darstellung temperierter vektorwertiger Distributionen durch holomorphe Funktionen II, Math. Ann. 198 (1972), 161-178.

[103] R. Meise, Sequence space representations for zero-solutions of convolution equations on ultradiferentiable functions of Roumieu type, Studia Math. 85 (1987), 203-227.

[104] R. Meise, B. A. Taylor, Whitney's extension theorem for ultradifferentiable functions of Beurling type, Ark. Mat. 26 (1988), 265-287.

[105] R. Meise, B. A. Taylor, D. Vogt, Equivalence of slowly decreasing functions and local Fourier expansions, Indiana Univ. Math. J. 36 (1987), 729-756.

[106] N. Ortner, P. Wagner, Sur quelques propriétés des espaces $\mathcal{D}'_{L^p}$ de Laurent Schwartz, Boll. Un. Mat. Ital. (6) 2-B (1983), 353-357.

[107] N. Ortner, P. Wagner, Applications of weighted $\mathcal{D}'_{L^p}$ spaces to the convolution of distributions, Bull. Polish Acad. Sci. 37 (1989), 579-595.

[108] J. N. Pandey, An extension of the Gelfand-Shilov technique for Hilbert transform, Journal of Applicable Analysis 13 (1982), 279–290.

[109] B. E. Petersen, Introduction to the Fourier Transform and Pseudo-differential Operators, Pittman, Boston, 1983.

[110] H. J. Petzsche, Die Nuklearität der Ultradistributionsräume und der Satz vom Kern I, Manuscripta Math. 24 (1978), 133-171.

[111] H. J. Petzsche, Generalized functions and the boundary values of holomorphic functions, J. Fac. Sci. Univ. Tokyo Sect. IA Math. 31 (1984), 391-431.

[112] S. Pilipović, Hilbert transformation of Beurling ultradistributions, Rend. Sem. Mat. Univ. Padova 77 (1987), 1-13.

[113] S. Pilipović, Boundary value representation for a class of Beurling ultradistributions, Portugaliae Math. 45 (1988), 201-219.

[114] S. Pilipović, Ultradistributional boundary values for a class of holomorphic functions, Comment Math. Univ. Sancti Pauli 37 (1988), 63-71.

[115] S. Pilipović, Tempered ultradistributions, Boll. Un. Mat. Ital. (7) 2-B (1988), 235-251.

[116] S. Pilipović, On the convolution in the space $\mathcal{D}^{(M_p)}_{L^2}$, Rend. Sem. Mat. Univ. Padova 79 (1988), 25-36.

[117] S. Pilipović, Some operations in $\sum_\alpha$, $\alpha > 1/2$, Radovi Mat. 5 (1989), 53-62.

[118] S. Pilipović, Beurling-Gevrey tempered ultradistributions as boundary values, Portug. Math. 48 (1991), 483–504.

[119] S. Pilipović, On the convolution in the space of Beurling ultradistributions, Comm. Math. Univ. St. Pauli 40 (1991), 15-27.

[120] S. Pilipović, Multipliers, convolutors and hypoeliptic convolutors for tempered ultradistributions, Proceedings International Symposium on Generalized Functions and their Applications, Banaras Hindu University, 1991.

[121] S. Pilipović, Characterization of bounded sets in spaces of ultradistributions, Proc. Amer. Math. Soc. 120 (1994), 1191-1206.

[122] S. Pilipović, Microlocal analysis of ultradistributions, Proc. Amer. Math. Soc. 126 (1998), 105-113.

[123] A. P. Robertson, W. Robertson, Topological Vector Spaces, Cambridge University Press, Cambrige, 1964.

[124] L. Rodino, Linear Partial Differential Operators and Gevrey Classes, World Scientific, Singapore, 1993.

[125] J. W. de Roever, Hyperfunctional singular support of distributions, J. Fac. Sci. Univ. Tokyo, Sect. IA Math. 31 (1985), 585-631.

[126] B. W. Ross, Analytic Functions and Distributions in Physics and Engineering, John Wiley and Sons, New York, 1969.

[127] C. Roumieu, Sur quelques extensions de la notion de distribution, Ann. Sci. École Norm, Sup. (3) 77 (1960), 41-121.

[128] W. Rudin, Lectures on the edge of the wedge theorem, Regional Conference Series in Mathematics (CBMS) No. 6, American Mathematical Society, Providence, 1971.

[129] M. Sato, Theory of hyperfunctions, I, J. Fac. Sci. Univ. Tokyo, Sect. IA Math. 8 (1959), 139-193.

[130] M. Sato, Theory of hyperfunctions, II, J. Fac. Sci. Univ. Tokyo, Sect. IA Math. 8 (1960), 398-437.

[131] L. Schwartz, Theórie des distributions, Hermann, Paris, 1950-51.

[132] L. Schwartz, Produits tensoriels topologiques et d'espaces vectoriels topologiques. Espaces vectoriels topologiques nucléaires, Séminaire Schwartz Faculté Sciences Paris, 1953–54.

[133] J. Sebastiõ e Silva, Les fonctions analytiques comme ultra-distributions dans le calcul opérationnel, Math. Ann. 136 (1958), 58-96.

[134] J. Sebastiõ e Silva, Les séries de multipôles des physiciens et la théorie des ultradistributions, Math. Ann. 174 (1967), 109-142.

[135] R. Shiraishi, On the definition of convolution for distributions, J. Sci. Hiroshima Univ. Ser. A 23 (1959), 19–32.

[136] O. P. Singh, J. N. Pandey, The n-dimensional Hilbert transform of distributions, its inversion and applications, Can. J. Math 42 (1990), 239–258.

[137] E. M. Stein, G. Weiss, Introduction to Fourier Analysis on Euclidean Spaces, Princeton University Press, Princeton, New Jersey, 1971.

[138] E. M. Stein, G. Weiss, M. Weiss, H^p classes of holomorphic functions in tube domains, Proc. Nat. Acad. Sci. U.S.A. 52 (1964), 1035-1039.

[139] H.-G. Tillmann, Distributionen als Randverteilungen analytischer Funktionen, Math. Z. 59 (1953), 61-83.

[140] H.-G. Tillmann, Distributionen als Randverteilungen analytischer Funktionen, II, Math. Z. 76 (1961), 5-21.

[141] H.-G. Tillmann, Darstellung der Schwartzschen Distributionen durch analytische Funktionen, Math. Z. 77 (1961), 106-124.

[142] H.-G. Tillmann, Darstellung vektorwertiger Distributionen durch holomorphe Funktionen, Math. Ann. 151 (1963), 286-295.

[143] E. C. Titchmarsh, Introduction to the Theory of Fourier Integrals, Oxford

Univ. Press, Oxford, 1937.

[144] J. Uryga, On a criterion of existence of convolution of generalized functions of the type $K\{M_p\}'$, Preprint 15, Ser. B, Institute of Mathematics, Polish Academy of Sciences, Warszawa, 1986 (in Polish).

[145] J. Uryga, On compatibility of supports of generalized functions of Gelfand-Shilov type, Bull. Acad. Polon. Sci. Ser. Sci. Math. Astronom. Phys. 36 (1988), 143-150.

[146] J. Uryga, On tensor product and convolution of generalized functions of Gelfand-Shilov type, Generalized Functions and Convergence, P. Antosik, A. Kamiński (editors), World Scientific, Singapore-New Jersey-London-Hong Kong, 1990, 251-264.

[147] V. S. Vladimirov, Methods of the Theory of Functions of Many Complex Variables, M.I.T. Press, Cambridge-Massachusetts, 1966.

[148] V. S. Vladimirov, Generalized functions with supports bounded from the side of an acute convex cone, Sibir. Math. J. 9 (1968), 930-937.

[149] V. S. Vladimirov, Equations of Mathematical Physics, Nauka, Moscow, 1968 (in Russian); English edition: Marcel Dekker, New York, 1971.

[150] V. S. Vladimirov, On the Cauchy-Bochner representations, Math. USSR Izv. 6 (1972), 529-535.

[151] V. S. Vladimirov, Generalized Functions in Mathematical Physics, Nauka Moscow, 1976 (in Russian); English edition: Mir Publishers, Moscow, 1979.

[152] D. Vogt, Vektorwertige Distributionen als Randvertielungen holomorpher Funktionen, Manuscripta Math. 17 (1975), 267-290.

[153] P. Wagner, Zur Faltung von Distributionen, Math. Ann. 276 (1987), 467-485.

[154] R. Wawak, Improper integrals of distributions, Studia Math. 86 (1987), 205-220.

[155] J. Wloka, Grundräume und verlagemeinerte Funktionen, Lecture Notes in Math. 82, Springer, Berlin-Heidelberg-New York, 1969.

[156] K. Yoshinaga, H. Ogata, On convolutions, J. Hiroshima Univ. Ser. A 22 (1958), 15–24.

[157] A. H. Zemanian, Generalized Integral Transformations, Dover Publications, New York, 1987.

[158] V. V. Zharinov, Distributive lattices and the "Edge of the Wedge" Theorem of Bogolybov, Proc. Conf. Generalized Functions and their Applications in Mathematical Physics - Moscow 1980, Acad. Sci. USSR, Moscow, 1981, 237-249.

[159] B. Ziemian, The modified Cauchy transformation with applications to generalized Taylor expansions, Studia Math. 102 (1992), 1–24.

Index